# THE SHAPING OF SOUTHERN ENGLAND

**Special Publications of the Institute of British Geographers**



No. 1 *Land Use and Resources: Studies in Applied Geography. A Memorial to Sir Dudley Stamp* (1968)

No. 2 *A Geomorphological Study of Post-Glacial Uplift with Particular Reference to Arctic Canada* (1970)
J. T. ANDREWS

No. 3 *Slopes: Form and Process* (1971)
Compiled for the British Geomorphological Research Group by D. BRUNSDEN

No. 4 *Polar Geomorphology* (1972)
Compiled for the British Geomorphological Research Group by R. J. PRICE and D. E. SUGDEN

No. 5 *Social Patterns in Cities* (1973)
Compiled for the Urban Study Group by B. D. CLARK and M. B. GLEAVE

No. 6 A contribution to the International Hydrological Decade—*Fluvial Processes in Instrumented Watersheds: Studies of Small Watersheds in the British Isles* (1974)
Edited for the British Geomorphological Research Group by K. J. GREGORY and D. E. WALLING

No. 7 *Progress in Geomorphology: Papers in Honour of David L. Linton* (1974)
Compiled and edited for the British Geomorphological Research Group by E. H. BROWN and R. S. WATERS

No. 8 *Inter-Regional Migration in Tropical Africa* (1975)
I. MASSER and W. T. S. GOULD with the assistance of A. D. GODDARD

No. 9 *Agrarian Landscape Terms: A Glossary for Historical Geography* (1976)
I. H. ADAMS

No. 10 *Change in the Countryside: Essays on Rural England, 1500-1900* (1979)
Compiled and edited for the Historical Geography Research Group by H. S. A. FOX and R. A. BUTLIN

All these publications are now available from Academic Press.

# THE SHAPING OF SOUTHERN ENGLAND

INSTITUTE OF BRITISH GEOGRAPHERS
SPECIAL PUBLICATION, No. 11

*Edited by*

DAVID K.C. JONES

*Department of Geography,*
*London School of Economics and Political Science*



1980

ACADEMIC PRESS
*A Subsidiary of Harcourt Brace Jovanovich, Publishers*
London New York Toronto Sydney San Francisco

ACADEMIC PRESS INC. (LONDON) LTD.
24/28 Oval Road,
London NW1

**United States Edition published by**
ACADEMIC PRESS INC.
111 Fifth Avenue
New York, New York 10003



**British Library Cataloguing in Publication Data**
The shaping of southern England. - (Institute of British Geographers. Special publications; Vol.11 ISSN 0073-9006).
1. Landforms - England
I. Jones, D K C    II. Series
551.4′09422    GB436.E5    80-40888

ISBN 0-12-388950-2



Text set in 11/13 Plantin
by DMB (Typesetting) Oxford
Printed in Great Britain by
Whitstable Litho Ltd., Whitstable, Kent

## CONTRIBUTORS

C. A. BAKER, St Lawrence College, Ramsgate, Kent CT10 2DT, England

G. H. CHEETHAM, Department of Geography, University of Reading, Whiteknights Road, Reading RG6 2AU, England

K. M. CLAYTON, School of Environmental Sciences, University of East Anglia, Norwich NR4 7TJ, England

A. S. GOUDIE, School of Geography, University of Oxford, Mansfield Road, Oxford OX1 3TB, England

C. P. GREEN, Department of Geography, Bedford College, University of London, Regent's Park, London NW1 4NS, England

D. T. JOHN, School of Geography, Kingston Polytechnic, Penrhyn Road, Kingston upon Thames, Surrey KT1 2EE, England

D. K. C. JONES, Department of Geography, London School of Economics and Political Science, Houghton Street, London WC2A 2AE, England

D. F. M. MCGREGOR, Department of Geography, Bedford College, University of London, Regent's Park, London NW1 4NS, England

R. J. SMALL, Department of Geography, The University, Southampton SO9 5NH, England

M. A. SUMMERFIELD, School of Geography, University of Oxford, Mansfield Road, Oxford OX1 3TB, England

R. B. G. WILLIAMS, The Geography Laboratory, University of Sussex, Falmer, Brighton, Sussex, England

# PREFACE

This collection of essays has been produced to celebrate the fortieth anniversary of the publication of *Structure, Surface and Drainage in South-East England* by S. W. Wooldridge and D. L. Linton. Originally published by the Institute of British Geographers in 1939 and later to appear in an updated form (1955), this important monograph marks a significant milestone in the study of the historical geomorphology of the British Isles. In it the authors presented a remarkable and commendably simple synthesis of previous work and outlined the first general model of landscape evolution for southern England. The widespread acceptance of this model is testimony to both the literary and field skills of the authors, and it is largely as a consequence of their labours that southern England came to be regarded as the 'heartland' for denudation chronology studies during the 1950s and 1960s. Thus the general evolutionary sequence established for the chalklands of this area came to be used as a blueprint for the interpretation of the harder rock terrains of upland Britain.

However, it is a feature of scientific enquiry that the passage of time results in established reconstructions being repeatedly tested in the light of new evidence, or through the re-interpretation of existing data because of new techniques of analysis or changing conceptual frameworks. This phenomenon has been particularly marked in the field of historical geomorphology, where the recent rapid development of Quaternary Studies has added new dimensions and vigour to the study of landform evolution. Thus the Davisian 'cyclical' framework and strongly morphological approach so characteristic of the denudation chronology school of two decades ago, has been replaced by a greater emphasis on biostratigraphic and lithostratigraphic evidence, and a growing appreciation that geomorphological landscapes, particularly those adjacent to the 'glacial limit', are the product of a variety of processes operating in response to complex patterns of environmental change. It is not surprising, therefore, that the resurgence of interest in landscape evolution studies should have resulted in criticisms of the Wooldridge and Linton reconstruction and its partial replacement by new models.

The essays which together make up *The Shaping of Southern England* seek to focus attention of these changing views, rather than to provide a complete and integrated guide to the current state of knowledge regarding the geomorphological evolution of the area. In order to achieve this, the authors have concentrated on those aspects of landform development discussed by Wooldridge and Linton in their monograph. Particular stress has been laid on Tertiary landscape evolution, and many of the central elements of their interpretation are re-evaluated in the light of recent research, especially that resulting from detailed studies into the nature and origin of the superficial deposits of the chalklands. Thus the classic ideas concerning a mid-Tertiary 'Alpine' folding phase, the evolution of a Mio-Pliocene peneplain, and the profound influence of a widespread Plio-Pleistocene marine incursion, are all challenged and arguments put forward in support of alternative models of landform and drainage evolution. Aspects of Quaternary development are examined in the second half of the book, including the effects of periglaciation on the Chalk and the significance of palaeohydrological variations in the evolution of river terraces. The results of recent lithostratigraphic investigations are stressed, particularly with respect to the glacial and fluvial sequences of the London Basin, and fully reasoned accounts given as to why new models have been proposed for both the patterns of glaciation and the evolution of the Lower Thames.

While many of the conclusions contained in this book differ from those of Wooldridge and Linton, the importance of their contribution is everywhere recognized and stressed. They, themselves, did not consider their interpretation to be necessarily the correct one, for in the preface to the 1955 edition of their monograph, they expressed the hope that it would prove of value to readers "despite its imperfections" and that "by criticism and further enquiry they correct these imperfections". It is hoped that these essays indicate the extent to which this has been achieved. Nevertheless, it is salutary to note that much of their work is still valid, despite the passage of four decades, and that many of the currently accepted ideas can be traced back to their roots in the writings of nineteenth century observers. While the process of re-interpretation will, no doubt, continue into the future under the influence of increasingly detailed and sophisticated studies, some of which may be of a radically different kind to those undertaken in the past, it is important that such developments should include the re-assessment of the legacy of previous work. This is contained in a huge and ever-increasing body of literature, within which occur highly significant and influential pieces produced by writers of capability and vision. *Structure, Surface and Drainage in South-East England* still stands pre-eminent in the former, Sidney William Wooldridge a clear example of the latter. Hopefully, this book will prove a valued addition.

*June 1980* *David K.C. Jones*

# CONTENTS

ONE

# The historical context of *Structure, Surface and Drainage in South-East England*

KEITH CLAYTON
*School of Environmental Sciences, University of East Anglia*

> "It would seem at first sight only natural to suppose that of all the river systems of Britain, that of south-eastern England might be singled out as the one whose history is most completely understood."
>
> David L. Linton (1932; p. 149).

In elucidating the history of a landscape, we must be aware that there is also a history of the ideas that have been advanced about that landscape. These ideas have been established on an intellectual gradient in which geomorphological concepts and various pieces of evidence have gradually increased in quantity and in reliability over time. The gradient is an irregular one, for from time to time syntheses of geomorphological concepts and of the field evidence promote a new level of understanding which may scarcely be improved upon for a generation. Looking back, there is no doubt that it is the successful combination of field evidence and geomorphological theory which has led to advances; the application of either in isolation has generally been insufficient to achieve much progress. Thus Prestwich brilliantly (and daringly) attributed the Lenham Beds to the Diestian in 1858, but it was not until 1927 that Wooldridge fully developed the concept of the Pliocene transgression of the London Basin, and a decade later that he and David Linton were persuaded to extend it throughout southern England. In much the same way, the strong theoretical arguments advanced by W. M. Davis in 1895 had little impact until they were incorporated into Linton's analysis of the Wessex rivers in 1932.

It is not my purpose here to review the antecedent literature at any length, but it does seem worthwhile setting the intellectual scene within which Wooldridge and Linton worked, and to explain the derivation of the ideas that were assembled to form their classic monograph. All twentieth-century work on south-east England has roots in the nineteenth century, some of them very strong, and this was certainly true of Wooldridge and Linton who were always very conscious of the long tradition of geological investigation of south-east England. The anticipation of modern ideas and arguments is no less true today as the following quotations will show:

> Whether the subaerial process has been continuous since early Oligocene times; whether it can be divided into two cycles, with a peneplain between; or whether the removal of the top of the dome has been assisted by marine planation, are questions so undecided that, ... geologists are perhaps in no closer agreement as to the physical history of the Wealden area than they were thirty years ago.

> This stage between the Chalk and the Eocene was a period of emergence, and consequently a period of subaerial denudation; the synclines were exposed to these agencies as well as the anticlines ... During the Eocene, the Weald was an island and attacked by subaerial agencies; this removed the Chalk from the central axis ... From that moment the rocks which outcropped in the centre of the Weald were less resistant than the Chalk which formed the rim, subaerial agencies eroded the area rapidly ... to form the great central Vale of the Weald. The Pliocene sea invaded and inundated the Weald, but failed to create a plain of marine denudation.

> The simplest explanation, therefore, of the Chalk Plateau seems to be to regard it as the product of the Diestian sea, while admitting that in some cases it differs but slightly from the old pre-Eocene plain, and that in others it has been modified by early consequent rivers as well (in all probability) by solution of the underlying Chalk.

Were I to set these unattributed in an examination question, they would as likely as not be accepted as Jones (1980; this volume p. 13). The first could as readily be attributed to Wooldridge (1927) and the last to Wooldridge and Linton (1939). In fact, the second is translated from Barrois (1876) and the other two are from Bury (1910, pp. 640, 652).

I am not old enough to have known directly of the work of Wooldridge and Linton in the late 1920s and 1930s, but like other geomorphologists of my generation, I have discussed that time with them both, and have at least the stimulus of remembering the tone of Wooldridge's voice as he spoke of the 'pink gravels of the geological map', or his widely relevant (though in application unique) concept of the '*pons asinorum*'. It is from that standpoint that I turn to the roots of *Structure, Surface and Drainage in South-East England*, which was first published as IBG Monograph No. 10 in 1939.

Reading accounts of the geomorphological evolution of a landscape, it is easy to forget just how difficult a topic we are tackling. We are attempting to use fragmentary and often obscure clues to reconstruct past landforms, most of which have subsequently vanished into thin air. The concomitant sedimentary record is best preserved in adjacent seas, and is hardly more accessible. Nevertheless, at any time since the early nineteenth century, widely accepted models of landform development have been available, and they have been used with remarkable confidence to interpret the fragmentary evidence. When we are not engaged in such reconstruction ourselves, we may contemplate the ease with which sparse evidence may be fitted to the fashionable model of the day. Furthermore, even the most determined advocate of the method of multiple working hypotheses (and there have always been too few of those) will find it difficult to collect enough evidence to throw out the majority of his explanatory models. If he does succeed, then it is difficult not to relax secure with the one explanation that does survive, rather than search for the evidence to find that one wanting and so force the erection of further explanatory models. More commonly, and south-east England illustrates this well, it will be left to geomorphologists of another school, or most commonly another generation, to find the surviving hypothesis wanting and to erect a new, and temporarily more attractive, model in its place.

In retrospect, although each major shift in interpretation may have required nothing more than a fresh viewpoint, a new hypothesis, or a new model to apply to the local landscape, it has in practice also involved the investigation of a new line of evidence. Inevitably some types of evidence become discredited (we no longer believe that chalk escarpments represent abandoned marine cliffs, the margins of glacial troughs, or fault scarps) but in general we are acquiring a more complex and sophisticated range of evidence in our search for understanding. At times the evidence itself has triggered the reopening of enquiry, or has demonstrated the inadequacy of a past interpretation. Most commonly, the feedback between theory and evidence is more complex: new evidence frequently requires consideration of new models (thus examination of sarsen stones led to consideration of models of climatic geomorphology, newly-discovered fossils imply a past transgression) and almost certainly the adoption of a new model (e.g. the concept of antecedence) will require investigation of new lines of evidence if earlier theories are to be rejected and the new ideas substantiated. At the same time there is no way in which evidence previously acquired may be easily disposed of in the interests of protecting a new hypothesis, although some authors during the last century have been a little too ready to discard well-established evidence as if it were but an invention. As Linton (1930) "with all due respect for official opinion" admonished the authors of the Guildford memoir (Dines and Edmunds, 1929) "although it is not possible to study them [the Dippenhall gravels] in section today, the authors of the Guildford memoir have given insufficient consideration to the trustworthy observations of former workers".

Wooldridge and Linton (1939) was a milestone, but the revolution in geomorphological understanding occurred in the decade 1857-1866. Until that time (and of course persisting after it despite the powerful arguments deployed) a confused, essentially catastrophist view of landscape evolution prevailed. It was as though Hutton and Playfair had not yet been published. The roots are scattered, but they were codified and entrenched by Lyell's advocacy of the marine theory from the first edition of his *Principles of Geology* in 1833. This theory has often since been confused with Ramsay's (1846) theory of a marine summit plane, but Lyell argued that all the essential elements of the landscape had been created by marine action, the rivers of today simply accepting and modifying them in the most minor ways. The core of the argument lay in the chalk scarps, which Lyell argued could only have been cut by the waves of the sea. The debris had (conveniently) been washed out by submarine currents through the gaps caused initially by transverse fractures and now occupied by grateful rivers, while any problems about the irregular levels at which these features might occur were explained not by warping, but by a slow process of uplift by which successive parts of the landscape (this rather suggests that some of the relief had prior existence) were brought into the action of the surf zone. Turning to the view westwards along the South Downs escarpment at the Devil's Dyke, Lyell wrote that any geologist "cannot fail to recognise in this view the exact likeness to a sea cliff"—a useful early example of the simplistic deduction of process from form.

Yet Lyell appealed less to catastrophic upheavals than did some of his contemporaries. Conybeare and Phillips (1822) could not understand how subsequent vales could be excavated upstream of the Chalk river gaps and with Buckland (1826) believed that "it is utterly impossible to explain the origin of any valleys of this description [inverted anticlinal vales] by denudation alone". Scrope and also Martin appealed with them for violent fracturing opening up the valleys, while Searles Wood Jr (1871) put forward even more complex patterns of tectonic movement, reversing the transverse rivers of the North Downs since he believed (cf. Geyl, 1976) the trumpet-shaped entries to the gaps in the North Downs to be abandoned river mouths. Murchison, too, favoured catastrophic denudation occurring during major and relatively recent episodes of folding. Prestwich (1890) and Osbert Fisher (1866) argued that the vales were glacial troughs, and the scarps created by glacial erosion, while others appealed to tidal waves to excavate the Weald.

This catastrophic mess was set right by the combined efforts of four men, each approaching the topic from a different standpoint and reasoning in different ways. They quote each other in reinforcement, but the trigger was undoubtedly Jukes' classic paper on Southern Ireland (1862). By showing that the rivers cross the ridges because they began at a higher level and have formed the entire landscape, he pushed 'rain and rivers' to its greatest possible extent. In a tentative postscript he wondered:

> My acquaintance with the Weald of Kent is too superficial to allow me to express an opinion; but perhaps I may venture to ask the question, whether the Chalk, when once bared by marine denudation, which perhaps removed it entirely from the centre of the district, has not been largely dissolved by atmospheric action; and whether the lateral river-valleys that now escape through ravines traversing the ruined walls of Chalk that surround the Weald may not be the expression of the former river-valleys that began to run down the slopes of the Chalk from the then-dominant ridge that first appeared as dry land during or after the Eocene period?

His views were taken up by Ramsay, who between 1862 and 1863 swung away from the marine hypothesis of Lyell as indicated by such statements as "of late I have begun to feel that it is not easily tenable, though it is perhaps not very easy completely to disprove it, especially to those who have long been accustomed to, and have never before doubted, the commonly received hypothesis". His book (Ramsay, 1864) is based on lectures, and it is fascinating to see the teacher teasing out the arguments and drawing on the evidence to present the case for subaerial denudation. The arguments he used included:

(1) the fact that the Chalk was 'cliffed' on one face only;

(2) that the fetch possible within the Weald seemed too small;

(3) the absence of surviving marine strata;

(4) the drainage pattern, (since it avoided the direct route eastwards to the Channel);

(5) the close adjustment of relief to lithology; and

(6) the work achieved by rivers in the same geological span in the Rhenish Uplands, the Jura and the Swiss Plain. The list has been drawn on by many later writers and has rarely been bettered. But it did not deter Wooldridge (1927) from dumping Ramsay into the 'marine' camp on the basis of his classic 1846 paper.

These new views were put forward with vigour and splendid clarity by Greenwood in his *Rain and Rivers; or Hutton and Playfair against Lyell and All Comers* (1857) which has special reference to the Weald. In the second edition (1866) he writes (pp. 57, 60):

> A stream running *through* ridges, large or small, is the simple consequence of the differing hardness of the ground through which it runs.... For a stream cannot run down where its bed is soft, and up again where it is hard. But the wash of rain digs down where the ground is soft, and leaves hills or ridges where it is hard. And as a stream cuts through a hard stratum, say the North or South Downs, the wash of rain is scooping out two lateral valleys *behind* it, that is, a valley behind each side of the gorge and ridge, as in the Weald Clay... . The ridges then, instead of being considered as barriers to the river, have been actually *formed* by the river by the abstraction of intervening masses.

Finally, Topley, both in the Memoir on the Weald (1875) and more particularly in Foster and Topley (1865), reviewed earlier theories, rejecting 'fracture' and 'marine action'. Their detailed geological mapping (especially of the river gravels) established the subaerial origin of the Medway basin, and with that the subaerial origin of the Chalk scarp, "it will be seen that we consider escarpments to be due to the difference of waste of hard and soft rocks under atmospheric denudation". They set out four reasons for rejecting Lyell's marine theory:

(1) the uneven height of the base of the scarp as compared with a wave-cut platform;

(2) the fact that most contemporary chalk cliffs cut across the strike;

(3) (Ramsay's points) the absence of marine shingle or fossils; and

(4) that the cliffing occurred only on the up-dip side of the Chalk ridge.

It is also perhaps worth recording that Topley (1875) with others took quite a detached view about the possibility of rather recent tectonic movements. He thought it possible that the marine gravels at Bourne Common above the Sussex Coastal Plain were to be correlated with the fossil beach exposed in the cliffs at Black Rock, Brighton. If so this represented displacement of 40 metres in 60 kilometres. Even if this flexure was a reality, he argued, it would be seriously felt only by the sluggish streams of the Weald Clay, "and even here it would only somewhat impede the drainage, not reverse it". The same readiness was shown by Bury (1910) who showed that Topley's suggestion of tilting would reverse the inclination of the crest of the South Downs, and who, more importantly, accepted the idea that the Wealden rivers were antecedent to the Oligocene/Miocene folds:

> The Wealden area seems to have been raised above the sea at the close of the Eocene period, but the principal longitudinal folds are generally attributed to Oligocene or Miocene times. In that case, assuming that there has been no more recent marine action, the river-system was antecedent to the folds, and their effect upon it would depend on so many factors - for instance, the degree of maturity of the rivers, the intensity and rapidity of the movements, etc. - that it seems impossible to lay down any definite rules which will guide us in distinguishing between such an antecedent river-system and one developed upon a marine plain after the folds were formed.
>
> Bury (1910; p. 659).

The progress made between 1857 and 1866 has not been matched since. The arguments rehearsed by Ramsay before his audience of working men in 1863 spelt out most of the points that were going to be chewed over again and again for the next one hundred years. Arguments put forward by Foster and Topley (1865), Bury (1910), and Wooldridge (1927) add remarkably little to the range of material so fully appreciated by Ramsay. One intervention that might have been expected to have had more effect was that by W. M. Davis (1895). As a visitor with well-developed ideas about the nature of subaerial erosion, he introduced a new theory that could stand as an alternative to the concept that only marine planation was capable of producing the summit plateau so characteristic of the Chalk. His proof of this reasonable idea was not found convincing by most of those who followed, but it did activate a new controversy about the reality of the summit plain (from which there were few dissenters) and its mode of formation, and hence its role in the evolution of the drainage pattern in particular. Until Wooldridge and Linton promoted the concept of partial submergence of a widespread subaerial plain, the Mio-Pliocene landsurface/Pliocene transgression division (to use the terms of 1939), workers on south-east England found themselves arguing either for a marine or for a subaerial origin for the summit plain. Davis also used the relationship between drainage and structure to elucidate the history of the area: this had been touched on from Buckland's (1826) 'valleys of elevation' onwards, but never before exploited so effectively. It came to form a crucial element in Linton's reasoning.

Davis's interpretation of south-east England was rapid and in many ways superficial; it demonstrated the power of his ideas, and the ease with which quite an elementary range of evidence could be used to construct the model. Yet it is a remarkable achievement and forms a clean break with the nineteenth-century geologists like Topley and Jukes-Browne, let alone the Kellaways of the nineteenth century like Searles Wood or Martin. What he argued was that the degree to which the drainage of south-east England was adjusted to structure, the extent to which the subsequent values had been excavated and the drainage integrated by capture, required two cycles of subaerial erosion, the first culminating in the production of the summit plane of south-east England, which he thus categorized as subaerial. Topley had regarded it as marine. Davis was able to erect his hypothesis on the evidence of relief, geology and drainage pattern alone, and he had no time for more than a most general look over the field evidence. That he was able to convince his audience of the power of rivers to reduce a landscape to a peneplain owes something to the case made out by the fluvialists in the marine-fluvial controversy of the second half of the nineteenth century. But his impact was reduced by the lack of detailed consideration of the evidence on the ground.

A masterly review (together with Davis greatly influencing Linton's later analysis) was provided by Henry Bury in 1910. From it we may extract his set of arguments on the nature of the summit plain. These were taken from the nineteenth-century literature but restated in the light of Bury's knowledge of the evidence on the ground, and necessarily reinterpreted in the light of Bury's understanding of geomorphological processes and relationships. His points were:

(1) the coincidence of the highest levels throughout the area (Ramsay, 1864; Topley, 1875);

(2) the even bevelling of the crest of the escarpment (the argument introduced by Davis in 1895);

(3) the distribution of flints (Foster and Topley, 1865);

(4) the Diestian deposits—including their distribution on the northern fringe of the Weald (Prestwich, 1858);

(5) the distribution of Lower Greensand chert (again Foster and Topley, 1865);

(6) signs of warping linked with the uplift of the summit plain (also Topley, 1875); this last dependent on more detailed consideration of summit heights, especially along the crest of the North Downs.

Bury's particular contribution in the second part of this long paper was to consider, in some detail, the river pattern and its relation to the folds, establishing some of the stages by which the present pattern had evolved by capture, and concluding that the changes required could readily be contained within the complex and long period of incision below a marine summit plain. The discussion was returned to by David Linton in 1932 in the report of his field excursion to the River Wey.

The consolidation of evidence achieved by Wooldridge and Linton in 1939 was a major achievement; they erected an edifice that remained virtually undisturbed for over thirty years, but as we are now beginning to see, it represented a reinterpretation of the record, not its establishment. But it will also be seen that their work has deeper roots than is commonly realized. As already noted, Prestwich established the Diestian age of the Lenham Beds in 1858, while the collection of evidence on other patches of presumed Pliocene shingle and their correlation by heavy mineral analysis can be traced back to Davies (1917). Linton's work on the Wessex drainage owed much to the mapping of the Geological Survey and particularly to the mapping of the Chalk zones by amateur geologists, while the detailed investigations of Henry Bury, J. F. N. Green, F. Gossling and others explored the structure, surface and drainage of many critical areas of the Weald and Hampshire Basin. They did this with care, and without too many misconceptions which might be imposed on the evidence, although some of the later work was a little too ready to reconstruct river profiles by extrapolating from sections preserved above knick-points. Thus Bury (1933) tackled the problem of the interpretation of a 'stairway' of terrace flats with the welcome comment "The truth seems to be that both in the Avon and Stour valleys there are numerous shelves out of which it is possible to select a few at 100 ft and 50 ft respectively above the river; they really occur at many other levels as well, and are seldom persistent for any distance."

Central to all this activity was Wooldridge's dedicated exploration of some rather dull-looking parts of the geology of south-eastern England. His early paper in 1923 began delightfully and accurately "The London Basin offers few attractions to the structural geologist". In this and subsequent papers (1926, 1927) he explored the evidence for the minor structures that do exist, and the stages of movement:

(1) post Bagshot Beds uplift;

(2) major response to Miocene stresses ('the outer ripples of the Alpine storm'—later shifted into the Oligocene);

(3) post-Diestian uplift, with no demonstrable E-W warping (first shown by Bury, 1922) and modest (if any) N-S warping.

He went on through painstaking field work backed up by laboratory investigation of heavy mineral and other lithological evidence (much of it in collaboration with others and inevitably none of it meeting today's standards of replication) to separate out a whole series of deposits of a sandy or pebbly nature. Some of these turned out to be outliers of various Upper Eocene beds, others matched the Headley-Lenham group of Davies (1917), some were classed as Pebble Gravel, and still others, by virtue of their erratic content, fell into a sequence of Pleistocene 'plateau' and 'terrace' gravels. Fossils were found at Netley Heath by Stebbing (1900) but they were poorly preserved and could only be matched with the Diestian genera. Davies (1917) kept these outcrops separate as the Headley Beds "pending definite evidence of their equivalence to the beds of East Kent", and re-examination of the Netley Heath casts (together with further finds) led to the Red Crag correlation, supporting Davies' caution.

In the London Basin a stairway of sediments of various ages (and fairly closely dated despite the absence of fossil evidence) from the Pebble Gravel and all the 'plateau' gravels down to the accepted river terraces, encouraged the erection of a denudational chronology into which the glacial incursions and related diversions of the Thames were fitted (Wooldridge, 1938). The sequence of events has never fitted neatly and frequently has been disturbed by the chance exposure of new lines of evidence. Wooldridge himself, then Clayton, Hey and many others, have found themselves revising their previous correlations and setting up new ones. The two-diversion model survived longest of all, but even it has been abandoned in recent years, while at the same time links with the East Anglian sequence have been usefully elaborated (see Baker and Jones, this volume p. 131).

An important part of geomorphological modelling over the past one hundred years has been the ability to think in terms not only of the present landscape, but of past landscapes on which the changes have been acted out. Explanation of the relative success of various branches of the Guildford Wey cannot simply be sought in the patterns of the published geological map—it is the outcrop pattern of one or more of the earlier high-level stages that accounts most convincingly for the capture pattern (Bury, 1910). Linton's (1930) field description of the Upper Wey leans heavily on the existence of major valley stages of low relief, which were the more readily disturbed as incision worked back along more favoured competing streams. It may have been presumptuous to correlate these (at a later time) with a regional 200-ft (60 m) Ambersham level, but they are real enough in the area around Selborne. In 1930 Linton was properly cautious; "the hypothesis that it is the local equivalent of the 200-ft platform described by Wooldridge in the regions to the north and east, would seem a reasonable one. On the other hand, if the gravels of Bury's 'Terrace C' of the Farnham Wey do truly belong to this period of stillstand, the evidence of the contained implements certainly suggests a more recent origin." In much the same way Wooldridge sought to educate his students in the message of the views north and south from Jack Straw's Castle on Hampstead Heath and by pointing to the (Lower Greensand) pin-hole chert at their feet to convince them of the reality of the time when the Thames III Valley simply was not there. As for the Thames at Stage I, it had flowed along the line of the Vale of St Albans, but at hill-top level, the Vale being formed at a later stage. Each student who could not grasp that sequence of events and sought to take the Stage I Thames through the Vale was castigated in stern tones, for they had failed to cross S. W. Wooldridge's '*pons asinorum*'.

In 1927 Wooldridge put forward five arguments for accepting the reality of the summit planation of the Weald. They make an interesting comparison with the subaerial-marine arguments already quoted from Ramsay (1846), Foster and Topley (1865), and Bury (1910). His points were:

(1) accordant summit levels;

(2) bevelled and straight scarps;

(3) scarcity of residual flint in the Central Weald;

(4) conversely the wide distribution of Lower Greensand chert along the North Downs;

(5) the relationship between the rivers and the longitudinal folds.

As he comments, "Not one of these arguments is final in itself, for all the features are capable of other explanations, although some of these are far from convincing. The collective testimony of the several arguments is strong, however ..."

What is missing from some of the most recent work, and it is missing because it is unfashionable, is the detailed knowledge of the ground held by so many of the older workers. It was the proud aim of Wooldridge and Linton that they should know all the ground, and that the name of a village would be enough to recall a site, a pattern of relief and geology, that was a piece of the jigsaw they had set out to reconstruct. In 1964, having recently moved, I remember telling David Linton that my nearest pub was at Stalisfield Green. "Oh yes", he replied, "there's a curious displacement in the structural contours of the Chalk there, they seem to be offset north by a minor monocline, and I never did fully understand how the dry valley pattern was related to it." They had both walked across every parish in south-east England, and Linton never crossed the Wessex Chalk without his one-inch map with the Chalk zones marked in on it. I rarely found a pit in southern Essex that Wooldridge could not recall and tell me his assignment of the deposits exposed. They were not alone in this, as the careful mapping, over very many years, of the Lower Greensand by J. F. Kirkaldy testifies: the result is an accumulation and integration of evidence that may be disturbed by some important new exposure, some change in correlation, but is not easily upset simply because it has been found to hang together so well.

This detailed knowledge of the ground, and the area we are concerned with is a large one, is also important in the testing of new working hypotheses. It is fully legitimate to establish a new model on the basis of local evidence of the evolution of an insequent river in relation to the dates of movement of a fold, to reinterpret a transgression on the basis of new fossils, or new correlations. Equally, improved knowledge of the East Anglian glacial succession or of the sedimentary hydraulics of river flood plains may allow quite new inferences about the Thames sequence. Yet if such ideas are to be seen as credible, if the hypothesis is to survive as a model, it must survive the test of application to the region as a whole. However the transgressive gaps of the South Downs may be explained, the concept should also work for the chalk-bound Meon, as well as for the small streams of the Isle of Wight, or the twin gaps at Corfe. Is warping to be limited to the explanation of the summit swell of the Chalk at Warden Hill, or does it also explain Borley Hill on the Kent-Surrey border? Where is the Hampshire Basin equivalent of the Pebble Gravel stage? As Wooldridge and Linton wrote in 1938, "only patient and long-continued re-examination of the ground can fully demonstrate the truth of what we have suggested".

As a synthesis of the geomorphology of a major region within the British Isles, *Structure, Surface and Drainage* survived its issue in an extended version in 1955 for perhaps half a dozen years. For the past twenty years it has had something of the style of a fossil, surviving from an age when a denudation chronology was the natural aim of a geomorphological synthesis, and where regional description carried on the long French tradition. The revision in 1955 left Linton's chapters untouched, and the additions by Wooldridge (especially on the former course of the Thames in Essex) lacked the quality of the original material. It would have helped if the original terminology of the Mio-Pliocene peneplain had been replaced by Linton's preferred version, the Pliocene landsurface. For such a surface can, by implication, be a palimpsest in a way that a peneplain may not. Yet if the style of the book is now out of fashion, it may at the very least be hailed as a classic work, and as yet hardly matched by other syntheses of regional geomorphology. It is a quirk of scientific progress that we often have learnt enough to question, or even to reject, past work of this kind, but somehow we then find ourselves insufficiently knowledgeable to erect something in its place. Comprehensive regional synthesis is not easy to achieve.

In retrospect, it may seem as though Wooldridge and Linton imposed their own model on the region with less hesitation than we might wish, given the paucity of the evidence outside a few favoured areas. It is worth placing on record that when Douglas Johnson visited England a little before the war (I believe it was in 1936) he was taken over the ground to find that David Linton was postulating a transgression in Wessex in order to explain the river patterns, although with no local evidence of marine deposits. It was Johnson who persuaded him that a ready explanation lay to hand in the 'Pliocene' episode Woodridge had established in the London Basin. As he put it, why invoke a local transgression when a regional one is to hand? Wooldridge and Linton were ready enough to accept the region-wide model as their IBG monograph and their less widely read paper (Wooldridge and Linton, 1938) in Johnson's journal suggests, but it was a readiness based on a detailed knowledge of the field relationships, and triggered by a visitor who regarded south-east England as a region you crossed by car in a day, not a region you walked over, field by field, over a decade or more. Knowledge and vision at both these scales has its part to play in the successful creation and testing of geomorphological explanation.

## References


Barrois, C. (1876). Recherches sur le terrain crétacé supérior de l'Angleterre et de l'Irelande, *Mém. Soc. Géol. du Nord* **1**, 1-232.

Buckland, W. (1826). On the formation of the valley of Kingsclere and other valleys by the elevation of the strata that enclose them, *Trans. Geol. Soc. (second series)* **2**, 119-30.

Bury, H. (1910). On the denudation of the western end of the Weald, *Q. J. Geol. Soc. Lond.* **66**, 640-92.

Bury, H. (1922). Some high-level gravels of North-east Hampshire, *Proc. Geol. Ass.* **33**, 81-103.

Bury, H. (1933). The plateau gravels of the Bournemouth area, *Proc. Geol. Ass.* **44**, 314-35.

Conybeare, W. D. and Phillips, W. (1822). "Outline of the Geology of England and Wales", William Phillips, London.

Davies, G. M. (1917). Excursion to Netley Heath, Newlands Corner and the Silent Pool, 23 September 1916, *Proc. Geol. Ass.* **27**, 48-51.

Davis, W. M. (1895). On the origin of certain English rivers, *Geogr. J.* **5**, 128-46.

Dines, H. G. and Edmunds, F. H. (1929). "The geology of the country around Aldershot and Guildford", Memoirs of the Geological Survey of England and Wales, Explanation of Sheet 285, p. 182.

Fisher, O. (1866). On the probably glacial origin of certain phenomena of denudation, *Geol. Mag.* **3**, 483-87.

Foster, C. Le Neve and Topley, W. (1865). On the superficial deposits of the valley of the Medway, with remarks on the denudation of the Weald, *Q. J. Geol. Soc. Lond.* **21**, 443-74.

Geyl, W. F. (1976). Tidal palaeomorphs in England, *Trans. Inst. Br. Geogr.* **1**, 203-24.

Greenwood, G. (1857). "Rain and rivers; or Hutton and Playfair against Lyell and all comers" (2nd edn 1866), London.

Jukes, J. B. (1862). On the mode of formation of some of the river valleys in the south of Ireland, *Q. J. Geol. Soc. Lond.* **18**, 378-400.

Linton, D. L. (1930). Notes on the development of the western part of the Wey drainage system, *Proc. Geol. Ass.* **41**, 160-74.

Linton, D. L. (1932). The origin of the Wessex rivers, *Scot. Geogr. Mag.* **48**, 149-66.

Lyell, Sir C. (1833). "Principles of Geology" (1st edn), John Murray, London.

Prestwich, Sir J. (1858). On the age of some sands and iron-sandstones on the North Downs, *Q. J. Geol. Soc. Lond.* **14**, 322-35.

Prestwich, Sir J. (1890). On the relation of the Westleton Beds or pebbly sands of Suffolk to those of Norfolk, and on their extension inland: with some observations on the final elevation and denudation of the Weald and of the Thames valley, etc. *Q. J. Geol. Soc. Lond.* **46**, Part I, 84-119, Part II, 120-54; Part III, 155-81.

Ramsay, A. (1846). On the denudation of South Wales and the adjacent counties of England, *Mem. Geol. Surv. U.K.*, pp. 326-8.

Ramsay, A. (1864). "The physical geology and geography of Great Britain" (2nd edn), Edward Stanford, London.

Stebbing, W. B. D. (1900). Excursion to Netley Heath and Newlands Corner, Saturday 11 August 1900, *Proc. Geol. Ass.* **16**, 524-6.

Topley, W. (1875). The geology of the Weald, *Mem. Geol. Surv. U.K.*, p. 503.

Wood, S. V. Jr (1871). On the evidence afforded by the detrital beds without and within the north-eastern part of the valley of the Weald as to the mode and date of the denudation of that valley, *Q. J. Geol. Soc. Lond.* **27**, 3-27.

Wooldridge, S. W. (1923). The minor structures of the London Basin, *Proc. Geol. Ass.* **34**, 175-93.

Wooldridge, S. W. (1926). The structural evolution of the London Basin, *Proc. Geol. Ass.* **37**, 162-96.

Wooldridge, S. W. (1927). The Pliocene period in the London Basin, *Proc. Geol. Ass.* **38**, 49-132.

Wooldridge, S. W. (1938). The glaciation of the London Basin and the evolution of the Lower Thames drainage system, *Q. J. Geol. Soc. Lond.* **94**, 627-67.

Wooldridge, S. W. and Linton, D. L. (1938). Influence of Pliocene transgression on the geomorphology of South-east England, *J. Geomorphology* **1**, 40-54.

Wooldridge, S. W. and Linton, D. L. (1939). Structure, surface and drainage in South-east England, *Trans. Inst. Br. Geogr.* **10**.

TWO

# The Tertiary evolution of south-east England with particular reference to the Weald

DAVID K. C. JONES
*Department of Geography, London School of Economics and Political Science*

## Introduction

The Weald must rank as one of the best-known denuded dome landscapes in the world, and yet great uncertainty still exists as to its geomorphological evolution. This is particularly true for the Tertiary, evidence for which has largely been destroyed by more recent erosion so that workers have been forced to extrapolate from surrounding chalklands and adjacent sedimentary basins. Almost a century of investigation and debate culminated in the publication of Wooldridge and Linton's remarkable monograph *Structure Surface and Drainage in South-East England* (1939) which synthesized and extended previous work (Linton, 1969; Kirkaldy, 1975) and presented the first comprehensive discussion of Tertiary landscape evolution. Their interpretation was so widely accepted that a second edition could be produced in 1955 with few significant modifications and continued to be employed unquestioningly until the early 1970s, even though a radically different explanation had been advanced by Pinchemel (1954). This situation is partly explicable by the fact that publication of the monograph's second edition failed to stimulate further research. The contemporary view appears to have been that the general sequence of events had been clearly defined and substantiated. As a consequence, the focus of 'denudation chronology' investigation switched to the harder rock terrains of Upland Britain, where the Wooldridge and Linton model was to prove a cornerstone for the interpretation of the South-west Peninsular (e.g. Brunsden, 1963), Wales (Brown, 1960a), and the southern Pennines (Linton, 1956a). The continued general lack of interest in the South-east led Clayton (1969) to remark, "It is disconcerting to consider how little has been accomplished in the last twenty years.... Any adequate new account must still lean heavily on pre-war material", a pronouncement that fortunately heralded a regeneration of research which was to be characterized by a move away from the Davisian approach adopted by earlier studies, together with a reduction in the emphasis placed on morphological evidence and its replacement by sedimentological analysis of soils and superficial deposits. As a consequence elements of the Wooldridge and Linton model have come to be so seriously questioned (Jones, 1974; Catt and Hodgson, 1976) that a radical reappraisal is now overdue.

The aim of this chapter is to examine available evidence for the Tertiary evolution of south-east England and particularly the Weald and adjacent Chalk cuestas. In order to do this it is necessary first to briefly describe the Wooldridge and Linton model prior to examining critically and rejecting the long-held views on the influence of the 'Calabrian' transgression, the former existence of a Mio-Pliocene peneplain and the occurrence of a brief mid-Tertiary 'Alpine storm'. To conclude, a model of landform evolution will be advanced which places greater emphasis on early Tertiary (Palaeogene) denudation.

## The Wooldridge and Linton model

The Wooldridge and Linton interpretation was dependent on the identification of remnants of three widely developed erosion surfaces; a warped marine-trimmed sub-Tertiary unconformity (Sub-Eocene Surface), a high-level unwarped subaerial late Tertiary (Neogene) peneplain and an unwarped end-Tertiary (Plio-Pleistocene) marine platform (Wooldridge and Linton, 1938a, 1939, 1955). The interrelationship of these three facets on the Chalk cuestas, with the inclined Sub-Eocene Surface fringing the present Tertiary outcrop and truncated upslope by the Plio-Pleistocene (Calabrian) platform, which in turn cut into the subaerial surface (Mio-Pliocene peneplain) preserved above 210 m O.D., appeared to establish convincingly the evolutionary sequence. In addition, the identification of two Neogene surfaces resolved the long-standing debate as to whether the high-level 'Summit Surface', first recognized by Prestwich (1858) on the evidence of widespread conformity of summit levels between 200 and 300 m, was the product of marine (Topley, 1875; Bury, 1910) or subaerial (Davis, 1895) denudation. The compound interpretation offered by Wooldridge and Linton satisfied both hypotheses, for it allowed the concordant drainage pattern of the Central Weald to be explained as the product of long-term subaerial evolution (Davis, 1895), while at the same time accounting for the widespread discordant relationships by invoking superimposition from an elevated marine shelf (Bury, 1910).

The fully embellished model, as applied in the early 1970s, can be briefly summarized as follows (Fig. 1):

(1) The progressive submergence of the area during the Cretaceous culminated in the deposition of a continuous sheet of Chalk more than 300 m thick in 200-300 m of water.

(2) Late Cretaceous uplift caused tilting towards the east and initiation of macro-structural features, including a subdued Weald-Artois Anticline. The onset of subaerial conditions saw the development of a consequent drainage pattern including two major eastward flowing trunk rivers within the ancestral London and Hampshire Basins (Linton, 1951; Brown, 1960b).

(3) A complex series of early Tertiary (Palaeogene) marine transgressions

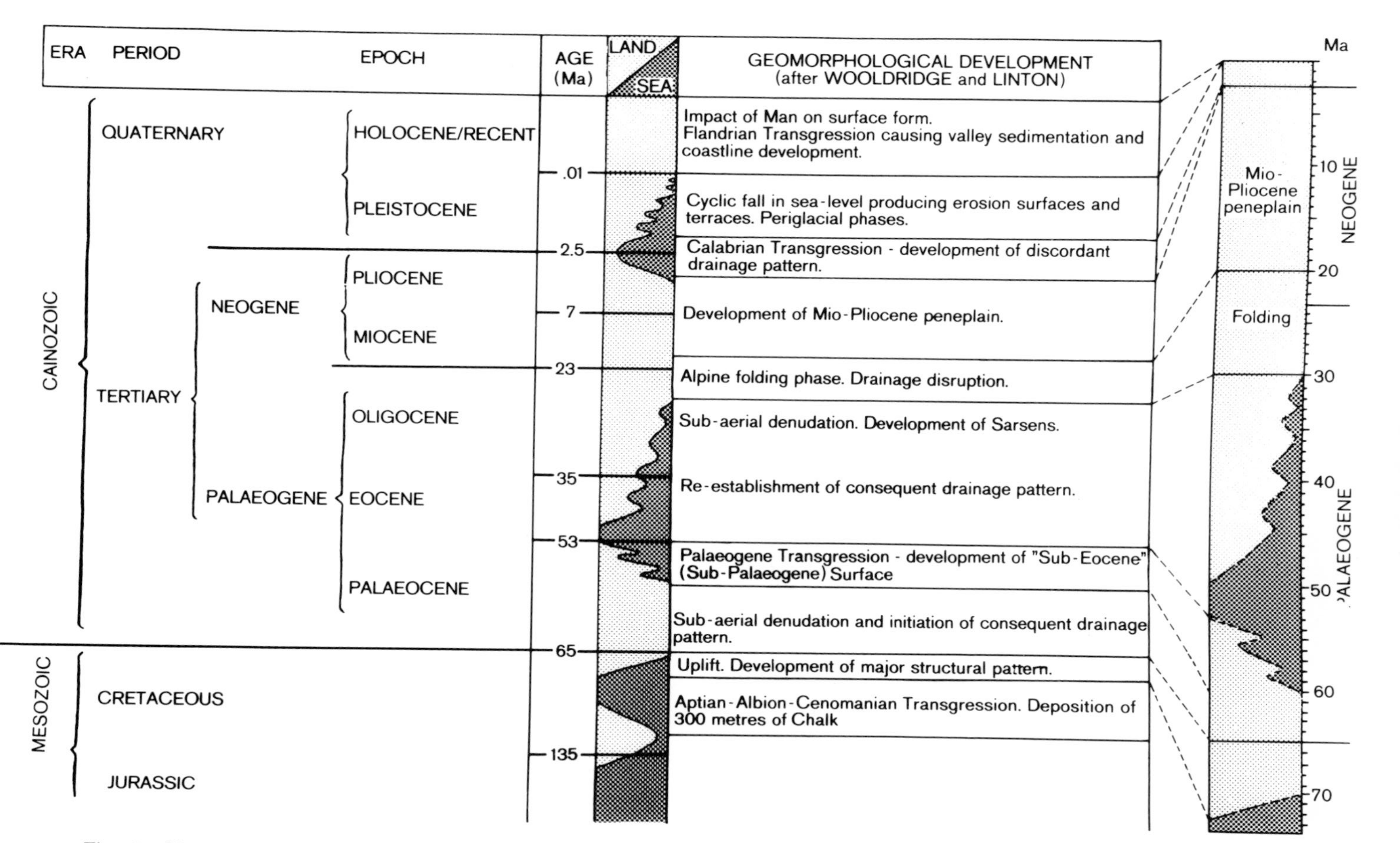


*Figure 1.* Chronological subdivision of the Tertiary and Quaternary showing the Wooldridge and Linton evolutionary interpretation.

culminated in the total submergence of the area by the London Clay (Eocene) sea, and resulted in the deposition of up to 800 m of fluvial, estuarine and marine sediments on a marine trimmed surface which was called the Sub-Eocene Surface but should, more correctly, be known as the Sub-Palaeogene Surface.

(4) The gradual emergence of the area during the Eocene and early Oligocene led to the progressive return of subaerial conditions and the re-establishment of a consequent drainage pattern, including an eastward flowing proto-Thames in the London Basin and the Solent River in the Hampshire Basin. Subaerial denudation was considered minimal under the prevailing conditions of tectonic stability and semi-arid climate—although Linton (1956b) argued for the local removal of the Eocene cover and erosion of the underlying Chalk—and resulted in the production of extensive low-relief surfaces with duricrusts (sarsens) (Clark *et al.*, 1967).

(5) Tectonic disturbance in the Upper Oligocene and early Miocene, due to the climax of the Alpine orogenic phase, greatly increased the amplitude of the macro-structures, while at the same time superimposing secondary folds and faults. Disturbance was considered violent enough to disrupt the existing consequent drainage pattern, thereby producing a new network dominated by east-west oriented synclinal segments (Linton, 1956b), and resulted in greatly accelerated denudation.

(6) Continuing subaerial denudation during the stable Miocene and Pliocene (Neogene) led to widespread erosion and the creation of a low relief surface (the Mio-Pliocene peneplain) extensively mantled by residual weathering deposits, and most particularly the Clay-with-Flints of the chalklands.

(7) The lower parts of this surface were inundated by the late Pliocene/early Pleistocene ('Calabrian') transgression. Widespread marine planation up to an elevation of 210 m eradicated much of the pre-existing drainage pattern, except for that on the unsubmerged 'Weald Island' located over the central and western Weald. Upon the withdrawal of this sea, a new system of trunk streams was established on the flanks of the Weald Arch which were consequent upon the slope of the erosion surface but discordant to the underlying structures.

(8) Generally falling base levels throughout the Pleistocene led to drainage incision and the production of flights of raised shorelines and river terraces, the most significant being the 60 m (200 ft) surface and the 30 m level (Hoxnian). There also occurred at least three phases of periglaciation with associated solifluction activity and loess deposition.

(9) Finally, the post-glacial eustatic rise in sea-level (Flandrian Transgression) from about − 100 m at the last glacial (Devensian) maxima (about 17 000 years B.P.) resulted in the re-inundation of the Dover Straits by 9600 B.P. (D'Olier, 1972) and the subsequent fashioning of the coastline.

The application of this evolutionary sequence to the Weald is portrayed in Fig. 2 and clearly illustrates the emphasis placed on Neogene denudation. In addition it shows how the fundamental surfaces decrease in age towards the axis of uplift, with the exhumed Sub-Eocene Surface on the lower backslopes of the Chalk being replaced by Neogene surfaces (Mio-Pliocene peneplain and 'Calabrian' Bench) on the higher parts of the Chalk cuestas, while the Weald *sensu stricto* is the product of Quaternary denudation.

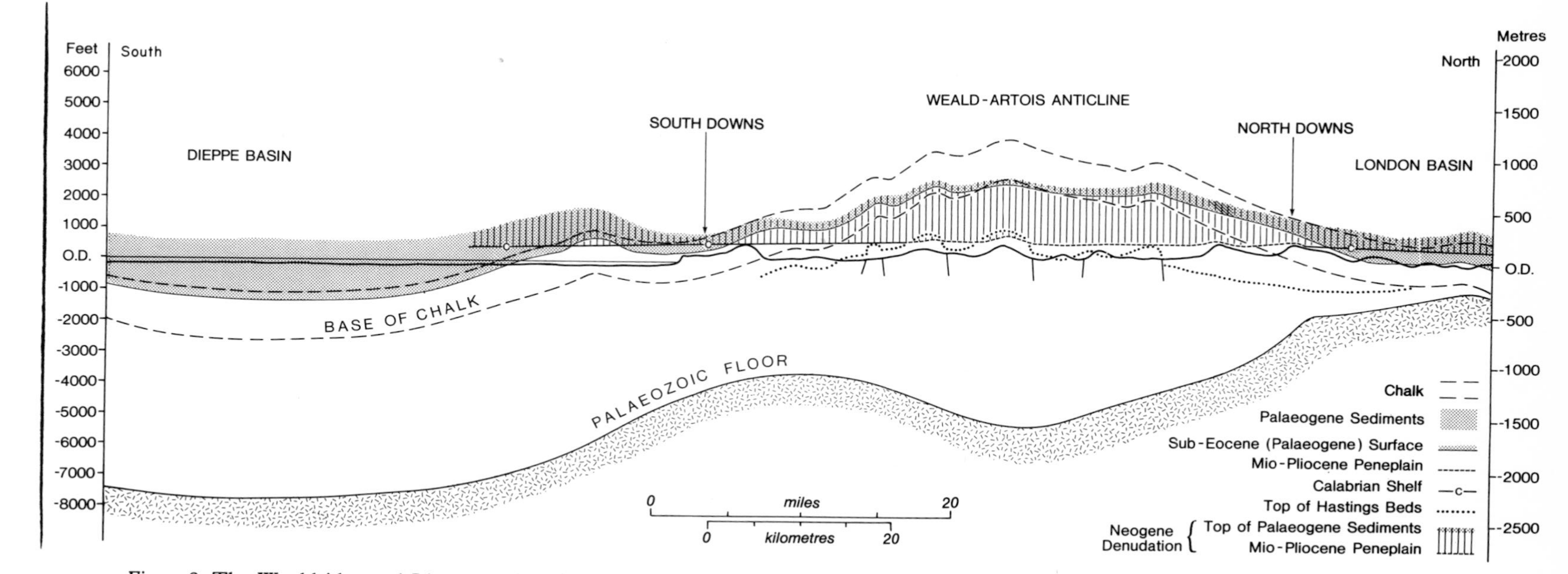


*Figure 2.* The Wooldridge and Linton model of landscape evolution as applied to the Weald-Artois Anticline and adjacent areas (based, in part, on Wooldridge and Linton (1955), Jones (1974) and Curry and Smith (1975)). Line of section runs N-S and passes close to Dartford, Sevenoaks and Beachy Head.

## Criticisms of the Wooldridge and Linton model of Tertiary landscape evolution

It is important to note at the outset that the Wooldridge and Linton reconstruction envisaged a sequence of well-defined and distinctive early, middle and late Tertiary evolutionary phases (Fig. 1). By ascribing most of the tectonic disturbance to a brief mid-Tertiary episode it proved possible to distinguish two phases of subaerial denudation characterized by relative stability. Their 'cyclical' interpretation emphasized the significance of the shorter, but more recent, Neogene phase as against the longer Palaeogene episode (Figs. 1 and 2). Contemporary research has questioned the existence of such 'distinctive' compartmented evolutionary episodes, preferring reconstructions characterized by more continuously changing environmental conditions, including pulsed tectonism (George, 1974).

Recent geomorpholgical investigations around the Weald have concentrated on examining the nature and interrelationship of the Sub-Eocene Surface and 'Calabrian' Bench, because of the restricted extent of ground above 210 m. Such studies have been undertaken in the South Downs (Hodgson *et al.*, 1967, 1974; Small and Fisher, 1970) and North Downs (Docherty, 1967; John, 1974), and are compatible with contemporary research in 'Wessex' (Waters, 1960; Clark *et al.*, 1967; Green, 1969, 1974) where the emphasis has been placed on the evolution of the higher chalklands and the significance of the Mio-Pliocene peneplain. As a consequence of these and related studies, numerous detailed criticisms of the Wooldridge and Linton model have emerged which have been extended by the regional syntheses of Worssam (1973), Jones (1974) and Catt and Hodgson (1976). The ideas that have recently gained favour centre on two main threads. First, the possibility that the early-Tertiary (Palaeogene) phase of erosion may have been considerably more important in gross landform sculpturing than had hitherto been generally recognized, a thesis originally advanced by Pinchemel (1954), and second, that the discordant drainage pattern may have been inherited from the early Tertiary and be explicable in terms of antecedence rather than superimposition.

The essence of the contemporary debate involves the following major questions:

(1) How much erosion was achieved during the Palaeogene and to what extent does the present chalkland macrotopography represent surfaces inherited from that time?

(2) Was the structural pattern created during a relatively short mid-Tertiary episode, or was it produced over a longer time span by much less violent disturbance?

(3) Did the Mio-Pliocene (Neogene) phase of subaerial denudation actually result in the formation of an extensive low relief peneplain?

(4) Did the Plio-Pleistocene sea transgress extensively and achieve sufficient erosion to eradicate much of the pre-existing drainage pattern, thereby causing the inception of the present discordant trunk streams?

These questions are examined in greater detail in the following sections.

## The validity of a widespread 'Calabrian' incursion

The concept of a Plio-Pleistocene marine incursion is based on apparently good morphological, sedimentological and faunal evidence preserved on the chalkland flanks of the London Basin (Wooldridge, 1927; Wooldridge and Linton, 1938a, 1939, 1955). Particularly impressive is the widely developed platform, or 'Calabrian Bench', which rises to a maximum elevation of 210 m along the presumed shoreline. However, uncertainty exists as to the age and origin of this feature because the evidence is spatially inconsistent. In Kent, the platform planes across the crest of the North Downs at 180-190 m and is patchily overlain by Lenham Beds, ferruginous pebbly sands intimately associated with Clay-with-Flints and often piped into the Chalk. These sands have been the subject of controversy since they were first described (Prestwich, 1858) and have been ascribed various ages from Miocene to late Pliocene (see Chatwin in Worssam, 1963; Curry *et al.*, 1978), with sea-level estimations ranging from 200 m to 265 m (Reid, 1890) and 270 m (Shotton, 1962). Elsewhere in the London Basin an early Pleistocene (Red Crag/Waltonian) incursion to 200-210 m is generally accepted, based on fossil evidence obtained from Netley Heath (Chatwin, 1927; Dines and Edmunds, 1933) and Rothamsted (Dines and Chatwin, 1930), the difference in age between the Lenham Beds (most frequently dated as Upper Pliocene ≡ Coralline Crag) and the deposits encountered elsewhere being explained by the sea transgressing from the east (Wooldridge, 1960).

The coastline established in the London Basin was extended to the west and south around the Weald on the basis of morphology, sediments, altitude and drainage pattern (Wooldridge and Linton, 1938a). The resultant palaeography has been severely criticized on four main grounds (Jones, 1974):

(1) Morphological evidence from the Hampshire Basin and southern Weald is best described as unconvincing. There are no clearly developed benches similar to those of the London Basin, which accounts for the variation in palaeogeographical reconstructions (see Jones, 1974, Fig. 1).

(2) The only 'Calabrian' sediments recorded outside the London Basin are those of Beachy Head (Edmunds, 1927) and Mersley Down, Isle of Wight. Neither of these deposits is *in situ* and the Isle of Wight example appears to be associated with the much lower 130 m marine level (Everard, 1954).

(3) The use of altitude (200 m *level*) as a basis for defining the former coastline is indefensible. The upper limit of the 'Calabrian Bench' has never been fixed accurately in the London Basin, so it is difficult to accept assertions that it is unwarped, especially as there is growing evidence that south-east England has suffered warping throughout the Quaternary (King, 1977).

(4) In many instances drainage pattern was the sole basis for defining the coastline, which was made to follow the boundary between supposedly concordant east-west dominated networks, such as in the north and centre of the Weald, and generally southward flowing discordant systems. Jones (1974) raised a number of serious objections to this approach and the resultant 'Gulf and Islands' palaeogeo-

graphy of the southern Weald, particularly noting (a) that discordant relationships are more widespread than envisaged by Wooldridge and Linton; (b) that further investigation of geological structure has revealed that the division into concordant and discordant networks is not as clear as was formerly envisaged; (c) that the use of drainage patterns as evidence for a former marine incursion, which was then used to explain the origin of the discordant rivers, is a most unsatisfactory circular argument if unsupported by sound morphological and sedimentological evidence, and (d) that no clear link has ever been established between the 'Calabrian' transgression and the development of discordant networks, for where 'physical' evidence for a 'marine' incursion appears well preserved, discordance is the exception, and vice versa.

In addition to these palaeogeographical arguments, a second group of objections has been raised (Jones, 1974) concerning the ability of the 'Calabrian' transgression to achieve the drainage modifications accredited to it by Wooldridge and Linton. These points relate particularly to the southern Weald and Hampshire Basin, where discordance is so widespread as to require the former existence of a broad and extensive planation surface, for which little evidence can be found. The enormous scale of erosion required not only to create such a surface, but also to obliterate a pre-existing drainage network occupying a landsurface with an estimated relative relief of up to 180 m, appears beyond the ability of what has generally been considered a brief marine incursion. Further, the postulated existence of a chain of islands along the South Downs, together with the recognition of the Western Rother as a relic of the pre-Calabrian drainage pattern (Wooldridge and Linton, 1955; Linton, 1956b) and the fact that many of the discordant streams do not flow orthogonally to the postulated coastline, all militate against the superimposition hypothesis.

Finally, the necessity for invoking a widespread marine incursion has been questioned. The superimposition hypothesis was developed in the belief that fold growth during the mid-Tertiary had been so rapid that the drainage pattern was disrupted to create an east-west concordant network, part of which survives in the Central Weald. There is no evidence to support such a view. In fact the division into concordant and discordant networks is seen to be increasingly artificial as more information is obtained on structure. The similarity of the drainage patterns on the northern and southern flanks of the Weald and their high degree of adjustment to macrostructure (Jones, 1974) points to the whole network having experienced a common evolutionary history.

Because of these criticisms the 'Calabrian' transgression can no longer be envisaged as having played a significant role in drainage evolution. It follows, therefore, that the main elements of the drainage net existed in the Neogene, a view supported by the evidence for southward flowing drainage across the South Downs prior to the supposed arrival of the 'Calabrian' sea (Small and Fisher, 1970). This, in turn, raises doubts concerning the impact of Tertiary folding. The most logical explanation of drainage development is that the trunk streams originated as consequents on the flanks of the proto-Weald Arch as it emerged from beneath the

Palaeogene sea in the Eocene (Jones, 1974). These streams developed on a cover of Palaeogene sediments and were repeatedly rejuvenated by the growth of the Weald Anticline, thereby enabling them to maintain their courses across small developing periclines. The formation of periclines in headwater areas, i.e. near the structural axis, led to the creation of generally concordant networks, such as in the Central Weald, either because the streams were too small to resist derangement or through drainage adjustments following superimposition from the thick and homogeneous Weald Clay. Elsewhere, however, the development of secondary folds resulted in discordance. As there is a markedly greater concentration of minor folds on the southern limb of the Weald Anticline, which occur as an almost continuous line of structures (Fig. 3), it is not surprising that the southward flowing rivers are

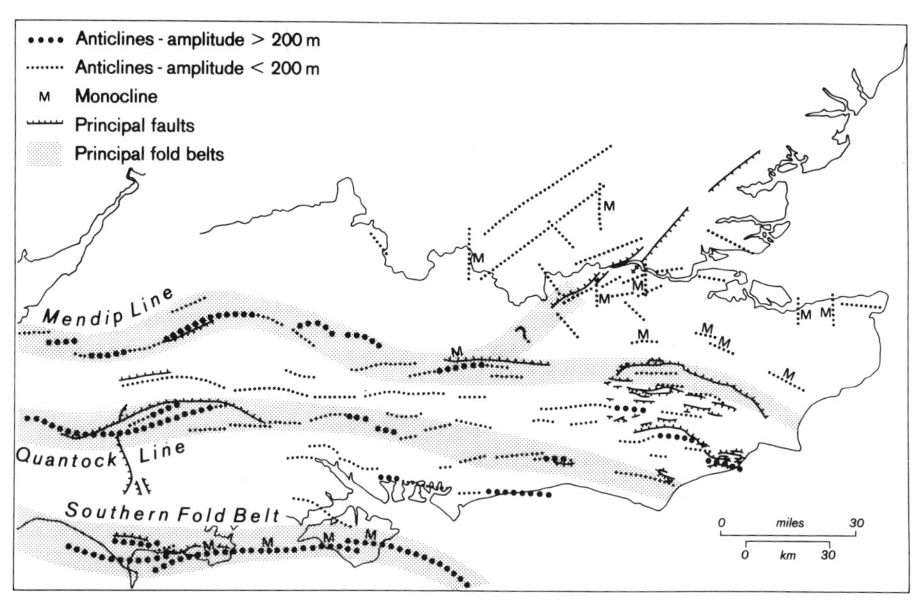


*Figure 3.* The secondary structures of the Weald and adjacent areas showing the three main structural belts (based on Wooldridge and Linton, 1955).

conspicuously discordant. This 'antecedent' model adequately explains both the distribution of discordant drainage as well as the apparent division into concordant and discordant networks which was the most important line of evidence used in invoking and delimiting a 'Calabrian' incursion in the southern Weald.

Rejection of the superimposition hypothesis still leaves unresolved questions as to the number, age and extent of late Cainozoic incursions, for problems exist concerning (1) the differing antiquity of the Plio-Pleistocene marine deposits identified on the North Downs and (2) why evidence for a marine episode should be so well preserved in the London Basin and yet virtually absent elsewhere in southern England. The three main palaeogeographical interpretations which have been proposed can be termed the 'single transgression', 'slow transgression' and 'double transgression' models. The 'single transgression' model was the original reconstruction and envisaged an extensive submergence in Pliocene times (Jukes-Browne, 1911; Wooldridge, 1927; Wooldridge and Linton, 1938, 1939). While there is still support for this view, the re-correlation of Red Crag as basal Pleistocene in 1948 resulted in the development of the 'slow transgression' model, in which the Pliocene sea is envisaged to have gradually spread westwards until the early Pleistocene (Wooldridge and Linton, 1955), thereby explaining why both the scale of erosion and age of fossils appear to diminish in this direction. Because of the supposed drainage evolution implications, attention soon came to be focussed on the maximum extent of submergence, so that the term 'Pleistocene sea' was introduced (Wooldridge, 1957) and led to the 'double transgression' model involving separate Pliocene and early Pleistocene incursions. This idea was first hinted at by Wooldridge and Linton (1955) and later amplified by Catt and Hodgson (1976), who argued that the early episode affected the eastern Weald and deposited the Lenham Beds, together with the Coralline Crag of East Anglia, and was followed, after an interval of over 2 Ma, by a brief but more widespread incursion which deposited the Red Crag sediments. The term 'Calabrian' only recently came to be associated with the early Pleistocene phase of maximum submergence (Wooldridge, 1960), following a suggestion by Baden-Powell (1955), but has subsequently come into general usage. It is clear that this term should now be discarded, based as it is on long distance correlation with southern Italy, and be replaced by the more applicable local names Red Crag transgression or Waltonian transgression.

All three hypotheses have suffered from the problem that deduced Pliocene sea-levels in Kent (up to 270 m) are higher than the upper limit of the 'Calabrian Bench' in the London Basin (210 m). In fact Jukes-Browne (1911) considered that most of the Weald had been submerged by the Pliocene sea. This discrepancy could be resolved if it were shown that the Kentish Downs had suffered uplift so that calculated Pliocene sea-levels appear artificially high but no evidence for such movements has been identified, the general consensus being that the Lenham Beds have probably been relatively lowered because of their proximity to the subsiding North Sea Basin. However, recent work at Headley Heath (John, 1974) has presented a possible solution. His analysis indicates that the locally developed 'Calabrian Bench' is not a product of early Pleistocene marine erosion but a structural feature produced by minor monoclinal flexures. It may, therefore, be either an element of the warped Sub-Palaeogene unconformity or a structural surface produced by the differential erosion of zones within the Upper Chalk. The marine 'Calabrian' deposits which rest upon this bench have, according to John, been let down through denudation, thereby indicating a sea-level in excess of the 200 m postulated by Wooldridge and Linton (1955), but for which no morphological evidence survives.

While there is still some support for a high-level early-Pleistocene marine transgression within the London Basin, this is no longer the case for the Hampshire Basin. Although Stevens (1959) mapped a 'Calabrian Bench' around the western Weald, subsequent work by Green (1969, 1974) has indicated that no morphological or sedimentological evidence for such an incursion can be identified either at, or above, the accepted 180-210 m level, the highest recorded marine feature being the 145 m surface (Everard, 1954). As it is difficult to explain why evidence for this incursion should be relatively down-warped into the Hampshire Basin, the obvious conclusions are either that the Red Crag transgression was restricted to the London Basin or that suitable conditions for the creation and/or preservation of evidence only occurred within the London Basin. The latter view appears to be supported by Catt and Hodgson (1976) who argue that much of the marine shelf was probably cut in Palaeogene sediments and has subsequently been destroyed by the ravages of Quaternary denudation. A third, more radical, explanation is that the concept of a major early Pleistocene transgression may be invalid and has resulted from the misinterpretation of the London Basin evidence, the most crucial elements of which were produced about fifty years ago (Chatwin, 1927; Wooldridge, 1927; Dines and Chatwin, 1930; Dines and Edmunds, 1933). Both the age and origin of the Red Crag ('Calabrian') deposits must now be considered in doubt, for no faunal remains have been described since 1933. Similarly, the pioneer sedimentological work of Wooldridge (1927), in which he used heavy mineral analysis to distinguish Plio-Pleistocene marine sediments from Palaeogene residuals, is in need of re-evaluation in the light of new analytic procedures and increased knowledge of the complex Palaeogene sequence. It is significant, therefore, that the Headley Heath sediments have been reinterpreted as Pleistocene aeolian cover sands (Letzer, 1973), while Docherty (1967) has indicated that certain of the 'Calabrian' deposits of north-west Kent are outliers of Blackheath Beds (upper Palaeocene/basal Eocene) resting on the Sub-Palaeogene Surface, an interpretation largely compatible with the conclusions of John (1974) and one that naturally leads to the suggestion that the 'Calabrian Bench' may well represent exhumed relics of one or more of the marine-trimmed surfaces produced in the early Palaeogene. Such an interpretation would adequately account for the apparent restriction of good morphological evidence to the London Basin.

Clearly, inundation was not the potent sculpturing event envisaged by Wooldridge and Linton, for re-appraisal of the platform and sediments (Docherty, 1967; Green, 1974; John, 1974), together with the general criticisms of the 'superimposition model' (Jones, 1974), indicates that *if* there was a widespread incursion then it must have been brief and impotent with a coastline that was determined by prevailing relief. By contrast, the conspicuous planation of the Kentish North Downs and associated Lenham Beds, has been taken to suggest a Pliocene (Chatwin in Worssam, 1963) or late Miocene (Curry *et al.,* 1978) marine episode of rather greater geomorphological significance, for the only river for which a superimposition origin can be cogently argued is the Kentish Stour which leaves the Weald between Chalk cuestas bevelled by the 'Lenham Surface'. However, as it does not display discordance, has a small catchment with no Central Weald headwaters, and appears the product of a marine event for which there is local morphological and sedimentological evidence, it cannot be used to support arguments for more general superimposition of drainage from an extensive hypothetical marine shelf. Furthermore, the occurrence of Clay-with-Flints *sensu stricto* (Loveday, 1962) underlying the Lenham Beds (Smart *et al.,* 1966) suggests that the 'Lenham Surface' may also represent an erosionally modified element of the Sub-Palaeogene unconformity (see later). This clearly poses difficulties for both the 'slow' and 'double' transgression models and adds to the growing scepticism concerning the geomorphological significance of late Cainozoic marine erosion, irrespective of arguments as to the number and timing of incursions.

Further analytical work is obviously required on the London Basin marine sediments to ascertain whether they are Pleistocene, Neogene or degraded remnants of the Palaeogene cover. Until such results are available it is possible to propose arguments in support of all three interpretations. However, on the basis of current evidence the weight of argument would appear to favour the 'single transgression' model with the Miocene/Pliocene sea briefly submerging much of the (then) lower-lying tracts of southern England. The shelf was probably cut mainly in Palaeogene deposits, as suggested by Catt and Hodgson (1976), and has been largely destroyed, while sedimentological evidence has been selectively preserved at those locations where the marine platform was underlain by gently inclined facets of the Sub-Palaeogene Surface (i.e. London Basin and Kent). There may well have been a later Red Crag incursion, but the evidence from East Anglia suggests that it was a shallow sea and probably restricted to the lower-lying eastern part of the London Basin. Such an interpretation is clearly compatible with observations on the character and distribution of evidence (Docherty, 1967; Green, 1969, 1974; John, 1974; Jones, 1974; Catt and Hodgson, 1976) and the view that marine action during the Plio-Pleistocene was of little significance in the geomorphological evolution of southern England (Pinchemel, 1954; Jones, 1974; Catt and Hodgson, 1976). Certainly the suggestion that the morphological and sedimentological evidence may not be related, and that maximum sea-level could have been well in excess of 200 m, indicate that the palaeogeographic reconstruction of Wooldridge and Linton (1938, 1939, 1955) should now be abandoned.

## The Mio-Pliocene peneplain

The identification of a high-level subaerial peneplain of late Tertiary (Neogene) age, preserved on the summit areas rising above 210 m, was based on evidence of summit conformity, morphology and superficial deposits (Wooldridge and Linton, 1939, 1955). The existence of such a surface is now disputed for a wide variety of reasons (Jones, 1974, 1980; Catt and Hodgson, 1976).

Much of the recent debate has centred on the origin and character of the residual weathering deposits known as the 'Clay-with-Flints' which were considered diagnostic of this surface when developed on Chalk (Wooldridge and Linton, 1955). This subject has already been exhaustively treated elsewhere (Hodgson *et al.,* 1974; Catt and Hodgson, 1976), the essence of the argument being as follows. First, Loveday (1962) indicated that the superficial cover of the chalklands can be subdivided into two groups, the 'Clay-with-Flints' *sensu stricto* and the 'Plateau Drift'. The 'Plateau Drift' is by far the more important in terms of thickness and extent. It is lithologically variable, often containing a significant component of aeolian material, and shows widespread signs of soliflunctional movements, thereby indicating a rather complex Pleistocene history. It occurs at all elevations above 100 m O.D. and is in no way diagnostic of Neogene subaerial denudation.

Clay-with-Flints *sensu stricto* is also found at all levels between the edge of the Palaeogene outcrop and the crest of the Chalk cuestas and usually occurs at the interface between Upper Chalk and a cover of Plateau Drift, Palaeogene debris or Lenham Beds (Smart *et al.*, 1966). Recent investigations have concluded that, while Clay-with-Flints may be formed beneath any suitable permeable cover (e.g. Plateau Drift) through dissolution of Chalk by percolating water, it is characteristically developed where Chalk is overlain by 'a thin and disrupted cover of (Eocene) Reading Beds' (Hodgson, 1967; Hodgson *et al.*, 1967) and thus reflects the former extent of the Reading Beds facet of the Sub-Palaeogene Surface (Hodgson *et al.*, 1974). Furthermore, its elevational distribution indicates that it cannot be viewed as a diagnostic weathering product of the warm and humid Neogene subaerial environment, but may have been formed at any time since the withdrawal of the Palaeogene sea, with many authors (e.g. Avery, 1964) favouring particularly active development in the Upper Palaeogene and warmer phases of the Pleistocene. It may even be forming slowly under contemporary conditions (Catt and Hodgson, 1976).

The presence of Clay-with-Flints and Plateau Drift on the higher parts of many Chalk backslopes, both often rich in Reading Beds material, is now taken to indicate the close proximity of the present topographic surface to the Sub-Palaeogene unconformity, an interpretation originally advanced by Jukes-Browne (1906). On this evidence the 'Summit Surface' of the chalklands can no longer be interpreted as a Neogene peneplain, but rather as the exhumed Sub-Palaeogene Surface largely stripped of its sedimentary cover and modified by weathering and subaerial denudation (Hodgson *et al.*, 1974; Catt and Hodgson, 1976; Jones, 1980).

Such a view is similar to that proposed by Pinchemel (1954), who suggested that the macro-form of the chalklands broadly reflects early Tertiary erosion, subsequent subaerial denudation being confined to landforms associated with drainage incision. Wooldridge and Linton also recognized a warped early Tertiary surface (the Sub-Eocene Surface) but confined its impact on landforms to the lowest Chalk backslopes adjacent to the present Palaeogene outcrop. They argued that if the surface seen emerging from under the Palaeogene deposits is projected over the Weald, it rapidly rises above the topographic surface, thereby indicating that it is of little significance in the modelling of the higher parts of the Chalk cuestas (Fig. 2). Such a view is obviously at variance with the conclusions of recent investigations, and two further lines of evidence can be advanced to militate against such an argument. First, mapping of the North Downs backslopes indicates that minor flexures are more widespread than previously envisaged and have imparted a hitherto unrecognized irregularity to the Sub-Palaeogene Surface (Docherty, 1967; John, 1974). Second, the cyclical nature of the Palaeogene sedimentary sequences preserved in the London and Hampshire Basins indicates several major fluctuations of sea-level, probably caused by tectonic movements. Under such circumstances, the resultant Sub-Palaeogene erosion surface would have been both polycyclic and diachronous (see later). Wooldridge and Linton's reconstruction, involving the projection of a simple planar surface, must, therefore, be considered increasingly inaccurate with distance from the surviving Palaeogene outcrop, for while they recognized the Eocene 'overstep' (i.e. the erosion of the Chalk) they failed to take sufficient note of the 'overlap'.

There is ample stratigraphic support for a diachronous interpretation of the Sub-Palaeogene Surface. In north Kent it is overlain by Thanet Sands (Palaeocene), while in north-west Kent, Thanet Sands, Woolwich Beds and Blackheath Beds—both the latter being considered as either Upper Palaeocene (Costa *et al.*, 1978; Fitch *et al.*, 1978) or basal Eocene (Odin *et al.*, 1978)—can all be found resting on Chalk (Docherty, 1967). Elsewhere around the Weald the Sub-Palaeocene Surface is overlain by Woolwich and Reading Beds. Although denudation has removed the Palaeogene cover from most of the Weald Arch, evidence from the opposite rim of the Hampshire Basin reveals Chalk directly overlain by London Clay near Cranborne and Bagshot Beds (Middle Eocene) at Hardy Monument (Dorset), while the variable dips visible at Creechbarrow suggest that both Bagshot Beds and Oligocene deposits (recently re-dated as mid-Eocene (Hooker, 1977)) overlapped the Lower Eocene to the south, so as to overlie the Sub-Palaeogene Surface above the Purbeck Downs (Arkell, 1947). It is reasonable to conclude from this evidence that the widely transgressing London Clay sea must have extensively trimmed the higher parts of the early Eocene land surface. Thus, stratigraphic arguments clearly support the existence of Sub-Palaeogene facets ranging in age from Palaeocene to mid-Eocene and possibly Oligocene.

In addition, there is geomorphological evidence for a compound Sub-Palaeogene Surface. Waters (1960) identified both subaerial Neogene and marine Palaeogene (Bagshot ≡ Middle Eocene) surfaces in south Dorset, while Green (1969, 1974) suggested the existence of Oligocene surfaces around the western rim of the Hampshire Basin. In the case of the Weald, Hodgson *et al.* (1974) have used the distribution of Clay-with-Flints to argue that the gross form of the South Downs backslopes is attributable to the Sub-Palaeogene Surface (Reading Beds facet). A similar explanation is emerging for the North Downs backslopes, including the 'Lenham' and 'Calabrian' benches (Docherty, 1967; John, 1974; Jones, 1980), and can be supported by evidence from the M20 excavations at Wrotham Hill (210-220 m), which revealed large sand bodies intimately associated with thick accumulations of Clay-with-Flints and Plateau Drift and resting on an irregular, piped Chalk surface. The survival of these sands indicates that the Sub-Palaeogene Surface lay at no great height above this relatively elevated part of the North Downs. Finally, Clark *et al.* (1967) have interpreted the gross morphology of the Marlborough Downs as largely the product of a 'composite Eogene cycle' on the basis of an examination of sarsens (relic blocks of sandstone usually composed of Reading Beds material and strongly cemented with silica and/or iron). Sarsens are widespread on the chalklands of southern England, including the North and South Downs, and are considered to be remnants of formerly extensive sheets of silcrete.

As this type of duricrust usually develops under tropical climates with a markedly seasonal rainfall, it is thought unlikely that it formed during the warm and moist Neogene. Most workers, therefore, support a Palaeogene origin, either penecontemporaneous with Reading Beds deposition, Middle Eocene (Bagshot), or Upper Eocene-Oligocene (Clark *et al.,* 1967). The widespread preservation of sarsens on the Marlborough Downs and Chiltern Hills thus indicates the former existence of Palaeogene landsurfaces—and therefore the Sub-Palaeogene Surface—at no great elevation above the present summits.

The increasingly numerous reports of the survival of Palaeogene sediments and surfaces on the higher chalklands casts severe doubt on both the efficacy of Neogene subaerial erosion and the concept of a Mio-Pliocene peneplain. In reality the elevated parts of southern England do not show the high degree of summit conformity that has so often been claimed, and even Wooldridge and Linton (1955) were forced to suggest that many areas over 250 m, such as Butser Hill (271 m), were residual monadnocks rising above the main surface. Examination of the South Downs (Small and Fisher, 1970) has indicated that the early Pleistocene landscape was already dissected to 90-120 m O.D., a conclusion that led Jones (1974) to suggest that the late Neogene landsurface may have had a regional relief of over 200 m and a local relative relief of 180 m. This is in no way compatible with the low relief surface envisaged by Wooldridge and Linton.

Morphological evidence in support of a Summit Peneplain is not well displayed around the Weald, the 'type area' being on the flanks of the Vale of Wardour. Here the Chalk appears to be extensively bevelled at 180-260 m, above which rise a number of monadnocks, such as Long Knoll (288 m). Green (1974) has reinterpreted the landscape components in this area, replacing the 'Monadnock Group', 'Mio-Pliocene Peneplain' and 'Calabrian Bench' (Wooldridge and Linton, 1955) with 'Watershed Residuals' (>260 m), 'Summit Peneplain: Higher Surface' (260-180 m) and 'Summit Peneplain: Lower Surface' (240-140 m), which he dates as Sub-Palaeogene (Oligocene?), Miocene and Pliocene respectively. The identification of two Neogene surfaces, whose creation was separated by a phase of uplift and warping in the late Miocene/early Pliocene, is further evidence against the late Tertiary peneplain model. Although there is still disagreement as to the extent to which Neogene erosion modified the Sub-Palaeogene Surface, with Green (1974) arguing for moderate erosion while Pinchemel (1954) and Small (1964) postulated minor incision, the important point is that the Sub-Palaeogene Surface is seen as having a significant influence on the present topography in what has hitherto been considered a classic area for demonstrating the character of the Mio-Pliocene peneplain.

It is now abundantly clear that stratigraphic, sedimentological and morphological arguments all support the abandonment of the Mio-Pliocene peneplain concept and its replacement by a model in which Neogene subaerial denudation led to the stripping away of the Palaeogene cover and incision into the underlying Mesozoic rocks. The recognition of Neogene surfaces in France (Cholley, 1957) and Wiltshire (Green, 1974) suggests that they may also have been developed around the Weald.

However, there is no evidence to support the former existence of an extensive low relief peneplain or 'Summit Surface'.

## Tectonic considerations

Wooldridge and Linton's interpretation of structural evolution envisaged the Tertiary as characterized by tectonic stability, except for two periods of disturbance associated with the Upper Cretaceous emergence and a brief mid-Tertiary (late Oligocene-early Miocene) 'Alpine' orogenic episode (Fig. 1). The early phase was considered to have resulted in the initial development of the macrostructural pattern of the London Basin, Weald-Artois Anticline and Hampshire-Dieppe Basin. There then followed a long period of relative quiescence until the mid-Tertiary, when new compressional movements enlarged the pre-existing flexures to their present amplitude (> 1600 m in the case of the Weald-Artois Anticline), while at the same time superimposing numerous minor east-west oriented periclines and faults (Fig. 3). This view can no longer be supported, for it is now recognized that southern England was located equidistant from three major disturbance zones—the expanding North Atlantic Basin, the subsiding North Sea Basin and the evolving Alpine mountain system—all of which have been active since the Lower Cretaceous (Read and Watson, 1975). Under such circumstances, it is logical to envisage that tectonic movements occurred continuously throughout the Tertiary, although the scale of activity probably varied depending on the interplay of the three causative mechanisms.

A second fallacy concerns the traditional over-emphasis of a brief 'Alpine' folding phase in the development of both macro- and microstructural patterns. This arose largely because of difficulties encountered in accurately dating the surface folds, a task normally accomplished by measuring the variable effect of flexuring on strata of differing ages and facies changes in sediments. The restricted occurrence of deposits dating from the last 40 Ma has therefore proved a major problem in southern England, especially as the classic 'Alpine' episode was envisaged as 30-20 Ma. This problem is particularly acute in the Weald where most flexures can only be dated as post-Lower Cretaceous or post-Chalk. Nevertheless, the belief that the periclinal folds were minor compressional structures created as the 'outer ripples of the Alpine storm' (Wooldridge and Linton, 1955, p. 14), together with their dating as mid-Tertiary, based on the flexuring of the Oligocene Hamstead Beds of the Isle of Wight, has led to a general tendency to ascribe fold development to this episode because of lack of evidence to the contrary. This view can now be discounted for three reasons:

(1) Studies of Alpine structural evolution have shown it to be exceedingly complex and both spatially and temporally variable, with movements occurring over the period late Jurassic-Pleistocene (Ager, 1975). In the western Alps, compression is thought to have led to pulsed activity between the mid-Cretaceous and Oligocene, followed by more local epeirogenetic movements in the Neogene and Quaternary (Read and Watson, 1975). This suggests that southern England could have been

affected by compressional stresses for a period of 70 Ma, including the whole of the Palaeogene, but ceasing by about 20 Ma. Thus, to emphasize a mid-Tertiary movement phase is, as George (1974, p. 117) remarked, "to be highly selective in the Cenozoic time-span".

(2) The fold pattern was explained as the product of northward directed compressional movements, thereby clearly implying an 'Alpine' origin. Variation in the size and trend of surface periclines was seen as a function of the differential influence of ancient structures in the buried Palaeozoic rocks, as determined by the thickness of the Mesozoic cover. Thus, the minor and variably oriented flexures of the London Basin (Fig. 3) were interpreted as the result of posthumous movements of faults in the shallowly covered Palaeozoic Floor, while the larger folds of the Weald and Hampshire Basin, with their similarity of trend (Fig. 3) and slight north-south asymmetry were seen as the product of deformation of the Mesozoic cover independent of Hercynian structures in the more deeply buried Palaeozoic rocks (see Fig. 4). This elegant analysis must now be discounted. The similarity of the

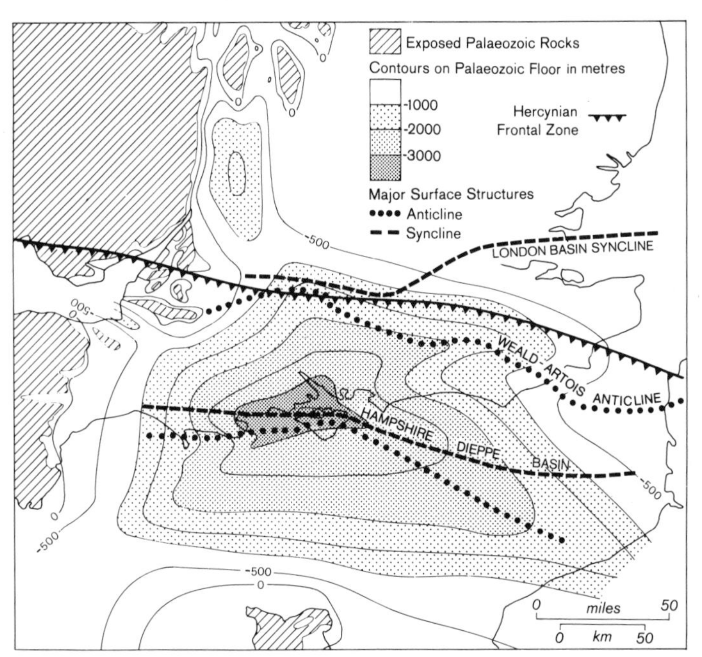


*Figure 4.* Relationship of surface macrostructures to the form of the Palaeozoic Floor and the Hercynian Front. Note that the Weald-Artois Anticline forms part of the 'tectonic divide' (Jones, 1974) which continues to the west in the 'Mendip Line'.

supposed 'Alpine' trend to the expected 'Hercynian' trend, the *en echelon* relationship of periclines, and their arrangement in three belts (Fig. 3) which continue to the west in observable Hercynian structures, all indicate the widespread influence of 'basement control'. In fact, the fault-bounded periclines of the Central Weald, which form a series of 'horsts' and 'grabens', can only be explained by invoking 'basement control' (Howitt, 1964; Gallois, 1965; Shephard-Thorn *et al.*,

1972; Lake, 1975) and has led to the suggestion (Lake, 1975, p. 554) that "offset horst structures may be represented at a higher structural level by offset open-fold structures". Further, the 'Mendip Line' (Fig. 3) appears intimately associated with the Hercynian Front (Fig. 4) while the relationship of the 'Quantock Line' and 'Southern Fold Belt' to the form of the Palaeozoic Floor (Fig. 4) indicates that they probably overlie Hercynian thrust zones. Thus the overall fold and fault pattern of southern England is increasingly recognized as reflecting deep-seated movements of ancient (Hercynian) structures which could have been activated at any stage of the Palaeogene.

(3) Exploration of the English Channel has revealed that the Dieppe Basin is a tectonically smooth structure (Curry and Smith, 1975; Smith and Curry, 1975). The location of this simple elongated flat syncline immediately south of the Weald-Artois anticlinorum (Fig. 2) strongly supports the view that secondary folds are the product of deep-seated movements of fault lines associated with the Hercynian Frontal Zone (Fig. 4) rather than compression of surface strata.

These arguments show that, while the development of the Alps was of importance in the structural evolution of southern England, it was only one of the tectonic influences that affected the area during the whole span of the Palaeogene. Further, the recognition of widespread 'basement control' indicates that surface flexuring could be the product of movements from a variety of directions and that firm evidence for the dating of a particular fold may be used to indicate the age of other folds on the same structural line. Arguments in favour of longer-term fold development have already been advanced by George (1974) and Jones (1974), and the presently available evidence for flexuring in the Weald and adjacent areas can be summarized as follows.

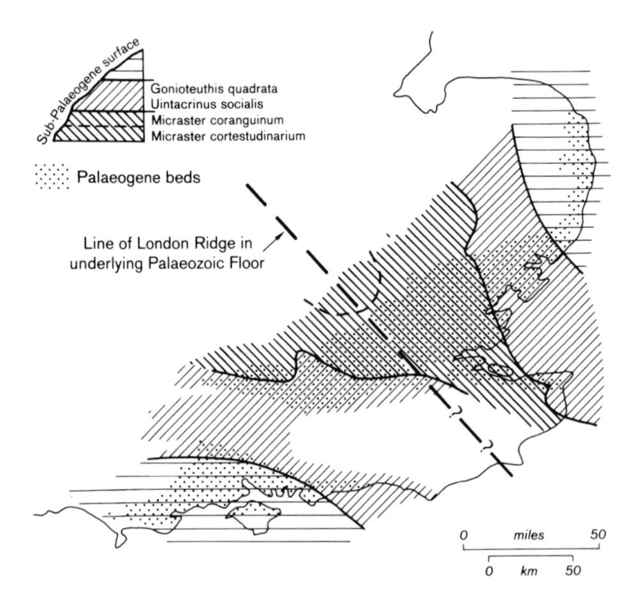


*Figure 5.* The pattern of early Palaeogene denudation as shown by the sub-crop of Upper Chalk zones onto clearly defined elements of the Sub-Palaeogene Surface (after Curry, 1965).

(1) Late Lower Cretaceous movements have been reported for the Boulogne area (Smith and Curry, 1975) as well as from the western Hampshire Basin (Arkell, 1947), where they appear to have been of considerable magnitude. The evidence from the Weald is less impressive, but nevertheless important movements took place during Lower Greensand times (Casey, 1961; Middlemiss, 1967, 1975), particularly along the Hog's Back Fault. During Gault times there was further vertical faulting along the 'Mendip Line' (Owen, 1971a) together with gentle NW-SE flexuring (Owen, 1971b, 1975).

(2) There is little evidence for differential movements during the early Upper Cretaceous, but every sign that an important phase of uplift and flexuring began in the Maastrichtian (70 Ma) (Hancock, 1975) and continued through the Palaeocene. Examination of the Sub-Palaeogene Surface in the London Basin reveals that Chalk, laid down beneath a sea 100-600 m deep (Hancock, 1975), was uplifted and eroded so that up to 250 m had been removed by the end of the Palaeocene (Wooldridge and Linton, 1938b; Curry, 1965). Taking 200 m as a generally accepted depth of the Chalk sea, this implies vertical movements of 450 m in 15 Ma along a NW-SE axis (Fig. 5).* Local flexuring also occurred, for it has been shown that the Purbeck Monocline (Arkell, 1947) and Dean Hill Anticline (Williams-Mitchell, 1956) began developing during the Palaeocene.

(3) Differential movements continued through the Palaeogene, as is testified by the unconformable relationships and thickness variations of the Palaeogene beds (Curry, 1965). The general lack of sediments of Oligocene age suggests that there was widespread uplift and broad-scale flexuring, while the further growth of the Purbeck fold (Arkell, 1947) indicates that minor structures continued to develop.

(4) Support for Neogene movements is more difficult to find because of the 'stratigraphical gap'. While there is certainly widespread evidence of both macro- and micro-flexuring in post-Palaeocene times, both on land and beneath the Channel (Smith and Curry, 1975), stratigraphic arguments that such movements took place in the Miocene, or later, depend entirely on the evidence of folded Oligocene beds in the Isle of Wight. Similarly, the geomorphological arguments for Neogene movements put forward by Clark *et al.* (1967) and Green (1974) must be recognized as having no firm stratigraphic basis. This is not to say that the Neogene was characterized by stability, but rather that there is little firm evidence for movement and reasonable grounds for questioning the relative importance of this episode. The most logical interpretation is that the macrostructural pattern continued to develop, possibly accompanied in the early Miocene by the further growth of minor structures along the southern rim of the Hampshire Basin.

(5) There is widespread support for Quaternary deformation but uncertainty as to the magnitude. It has been argued (McGinnis, 1970; Worssam, 1973; Jones, 1974) that denudational unloading of the Weald could have resulted in isostatic movements, or what Jones (1974) termed 'post-orogenic flexuring'. Such movements

* The interpretation adopted in this chapter is that the Chalk suffered subaerial denudation prior to marine trimming in the Palaeogene. Curry *et al.* (1978) favour marine erosion and would reduce the scale of uplift.

could have operated at both regional and local scales, possibly assisted by diapiric movements of evaporites contained in the Mesozoic sequence. Their impact would have been superimposed on the general north-eastward tilt produced by the continued subsidence of the North Sea Basin (West, 1972; George, 1974), which has deformed the onshore Sub-Pliocene Surface by over 250 m. Most estimates have been based on the elevation of Plio-Pleistocene marine deposits, and suggest uplift of the North Downs by 200 to 270 m; Jukes-Browne (1911) went further and postulated a relative uplift of the central Weald by an additional 172 m, a view supported by Worssam (1973). However, uncertainty as to the age and height of this incursion, together with a lack of knowledge as to the eustatic component of change, all make computation of isostatic deformation difficult.

It is possible to draw three general conclusions concerning the structural evolution of southern England. First, flexuring and warping have occurred over the last 90 Ma, mainly through re-activation of pre-existing fault systems in the Palaeozoic Floor by stresses generated by the Atlantic Ocean Rift and Alpine orogeny, with some warping caused by diastrophism of the North Sea Basin. The lower Palaeogene flexuring increases significantly to the west, both in Dorset (Arkell, 1947) and the English Channel (Smith and Curry, 1975), and indicates that early movements were probably caused by the Atlantic Rift.

Second, the structural pattern has clearly changed over time in response to variations in stress generation. The observed pattern of pre-Eocene denudation (Fig. 5) clearly shows a dominant early Tertiary flexure with a NW-SE 'Charnian' trend—apparently related to the 'London Ridge' or 'Nuneaton Line' feature of the Palaeozoic Floor, noted by Wooldridge and Linton (1938b)—which transversely crosses the present macrostructural pattern. Owen (1975) has produced evidence for similarly oriented flexuring in the Lower Cretaceous, which suggests that late Mesozoic movements, including the highly significant Upper Cretaceous emergence, were dominated by the Atlantic Rift System and most particularly that branch of the developing Oceanic Fracture Zone which led to the opening of the Bay of Biscay (the Biscay Rift). The impact of this lineation can still be distinguished, for the Weald-Artois Anticline is not a simple structure but can be divided into three segments (Fig. 4), the two extremities having an E-W 'Alpine' (Hercynian?) trend while the central section, east of a line Lewes-Medway Gap, is oriented NW-SE in sympathy with the continuation of the 'London Ridge' feature (Fig. 6). This latter trend is distinguishable in the outcrop patterns (especially the Gault), in the orientation of minor structures in the Central Weald and further north, as well as in the buried structures of the Palaeozoic Floor revealed in the Kent coalfield (Lake, 1975). However, the stratigraphy of the early Palaeogene indicates that this lineation was superseded, in the Palaeocene, by E-W flexuring, for the Thanet Sand and later incursions appear to have invaded an ancestral London Basin which had developed across the earlier 'Charnian' arch. Such a change suggests that 'Alpine' generated stresses became dominant at this time.

The third conclusion relates to the significance of Palaeogene folding. It is clear from inspection of the Sub-Palaeogene unconformity that the period 70-60 Ma was

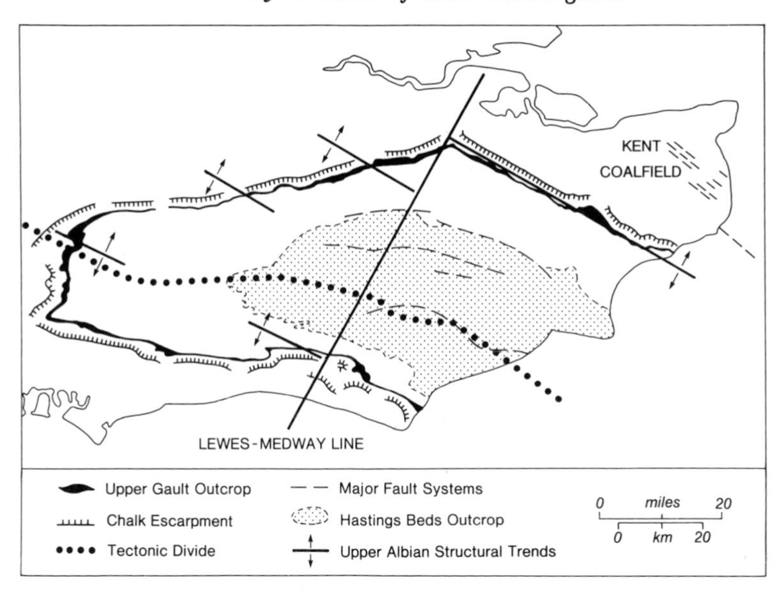


*Figure 6.* Selected structural lines in the Weald to show the importance of a NW-SE trend and the subdivision based on the Lewes-Medway line.

characterized by massive uplift and macro-flexuring along NW-SE and W-E axes. It is also clear that the growing influence of 'Alpine' movements in the Palaeocene led to the development of small-scale surface periclines (Arkell, 1947; Williams-Mitchell, 1956). Thus, there is no reason why both micro- and macrostructures should not have continued to develop during the remainder of the Palaeogene; in fact the re-dating of the 'Alpine' orogenic phase as mainly pre-Miocene, together with the recognition of the tectonic significance of the Atlantic Rift in the late Cretaceous and the relatively great length of the Palaeogene as compared with Neogene, all indicate that much of the present flexure pattern could have developed by the end of the Oligocene, with later movements largely restricted to broad-scale flexuring, possibly assisted by isostatic movements, and downwarping towards the North Sea Basin.

This interpretation of tectonic evolution clearly differs from that of Wooldridge and Linton, in that fold growth is seen as a long-term process, which took place earlier in the Tertiary (Palaeogene) through the complex interaction of stresses from two sources. No brief 'Alpine storm' can be recognized and thus the hypothesis that mid-Tertiary folding led to drainage derangement can no longer be supported.

## Towards a new model of Tertiary landscape evolution

It should be clear that the long-favoured Wooldridge and Linton model must now be abandoned in favour of a reconstruction that places greater emphasis on the creation of a compound Sub-Palaeogene Surface—here defined as the eroded surface of Mesozoic strata created during the Palaeogene. The view that this surface lay slightly above the present Chalk summits requires both a dramatic increase in the scale of denudation attributable to the Palaeogene, as well as the development of a new model in which the Sub-Palaeogene Surface bends over the Chalk cuestas so as to pass above the Weald at a lower elevation than the 800-900 m envisaged by Wooldridge and Linton (Fig. 2).

## Models of Palaeogene denudation

The Palaeogene deposits of southern England are the main evidence for arguing that the Sub-Palaeogene unconformity was a multi-faceted polygenetic diachronous surface which had an initial variability of slope that has subsequently been exacerbated by differential flexuring and warping. These complex sequences of inter-digitated fluvial and marine sediments were originally deposited on the western margin of the main Palaeogene sedimentary basin (the Anglo-Paris-Belgium Basin). The recognition of seven sedimentary cycles (Stamp, 1921) clearly indicates repeated fluctuations in sea-level, now mainly attributed to tectonic causes (Curry, 1965), which resulted in considerable shifts in coastline position. However, as palaeogeographical reconstructions show that marine influences were largely confined to the London and Hampshire-Dieppe Basins (see Hester, 1965), it follows that the present macrostructural pattern had been outlined by the mid-Palaeocene. Thus the Palaeogene sea first invaded the London Basin in the Thanet Sand (Palaeocene) transgression (60 Ma, Fitch *et al.,* 1978), the Dieppe Basin and the floor of the Hampshire Basin not being affected until the subsequent Woolwich/Reading Beds incursion (55 Ma) (Hester, 1965), while the Hampshire Basin was not wholly submerged until London Clay (early Eocene) times (53 Ma). The Palaeogene deposits not only overlie eroded Chalk but also display internal discontinuities and overlap relationships indicative of alternating phases of erosion and deposition. These characteristics are central to the polycyclic interpretation of the Sub-Palaeogene Surface, for they point to the continuation of differential movements (Curry, 1965) and spatial variation in the balance between erosion and deposition. The thickness of the Palaeogene sequences preserved in the London, Hampshire and Dieppe Basins (240 m, 730 m and 380 m respectively) clearly indicates that deposition exceeded erosion in these subsiding areas. This conclusion is supported by sedimentation rate calculations for the thick shallow-water deposits of the Hampshire Basin, which suggest continuous or intermittent subsidence from the late Palaeocene to early Oligocene (Odin *et al.,* 1978). However, stratigraphic and lithological characteristics indicate a complex situation in which sediment accumulation in basins occurred contemporaneously with subaerial denudation of adjacent upstanding areas undergoing uplift. Such differential movements resulted in repeated marine incursions into low-lying areas, the sea planing across recently deposited sediments on basin flanks—thereby producing the disconformities—so as to erode adjacent outcrops of Mesozoic strata. Thus a diachronous Sub-Palaeogene Surface continued to evolve as long as Mesozoic rocks were exposed to denudation. That this was indeed the case for a prolonged time period is shown by the identification of Thanet Sands, Blackheath Beds and Woolwich/Reading Beds resting on Chalk in the London Basin, while in the Hampshire Basin there is overlap by the London Clay and Bagshot Beds (middle Eocene) and indications that the same was formerly true of the Creechbarrow Limestone, for long considered of Oligocene (Arkell, 1947) but recently re-dated as mid-Eocene (Hooker, 1977). The Sub-Palaeogene Surface is thus clearly diachronous, with firm stratigraphic evidence for an age range of 60 Ma to 43 Ma and possibly 33 Ma (Fitch *et al.*, 1978), a span of 17-27 Ma. Even the more condensed time-scale of Odin *et al.* (1978) gives figures of 14-24 Ma (Curry *et al.*, 1978).

Although this interpretation fits the evidence in western areas, its application to the Weald depends on evidence of the early initiation of a flexure that evolved into the Weald-Artoise Anticline. Opinions differ markedly on this matter. Early doming was advocated by Prestwich (1852), Stamp (1921), Wooldridge and Linton (1938b, 1939, 1955) and George (1974), but rejected by Davis and Elliot (1957) and Curry (1965) on sedimentological grounds, while recent structural studies (Terris and Bullerwell, 1965; Lake, 1975) have been equivocal. On balance, arguments in favour of an early Palaeogene upwarp appear in the ascendency, but depend heavily on the evidence of variations in thickness of preserved Chalk and the zonal subcrop pattern onto clearly defined elements of the Sub-Palaeogene Surface.

The thickness of Chalk varies in a general north-east direction, declining from 420-500 m along the southern rim of the Hampshire-Dieppe Basin, to 300-350 m on the South Downs and 170-300 m in the London Basin, before increasing again to about 400 m in Norfolk (Curry and Smith, 1975; Hancock, 1975). Such variation cannot be explained by invoking differences in sedimentation rates or contemporaneous erosion and appears largely the product of post-Cretaceous denudation (Hancock, 1975). This view is supported by the distribution of Chalk zones (Fig. 5), which shows a similar pattern, with the oldest zones underlying the Sub-Palaeogene unconformity in the London Basin and then decreasing in age to both north-east and south-west (Curry, 1965). Thus, up to 250 m of Chalk appears to have been removed *by* the Upper Palaeocene (*c.* 55 Ma) along an axis of uplift with a Charnian (NW-SE) trend. While this feature was recognized by Wooldridge and Linton, it can now be extended south-eastwards on the basis of evidence from Chatham and Warlingham (Curry, 1965; Gallois, 1965) and the relatively small thickness of Chalk (250 m) reported beneath the Channel south-east of Hastings (Curry and Smith, 1975).

Two further lines of evidence support an early Palaeogene flexure. First, although the angular discordance between the Palaeocene and Chalk is small on the southern flank of the London Basin, the overstep rate of 6 m km$^{-1}$ (Wooldridge and Linton, 1955) indicates breaching of the Chalk above the Central Weald. Second, the Palaeogene sedimentary sequence is characterized by sands, gravels and clays, with limited limestone development restricted to the Hampshire Basin, a remarkable situation considering that it was a Chalk landsurface that was being denuded. While it is probable that the calcium carbonate was removed in solution and deposited elsewhere, it is still necessary to find a source for the arenaceous and argillaceous sediments. Mineralogical analysis indicates supply from the west, thereby implying rapid and widespread denudation of the Chalk cover. However, the occurrence of Lower Greensand chert in the Bagshot Beds of the London Basin (Dines and Edmunds, 1929) suggests unroofing of the Weald by the late Palaeocene, and certainly by the mid-Eocene, thereby implying the existence of an upwarp.

These arguments, supported by palaeogeographical reconstructions of the Palaeogene and the evidence for long-term pulsed tectonism, all suggest the early creation and continued development of a structural 'high' in the Weald area. However, there are indications that this feature changed shape over time as the structural framework evolved. Thus the late Cretaceous 'Charnian' arch was disturbed by complex vertical movements, beginning in the Palaeocene, which resulted in the present macrostructural pattern. Evidence for such movements is preserved in the London Basin where differential subsidence led to the creation of basins on either side of the 'Charnian axis', as revealed by the form of the Sub-Palaeogene Surface and isopachytes of the London Clay (Wooldridge and Linton, 1938b). A similar situation may be envisaged for the Weald where a portion of the original arch was modified by later E-W oriented uplifts.

The stratigraphy of the Upper Palaeocene and Eocene shows progressive inundation westwards, culminating with submergence of all areas east of a line Lyme Bay-Wash by the London Clay sea (Davis and Elliot, 1957). Although London Clay has only been found resting directly on Chalk near Cranborne, there is little doubt that a marine-trimmed surface was widely developed on the Mesozoic rocks forming the relatively elevated rims of the structural basins and Weald. The London Clay itself was deposited extensively, possibly as a continuous sheet, over the whole area. Up to 180 m has been recorded in the eastern London Basin but it thins and becomes sandier westwards. While the existence of a Weald Island in the London Clay sea (Prestwich, 1852; Stamp, 1921) has been refuted by sedimentological studies, which have shown the formation to be lithologically uniform and deposited in 180-360 m of water (Wrigley, 1940; Davis and Elliott, 1957), it seems probable that only a thin cover was laid on the developing Weald Arch and was easily removed on emergence. The subsequent Bagshot incursion is of uncertain extent, after which subaerial conditions progressively returned to the whole area, except for the Hampshire-Dieppe Basin where marine episodes continued for another 10 Ma. Although Bagshot Beds rest on Chalk in Dorset and the incursion appears of landscape significance further to the west (Waters, 1960), nowhere in the London Basin is this formation seen to overstep the London Clay; the inclusion of Lower Greensand chert being taken here to indicate a land mass located over the Weald.

The Sub-Palaeogene Surface thus underwent two distinct evolutionary phases. During the Lower Palaeogene, and culminating with the maximum extent of the London Clay (Ypresian) sea at about 49 Ma (Fitch *et al.*, 1978), a diachronous marine-trimmed unconformity was developed over the whole of southern England which decreased in age westward as well as towards the crest of the Weald Arch. Although this marine unconformity continued to develop in Dorset until Bagshot times and possibly into the Oligocene, elsewhere, the return of subaerial conditions led to the exposure of a blanket of unconsolidated Palaeogene sediments and their subsequent denudation, so that higher parts of the previously created marine unconformity were revealed and then destroyed during the development of a late Palaeogene land surface. Thus, the Sub-Palaeogene Surface continued to evolve over the Weald while in the adjacent basins it was buried and fossilized. The London Clay marine incursion (53-49 Ma) is therefore a fundamental dividing point in the Tertiary evolution of the Weald, for it not only separates the two main phases of denudation but also marks the point where the Sub-Palaeogene Surface ceased to evolve through marine agencies and began to develop under subaerial conditions—conditions which were to dominate the next 49 Ma except for a possible brief interruption in the Pliocene.

The stripping away of the thin and unconsolidated Palaeogene cover from the Weald proceeded rapidly, assisted by flexuring, thereby exposing the underlying rocks to denudation. Certainly the presence of Lower Greensand chert in the Bagshot Beds of the London Basin indicates that the London Clay cover was indeed quickly removed and erosion had cut deeply into the Mesozoic sequence. The elevation of the Sub-Palaeogene Surface over the Weald axis will probably never be known accurately and estimates will vary depending on calculations as to Palaeogene denudation rates, the thickness of early Palaeogene deposits and the degree of flexuring accredited to mid-Tertiary movements. Nevertheless it appears reasonable to suggest that the remaining 26 Ma of the Palaeogene witnessed considerable sculpturing of Mesozoic strata so as to yield a surface with an elevation below that postulated by Wooldridge and Linton based on the projection of marine-trimmed facets (Fig. 2). Comparative evidence from the western rim of the Hampshire Basin, where Oligocene surfaces have been tentatively recognized at 250-300 m (Arkell, 1947; Green, 1969, 1974), together with the recognition of the increased scale of Palaeogene tectonic movements, suggest that the Sub-Palaeogene Surface probably lay at 350-450 m over the Central Weald and was largely

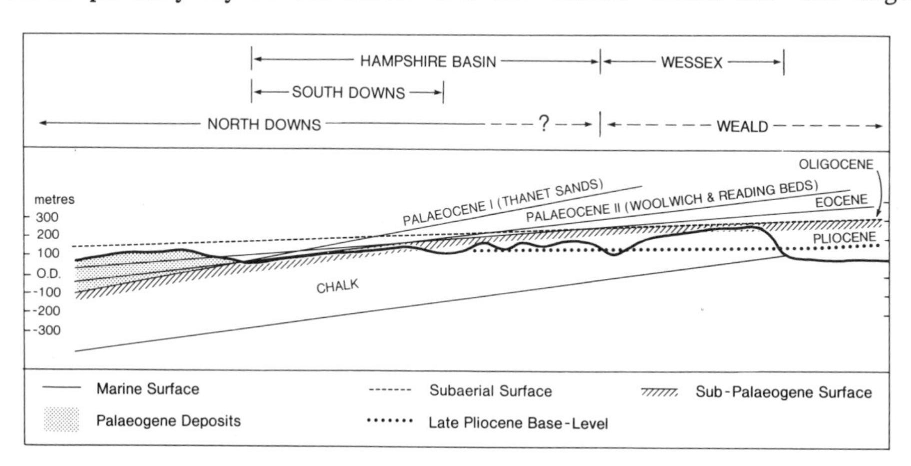


*Figure 7.* Diagrammatic representation of the multi-faceted polygenetic diachronous Sub-Palaeogene Surface. The bars at the top of the diagram show how recently advanced erosion surface interpretations of the chalklands can be interrelated by employing this model.

developed on relatively weak Lower Cretaceous strata (e.g. Weald Clay) (see Fig. 9).

The nature of the envisaged diachronous multi-facet Sub-Palaeogene Surface is illustrated in Fig. 7. It is clear that the number, inclination and age of facets will vary from area to area depending on the interplay between the pattern of Palaeogene erosion and the subsequent tectonic and denudation history. Nevertheless, use of the general model makes it possible to relate the three apparently contradictory erosion surface interpretations that have recently been advanced for the chalklands. Thus the view (Hodgson *et al.,* 1974) that the South Downs backslopes represent the Reading Beds facet is compatible with both the more complex interrelationship reported for the North Downs (Docherty, 1967) and the reinterpretation of the Wardour area as Miocene and Pliocene surfaces with residual remnants of a higher Oligocene level (Green, 1974).

Employment of this model also resolves the fundamental dichotomy between the view that the Sub-Palaeogene Surface lay close above the chalkland summits (Pinchemel, 1954; Green, 1969; Hodgson *et al.,* 1974; Catt and Hodgson, 1976) and the Wooldridge and Linton interpretation so aptly expressed in the statement (Linton, 1964, p. 115) that "the recent proximity of the Eocene base is readily disproved by the fact that these crests fall much lower in the Senonian zonal sequence than that outcropped locally on the Sub-Eocene plane". It can now be argued that the classical interpretation was based on the projection of clearly identifiable erosional planes adjacent to Palaeogene outcrops and so utilized some of the oldest and most markedly tilted elements of this compound surface (Fig. 7).

This model can also be extended to incorporate the view that the general flattening of the upper parts of many cuesta backslopes, together with the conformity of summit and ridge levels, may be due, in part, to solutional lowering following the removal of the Palaeogene cover. The possible effect of solution on Chalk surfaces has received scant attention, except in the case of the Lenham Surface (Bisson in Smart *et al.,* 1966), even though work elsewhere (Pitty, 1968) has shown that differential lowering can account for the creation of conformable summits previously interpreted as dissected subaerial peneplains. The same general argument can be adopted here by postulating that progressive stripping of the Palaeogene cover revealed Chalk which then suffered solutional lowering. As the magnitude of lowering would be related to length of exposure, it follows that the higher parts of the Chalk cuestas would suffer the greatest reduction, especially as the rate of exhumation would diminish towards basin areas because of the increasing thickness of Palaeogene sediments. This model is illustrated in Fig. 8 and clearly shows how the overall convex form of many cuesta backslopes, the often noted increasing thickness of 'Clay-with-Flints' towards the crestline (Avery, 1964) and the apparent departure of the topographic and Sub-Palaeogene surfaces noted in Fig. 7 and focused upon by Wooldridge and Linton (1955), can all be explained without recourse to Neogene peneplanation.

It must be noted here that not all the major features of chalkland physiography are explicable by this model. Differential erosion of Upper Chalk zones is obviously essential in the production of the secondary escarpments of the South Downs and

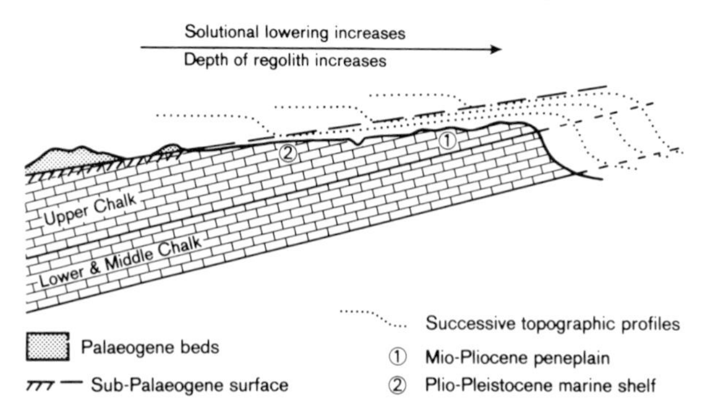


*Figure 8.* Diagrammatic representation of hypothetical chalkland area to show how summit conformity and the departure of topographic and Sub-Palaeogene surfaces can be explained by retreat of Palaeogene beds and solutional lowering of exposed Chalk.

Hampshire Basin (Small and Fisher, 1970; Hodgson *et al.,* 1974) and the related stripped stratum planes or *Micraster* surfaces so excellently developed on Salisbury Plain (Small, 1964). The distribution of highest ground, including the residual hills, also poses a problem, for many cuesta backslopes display a marked topographic rise near the crestline. Some of these were interpreted as 'Calabrian' clifflines and may now be explained as the intersection of facets of the Sub-Palaeogene Surface (Hester, 1965) or the result of monoclinal folds (Docherty, 1967; John, 1974). In other areas, such as Wardour, similar features have been interpreted as the junction between Sub-Palaeogene and Neogene surfaces (Green, 1974). A further explanation of this 'terminal rise' is an upward bending of strata near escarpments. The association of minor monoclinal flexures and Chalk escarpments is a common feature of geological interpretations, particularly around the Weald (Jones, 1974). Although invariably attributed to tectonic deformation in the Miocene, their distribution and variable trend indicate that they may be a product of Pleistocene isostatic movements caused by erosional unloading, a mechanism which has been described as 'non-uniform uplift' (Worssam, 1973) and 'post-orogenic flexuring' (Jones, 1974). These arguments suggest that the presence of high crestlines does not pose a major obstacle to the model's general applicability.

Acceptance of this model depends upon support for the view that Palaeogene denudation was significantly greater than has hitherto been generally appreciated. The subaerial interval separating the late Cretaceous emergence (70 Ma) from the deposition of the Thanet Sands is calculated as 10 Ma (Fitch *et al.,* 1978) to 15 Ma (Curry, 1965; Odin *et al.,* 1978). The diachronous nature of the Sub-Palaeogene Surface indicates that the period during which the sculpturing of Mesozoic rocks *could* have occurred ranges from 10-20 Ma within the structural basins to about 47 Ma on elevated areas, such as the Weald. This is a substantial interval and suggests severe denudation under the prevailing conditions of instability, but it is difficult to verify because of the lack of sediments dating from the Upper Palaeogene and Neogene. However, the Chalk cover is known to have been breached over much of south-western England (George, 1974) and in Purbeck

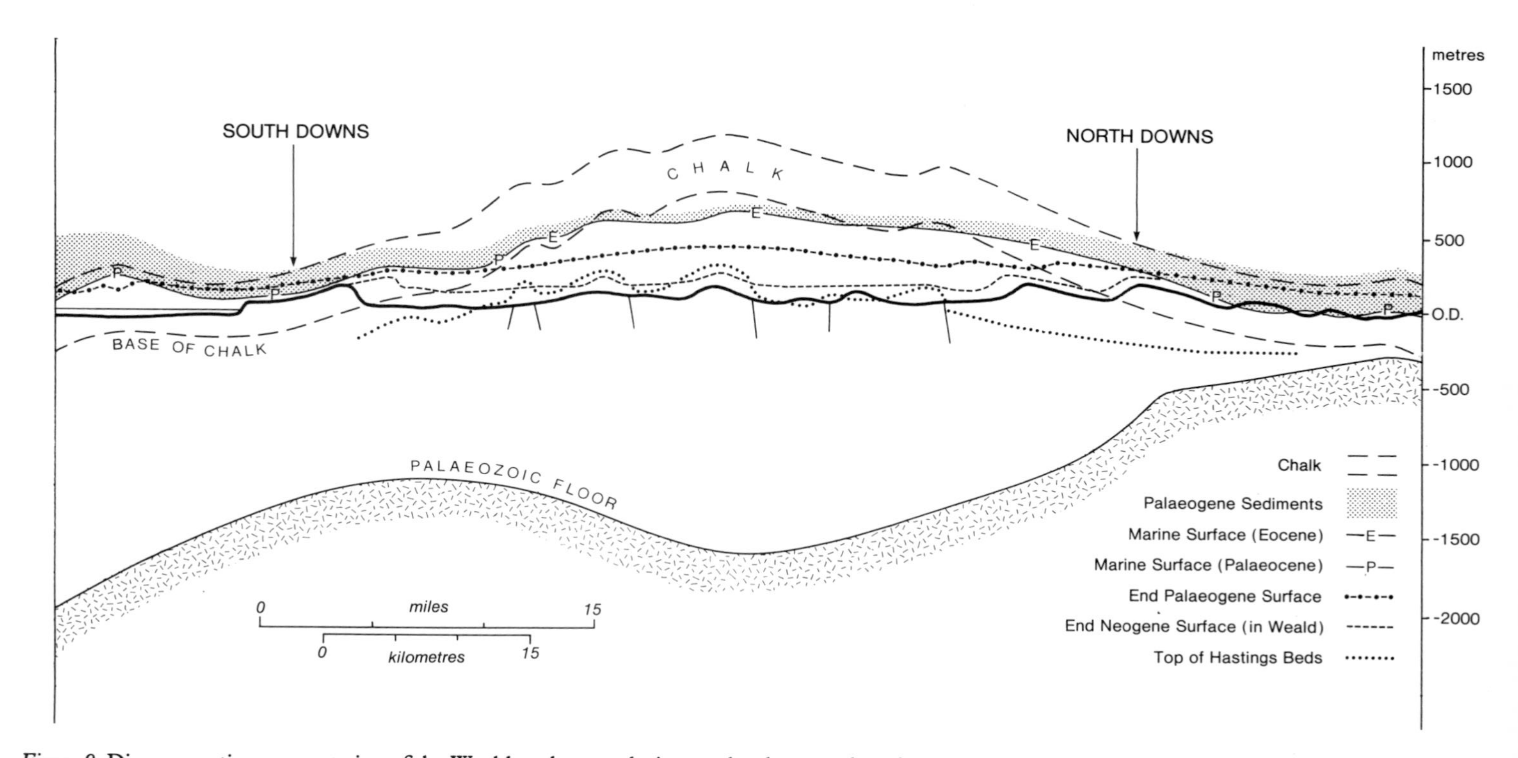


*Figure 9.* Diagrammatic representation of the Weald to show evolutionary development based on arguments presented in this chapter. Line of section runs N-S and passes close to Dartford, Sevenoaks and Beachy Head.

(Arkell, 1947) by the mid-Eocene. The same appears to be true of the Weald, for the evidence of the Sub-Palaeogene Surface (Fig. 5) and the rate of Palaeogene overstep both indicate breaching of the Chalk by the Eocene. While up to 250 m of Chalk was removed from the London Basin as a result of pre-Eocene denudation, the figure probably rose to 450 m over the Weald because both the Late Cretaceous and Palaeogene uplift axes crossed in this area. This figure is supported by the evidence of the Bagshot Beds which indicates denudation to 550 m by the mid-Eocene.

Assuming that the crest of the Weald-Artois structure was shaved across by the London Clay sea, that the London Clay cover was thin, and that denudation continued for the remaining 26 Ma of the Palaeogene, it is not difficult to envisage the removal of the additional 250 m of strata required by the interpretation shown in Fig. 9. This works out at about 10 mm per 1000 years, a relatively low figure compared with recent estimations of mean denudation (Schumm, 1963; Young, 1974) and less than most calculated rates for solutional lowering of limestone terrains (Pitty, 1968; Young, 1974).

## Conclusion

The model proposed in this paper is a response to the growing view that the chalklands are a palimpsest of surfaces developed over the whole of the Tertiary. It represents a viable alternative to the Neogene dominated model of Wooldridge and Linton, now largely rejected, from which it differs by emphasizing the importance of Palaeogene denudation in the creation of a variably warped surface, here termed a multi-faceted diachronous polygenetic Sub-Palaeogene Surface, as well as by stressing the significance of the London Clay (basal Eocene) transgression. The reinterpretation of the late Neogene incursion as a relatively insignificant geomorphological episode of uncertain extent, together with the view that tectonic activity was characteristic of the Tertiary, and particularly the Palaeogene, indicates that the drainage pattern is inherited largely from the network that developed on the flanks of the Weald upon the withdrawal of the Palaeogene sea. Similarly, the recognition of the broadscale importance of the Sub-Palaeogene Surface and the rejection of the idea of a Mio-Pliocene peneplain suggest, in turn, that the concept of a 'Summit Surface' (Ramsay, 1864; Topley, 1875; Davis, 1895; Mackinder, 1902; Wooldridge and Linton, 1938a, 1955) should also be abandoned, for it implies a single quasi-horizontal erosional feature developed over a particular period of time. The often noted summit conformity is here considered an accidental product of denudation, including solutional lowering, acting on Mesozoic rocks revealed through the progressive exhumation of a warped diachronous surface.

The application of this model to the Weald results in the evolutionary sequence summarized below and portrayed in Fig. 9:

(1) The Upper Cretaceous (Maastrichtian; 70 Ma) saw considerable uplift (300-450 m) and warping. The newly-created subaerial environment was dominated by a major NW-SE upwarp, which probably extended from the Midlands across the

present London Basin and Weald to near the French coast. Rapid denudation of the emergent Chalk along this axis resulted in the removal of up to 200 m of strata by the early Palaeocene.

(2) Tectonic deformation continued through the Palaeocene, resulting in the demise of the 'Charnian Arch' and the initiation of the present macro-flexure pattern and pericline formation. These movements probably caused drainage adjustments, the original structurally concordant pattern, which must have focused on the site of the Hampshire Basin, being partially redirected to the evolving London Basin so as to produce a network dominated by major eastward flowing rivers similar to that envisaged by Linton (1951) and Brown (1960b). Denudation proceeded on structurally elevated areas while the basins were progressively submerged by the Palaeogene sea.

(3) The early Palaeogene transgressive phase (Fig. 1) culminated with the complete submergence by the London Clay (Ypresian) sea (53-49 Ma). As a result, a blanket of sediments was deposited on a compound marine-trimmed surface (Sub-Palaeogene Surface) which overstepped the Chalk so as to plane across Lower Cretaceous rocks above the Central Weald (Fig. 9).

(4) Continued flexuring in the early Eocene saw the emergence of the Weald and initiation of a concordant radial drainage pattern on the mantle of Palaeogene deposits (Jones, 1974). Denudation quickly stripped away the thin Palaeogene cover from the crest of the growing Weald-Artois Anticline, thereby revealing the Mesozoic rocks underlying the marine unconformity which then underwent subaerial erosion.

(5) The remainder of the Palaeogene was characterized by further growth of the Weald-Artois Anticline which stimulated subaerial denudation. The progressive withdrawal of the sea from the structural basins led to the eastward extension of the proto-Thames and Frome-Solent and the integration of the drainage net. Examination of sarsens (Clarke *et al.*, 1967) suggests that extensive surfaces developed in the Oligocene which lay a little above the highest chalkland summits, i.e. 250-300 m O.D. The existence of duricrust relics indicates that the End Palaeogene Surface was one of low relief developed under semi-arid conditions. As Weald Clay would have been exposed over the Weald it is unlikely that this surface rose much towards the axis of uplift, although subsequent warping has probably deformed it so that it now lies at 350-450 m O.D. (Fig. 9). Differential movements maintained the erosional potential of the main rivers, thereby enabling them to resist derangement by the growth of periclines, and so resulted in the development of the widespread discordant relationships (Jones, 1974).

(6) The Neogene saw continued up-arching, but at a slower rate, possibly accompanied by localized small-scale flexuring during the early Miocene. Fluvial incision continued so that by the close of the Neogene the river valleys may have been cut down to 100-150 m O.D. around the Weald (Small and Fisher, 1970). The widespread removal of the Palaeogene cover and Chalk led to the exposure of an extensive Lower Cretaceous outcrop (Fig. 9). Lithological controls became important and resulted in drainage net extension through subsequent stream

development and the creation of a landscape of moderate relief characterized by scarps and vales (Fig. 9). During the Miocene/Pliocene there was a marine incursion of uncertain extent which may have planed across the Kentish Downs but failed to leave any other morphological evidence of its occurrence. While the nature and significance of the Lewes-Medway axis (Fig. 6) has yet to be evaluated, it is important to note that it separates the Lenham marine deposits from those of supposedly Red Crag age.

(7) The short span of the Quaternary saw widespread incision in response to glacially lowered sea-levels and against a background of local deformation. This resulted in further removal of Palaeogene sediments, slow retreat of Chalk escarpments and lowering of clay vales so that relative relief increased through time (Fig. 9). Lithological controls continued to influence drainage development with expansion of certain catchments leading to the creation of wind-gaps through the Chalk cuestas (e.g. Washington Gap). The gradual emergence of the structurally complex Hastings Beds outcrop from beneath a thick cover of Weald Clay (Fig. 9) probably resulted in further drainage adjustments so as to produce the predominantly concordant networks of the Central Weald.

The adoption of this interpretation involves some major conceptual changes regarding the palaeogeomorphology of southern England. First, it envisages continuity of landscape and drainage development for 50 Ma. Second, it greatly increases the level of inheritance from the early Tertiary and therefore extends the relevant time-span for evolutionary studies. Third, it indicates that Pleistocene denudation was mainly concerned with accentuating pre-existing landform patterns. Finally, it raises questions concerning the reality and age-range of the various 'geomorphological staircases' that have been proposed, both for areas adjacent to the Weald and also the harder rock terrains of Upland Britain.

## Acknowledgements

The author wishes to thank Dr C. P. Green for his helpful and constructive comments. In addition he is grateful to Pat Farnsworth for typing and Jane Shepherd, Barbara Glover and Alison Fisher for drawing the maps and diagrams.

## References

Ager, D. V. (1975). The geological evolution of Europe, *Proc. Geol. Ass.* **86**, 127-154.

Arkell, W. J. (1947). The geology of the country around Weymouth, Swanage, Corfe and Lulworth, *Mem. Geol. Surv. U.K.*

Avery, B. W. (1964). Soils and land use of the district round Aylesbury and Hemel Hempstead, *Mem. Soil. Surv. England and Wales.*

Baden-Powell, D. F. W. (1955). The correlation of the Pliocene and Pleistocene marine beds of Britain and the Mediterranean, *Proc. Geol. Ass.* **66**, 271-92.

Brown, E. H. (1960a). "The Relief and Drainage of Wales". University of Wales Press, Cardiff.

Brown, E. H. (1960b). The building of southern Britain, *Z. Geomorph.* **4**, 264-74.

Brunsden, D. (1963). The denudation chronology of the River Dart, *Trans. Inst. Br. Geogr.* **32**, 49-64.

Bury, H. (1910). On the denudation of the western end of the Weald, *Q. J. Geol. Soc. Lond.* **66**, 640-92.

Casey, R. (1961). The stratigraphical Palaeontology of the Lower Greensand, *Palaeontology* **3**, 487-621.

Catt, J. A. & Hodgson, J. M. (1976). Soils and geomorphology of the Chalk in south-east England, *Earth Surface Processes* **1**, 181-193.

Chatwin, C. P. (1927). Fossils from the ironsands on Netley Heath (Surrey), *Mem. Geol. Surv. Summ. Prog.* (1926) 154-7.

Cholley, A. (1957). "Recherches Morphologiques". Colin, Paris.

Clark, M. J., Lewin, J. and Small, R. J. (1967). The sarsen stones of the Marlborough Downs and their geomorphological implications, *Southamp. Res. Ser. Geogr.* **4**, 3-40.

Clayton, K. M. (1969). Post-war research on the geomorphology of south-east England, *Area* **1** (2), 9-12.

Costa, L., Denison, C. and Downie, C. (1978). The Palaeocene/Eocene boundary in the Anglo-Paris Basin, *J. Geol. Soc. Lond.* **135**, 261-4.

Curry, D. (1965). The Palaeogene beds of South-east England, *Proc. Geol. Ass.* **76**, 151-73.

Curry, D. and Smith, A. J. (1975). New discoveries concerning the geology of the central and eastern parts of the English Channel, *Phil. Trans. Roy. Soc. Lond.* **A279**, 155-67.

Curry, D. *et al.* (1978). A correlation of Tertiary rocks in the British Isles, *Geol. Soc.* Special Report No. 12.

D'Olier, B. (1972). Subsidence and sea-level rise in the Thames Estuary, *Phil. Trans. Roy. Soc. Lond.* **A272**, 121-30.

Davis, A. G. and Elliot, G. F. (1957). The Palaeogeography of the London Clay Sea, *Proc. Geol. Ass.* **68**, 255-77.

Davis, W. M. (1895). On the origin of certain English rivers, *Geog. J.* **5**, 128-46.

Dines, H. G. and Chatwin, C. P. (1930). Pliocene sandstone from Rothamsted (Hertfordshire), *Mem. Geol. Surv Summ. Prog.* (1929) 1-7.

Dines, H. G. and Edmunds, F. H. (1929). The Geology of the Country around Aldershot and Guildford, *Mem. Geol. Surv. U.K.*

Dines, H. G. and Edmunds, F. H. (1933). The geology of the country around Reigate and Dorking, *Mem. Geol. Surv. U.K.*

Docherty, J. (1967). The exhumed sub-Tertiary surface in north-west Kent, *South-east Nat.* **70**, 19-31.

Edmunds, F. H. (1927). Pliocene deposits on the South Downs, *Geol. Mag.* **64**, 287.

Everard, C. E. (1954). The Solent river: a geomorphological study, *Trans. Inst. Br. Geogr.* **20**, 41-58.

Fitch, F. J., Hooker, P. J., Miller, J. A. and Brereton, N. R. (1978). Glauconite dating of Palaeocene-Eocene rocks from East Kent and the time-scale of Palaeogene volcanism in the North Atlantic region, *J. Geol. Soc. Lond.* **135**, 499-512.

Gallois, R. W. (1965). The Wealden District, *Mem. Geol. Surv. U.K.*

George, T. N. (1974). Prologue to the geomorphology of Britain, *in* E. H. Brown and R. S. Waters (eds), "Progress in Geomorphology", *Inst. Br. Geogr. Sp. Pub.* **7**, 113-125.

Green, C. P. (1969). An early Tertiary surface in Wiltshire, *Trans. Inst. Br. Geogr.* **47**, 61-72.

Green, C. P. (1974). The Summit surface on the Wessex Chalk, *in* E. H. Brown and R. S. Waters (eds), "Progress in Geomorphology", *Inst. Br. Geogr. Sp. Pub.* **7**, 127-138.

Hancock, J. M. (1975). The petrology of the Chalk, *Proc. Geol. Ass.* **86**, 499-535.

Hester, S. W. (1965). Stratigraphy and Palaeontology of the Woolwich and Reading Beds, *Bull. Geol. Surv. U.K.* **23**.

Hodgson, J. M. (1967). Soils of the West Sussex Coastal Plain, *Mem. Soil Surv. England and Wales* **3**.

Hodgson, J. M., Catt, J. A. and Weir, A. H. (1967). The origin and development of Clay-with-Flints and associated soil horizons on the South Downs, *J. Soil Sci.* **18**, 85-102.

Hodgson, J. M., Rayner, J. H. and Catt, J. A. (1974). The geomorphological significance of the Clay-with-Flints on the South Downs, *Trans. Inst. Br. Geogr.* **61**, 119-129.

Hooker, J. J. (1977). The Creechbarrow Limestone—its biota and correlation, *Tertiary Res.* **1**, 139-45.

Howitt, F. (1964). The stratigraphy and structure of the Purbeck inliers of Sussex, *Q. J. Geol. Soc. Lond.* **119**, 77-114.

John, D. I. (1974). A study of the soils and superficial deposits on the North Downs of Surrey. Ph.D. thesis, University of London (unpublished).

Jones, D. K. C. (1974). The influence of the Calabrian transgression on the drainage evolution of south-east England, *in* E. H. Brown and R. S. Waters (eds), "Progress in Geomorphology", *Inst. Br. Geogr. Sp. Pub.* **7**, 139-158.

Jones, D. K. C. (1980). "Southeast and Southern England". Methuen, London (in press).

Jukes-Browne, A. J. (1906). The Clay-with-Flints; its origin and distribution, *Q. J. Geol. Soc. Lond.* **62**, 132-164.

Jukes-Browne, A. J. (1911). "The Building of the British Isles". Edward Stanford, London.

King, C. A. M. (1977). The early Quaternary landscape with consideration of neotectonic matters, *in* F. W. Shotton (ed.), "British Quaternary Studies". Oxford University Press.

Kirkaldy, J. F. (1975). William Topley and The Geology of the Weald, *Proc. Geol. Ass.* **86**, 373-388.

Lake, R. D. (1975). The structure of the Weald—a review, *Proc. Geol. Ass.* **86**, 549-57.

Letzer, J. M. (1973). The nature and origin of superficial sands at Headley Heath, Surrey, *Proc. Trans. Croydon Nat. Hist. Sci. Soc.* **14**, 263-8.

Loveday, J. A. (1962). Plateau deposits of the southern Chiltern Hills, *Proc. Geol. Ass.* **73**, 83-101.

Linton, D. L. (1951). Problems of Scottish scenery, *Scot. Geogr. Mag.* **67**, 65-85.

Linton, D. L. (1956a). Geomorphology, *in* D. L. Linton (ed.), "A survey of Sheffield and its region", pp. 24-43. British Association, Sheffield.

Linton, D. L. (1956b). The Sussex rivers, *Geography* **41**, 233-247.

Linton, D. L. (1964). Tertiary landscape evolution, *in* J. W. Watson and J. B. Sissons (eds), "The British Isles", pp. 110-113. Nelson, London.

Linton, D. L. (1969). The formative years in geomorphological research in south-east England, *Area* **1** (2), 1-8.

Mackinder, H. J. (1902). "Britain and the British Seas". Heinemann, London.

McGinnis, L. D. (1970). Tectonics and the gravity field in the continental interior, *J. Geophys. Res.* **75**, 317-31.

Middlemiss, F. A. (1967). Analysis of structure in a region of gentle enechelon folding, *Neues Jb. Miner. Geol. Paläont. Abh.* **129**, 137-156.

Middlemiss, F. A. (1975). Studies in the sedimentation of the Lower Greensand of the Weald, 1875-1975: a review and commentary, *Proc. Geol. Ass.* **86**, 457-73.

Odin, G. S., Curry, D. and Hunziker, J. C. (1978). Radiometric dates from N.W. European glauconites and the Palaeogene time-scale, *J. Geol. Soc. Lond.* **135**, 481-97.

Owen, H. G. (1971a). Middle Albian stratigraphy in the Anglo-Paris Basin, *Bull. Br. Mus. Nat. Hist. (Geol.) Suppl.* **8**, 1-164.

Owen, H. G. (1971b). The stratigraphy of the Gault in the Thames Estuary and its bearing on the Mesozoic tectonic history of the area, *Proc. Geol. Ass.* **82**, 187-207.

Owen, H. G. (1975). The stratigraphy of the Gault and Upper Greensand of the Weald, *Proc. Geol. Ass.* **86**, 475-98.

Pinchemel, P. (1954). "Les plaines de craie du nord-ouest du Bassin Parisien et du sud-est du Bassin de Londres et leur bordures. Etude de géomorphologie". Colin, Paris.

Pitty, A. (1968). The scale and significance of solutional loss from the limestone tract of the Southern Pennines, *Proc. Geol. Ass.* **79**, 153-78.

Prestwich, J. (1852). On the structure of the strata between the London Clay and the Chalk in the London and Hampshire Tertiary systems, Part III, *Q. J. Geol. Soc. Lond.* **8**, 235-64.

Prestwich, J. (1858). On the age of some sands and ironstones on the North Downs, *Q. J. Geol. Soc. Lond.* **14**, 322-35.

Ramsay, A. E. (1864). "The Physical Geology and Geography of Britain". Stanford, London.

Read, H. H. and Watson, J. (1975). "Introduction to Geology: Vol. 2 Earth History". Macmillan, London.

Reid, C. (1890). The Pliocene deposits of Britain, *Mem. Geol. Surv. U.K.*

Schumm, S. A. (1963). Disparity between present rates of denudation and orogeny, *U.S. Geol. Surv. Prof. Paper,* 454.

Shephard-Thorn, E. R., Lake, R. D. and Atitullah, E. A. (1972). Basement control of structures in the Mesozoic rocks in the Strait of Dover region, and its reflexion in certain features of the present land and submarine topography, *Phil. Trans. Roy. Soc. Lond.* **A272**, 99-113.

Shotton, F. W. (1962). The physical background to Britain in the Pleistocene, *Adv. Sci.* **19**, 193-206.

Small, R. J. (1964). Geomorphology, *in* F. J. Monkhouse (ed.), "A survey of Southampton and its region", pp. 37-50. British Association, Southampton.

Small, R. J. and Fisher, G. C. (1970). The origin of the secondary escarpment of the South Downs, *Trans. Inst. Br. Geogr.* **49**, 97-107.

Smart, J. G. O., Bissom, G. and Worssam, B. C. (1966). The geology of the country around Canterbury and Folkestone, *Mem. Geol. Surv. U.K.*

Smith, A. J. and Curry, D. (1975). The structure and geological evolution of the English Channel, *Phil. Trans. Roy. Soc. Lond.* **A279**, 3-20.

Stamp, L. D. (1921). On cycles of sedimentation in the Eocene strata of the Anglo-Franco-Belgium Basin, *Geol. Mag.* **63**, 108-14.

Stevens, A. J. (1959). Surfaces, soils and land use in north-east Hampshire, *Trans. Inst. Br. Geogr.* **26**, 51-66.

Terris, A. P. and Bullerwell, W. (1965). Investigations into the underground structure of Southern England, *Adv. Sci.* **22**, 232-252.

Topley, W. (1875). The Geology of the Weald, *Mem. Geol. Surv. U.K.*

Waters, R. S. (1960). The bearing of superficial deposits on the age and origin of the upland plain of east Devon, west Dorset and south Somerset, *Trans. Inst. Br. Geogr.* **28**, 89-95.

West, R. G. (1972). Relative land-sea-level changes in south-eastern England during the Pleistocene, *Phil. Trans. Roy. Soc. Lond.* **A272**, 87-98.

Williams-Mitchell, E. (1956). The stratigraphy and structure of the chalk of Dean Hill anticline, Wiltshire, *Proc. Geol. Ass.* **26**, 221-7.

Wooldridge, S. W. (1927). The Pliocene period in the London Basin, *Proc. Geol. Ass.* **38**, 49-132.

Wooldridge, S. W. (1957). Some aspects of the physiography of the Thames in relation to the Ice Age and Early Man, *Proc. Prehist. Soc.* **23**, 1-19.

Wooldridge, S. W. (1960). The Pleistocene succession in the London Basin, *Proc. Geol. Ass.* **71**, 113-29.

Wooldridge, S. W. and Linton, D. L. (1938a). Influence of the Pliocene transgression on the geomorphology of south-east England. *J. Geomorphology* **1**, 40-54.

Wooldridge, S. W. and Linton, D. L. (1938b). Some episodes in the structural evolution of south-east England considered in relation to the concealed boundary of Meso-Europe *Proc. Geol. Ass.* **49**, 264-291.

Wooldridge, S. W. and Linton, D. L. (1939). Structure, surface and drainage in south-east England, *Inst. Br. Geogr.* Publication No. 10.

Wooldridge, S. W. and Linton, D. L. (1955). "Structure, surface and drainage in south-east England". George Philip, London.

Worssam, B. C. (1963). The geology of the country around Maidstone, *Mem. Geol. Surv. U.K.*

Worssam, B. C. (1963). The geology of the country around Maidstone, *Mem. Geol. Surv. Inst. Geol. Sci.* Report No. 73/17.

Wrigley, A. G. (1940). The faunal succession in the London Clay, etc. *Proc. Geol. Ass.* **51**, 230-55.

Young, A. (1974). The rate of slope retreat, *in* E. H. Brown and R. S. Waters (eds), "Progress in Geomorphology", *Inst. Br. Geogr. Sp. Pub.* **7**, 65-78.

THREE

# The Tertiary geomorphological evolution of south-east England: an alternative interpretation

R. J. SMALL
*Department of Geography, University of Southampton*

## Introduction

It is now thirty years since *Structure, Surface and Drainage in South-East England* appeared in the Transactions of this Institute. When republished in a modified form in 1955, the book's main conclusions had faced no serious challenge; rather, most researchers of the 1940s and 1950s were seemingly concerned to elaborate, not to redraw, Wooldridge and Linton's 'Grand Design'. The reasons for this are not hard to find. Not only was *Structure, Surface and Drainage* the most persuasive and masterful piece of writing that British geomorphology had produced in half a century; its authors were still dominating figures, influencing the ideas and methods of many younger workers who came under their sway. During the 1960s and 1970s, with the decline of interest in denudation chronology, *Structure, Surface and Drainage* inevitably lost much of its impact, and investigation into the erosional history of south-east England in the pre-Pleistocene period became increasingly unfashionable. Fortunately there were exceptions, such as C. P. Green's study of surfaces and drainage evolution in the Vale of Wardour area (1969, 1974) and D. K. C. Jones's reconsideration of the influence of the Calabrian transgression on drainage evolution in south-east England (1974). During this period my own researches, in collaboration with colleagues at Southampton University, were centred on aspects of chalkland geomorphology, notably chalk escarpments, dry valleys and sarsen stones. As my familiarity with the Wessex area and the South Downs grew, I became aware of certain inconsistencies in the Wooldridge and Linton synthesis, and of evidence that they did not seem to take sufficiently into account. It is the aim of this review paper to present a reinterpretation of the Tertiary evolution of south-east England, in which the views of Wooldridge and Linton are revised rather than rejected (with the exception of the influence of the Calabrian transgression, which they may seriously have overestimated). In presenting this discussion I write from a standpoint of admiration of what must still be regarded as a geomorphological classic.

It is convenient to present the discussion under three headings: the Alpine folding, the erosion surfaces, and the drainage evolution.

## Alpine folding

The most intractable problem posed by the Tertiary evolutionary history of south-east England is arguably that of the nature and age of the so-called Alpine structures. These are considered in detail by Wooldridge and Linton, who conclude (p.14) that it was the "Oligo-Miocene or mid-Tertiary movements" that imposed the existing major structures on the region. "In some cases they augmented, emphasized or brought to completion structures which had been growing intermittently for a long earlier period, but to a considerable extent they also imposed upon the region important new lineaments, of whose earlier growth we can discern hardly any indications". The assumption of a mid-Tertiary 'Alpine storm' is convenient, in that it allows a clear distinction to be drawn between pre-Alpine surfaces (necessarily deformed by the folding) and post-Alpine surfaces (unwarped and in some instances eroded across the Alpine structures). However, it is important to consider the possibility that *all or most* folds underwent significant development prior to the final tectonic phase, the date of which may also be open to question.

That the early Tertiary period saw some crustal instability is accepted by Wooldridge and Linton (pp. 10-11). In addition to a pre-Eocene tilting of the Chalk, evidenced by the overstepping of the fossil-zones of the Senonian Chalk by the lowest Eocene strata, there are indications of further movements occurring during the Eocene and Oligocene. For example, the London Basin "grew spasmodically", the periods of deepening ushering in marine transgressions, whilst complementary uparching of the Wealden anticlinorium is implied by the inference that the central Wealden area "projected as an island above the Eocene sea". Nevertheless, Wooldridge and Linton write of these early phases of folding: "What happened in the major upwarped areas between the Eocene depressions must for ever remain a subject for surmise or, at best, for argument by analogy".

It is true that much valuable evidence has been removed from anticlinal areas by late-Tertiary denudation, but the little that remains points towards important conclusions. Williams-Mitchell (1956) has studied in detail the stratigraphy of the Dean Hill anticline, a minor east-west flexure between the Avon and Test valleys. He shows that the lowermost Palaeocene strata (Reading Beds) overstep the chalk fossil-zones to the north and south of the fold axis. They rest on the outcrop of the lower *mucronata* Zone everywhere to the north and east except near West Dean, where they lie on the *quadrata* Zone. At the western end of the fold the Eocenes transgress the *quadrata* to the *pilula* Zone. Williams-Mitchell concludes that the Dean Hill anticline "can be dated to the interval between the lower *mucronata* Zone and the earliest Eocene strata ... deposited in the area". Later development of the anticline is indicated by deformation of the Tertiary rocks, but it is incontrovertible that the fold experienced appreciable growth *and* crestal denudation (involving the removal of some 100 m of chalk) in the pre-Eocene period. A crucial point is that the Dean Hill anticline is but one component in a series of east-west folds traceable from the Meon valley in east Hampshire, past Winchester to Bower Chalke in Wiltshire. It is tempting to suppose that these folds experienced equivalent pre-Eocene development.

More striking evidence of early Tertiary folding and erosion is displayed by the important Weymouth-Purbeck anticlinal structure. To the north-west of Weymouth, at the Hardy Monument, deposits of gravels (predominantly of broken flint, but also containing quartz pebbles and Palaeozoic detritus of westerly provenance) cap the steeply dipping chalk at up to 237 m O.D. These 'Blackdown Beds', contrasting in height and composition with other plateau-gravels farther to the east, are widely accepted by geologists as of Eocene age. Their very existence and their relationships to equivalent formations elsewhere in the western Hampshire Basin, point to (a) significant early Tertiary development of the Weymouth anticline, and (b) an early Tertiary phase of subaerial erosion, probably quite distinct from that affecting the newly-formed Dean Hill anticline.

The Creechbarrow Beds, west of Corfe, have been discussed by Arkell (1947). These comprise sands, clays, flints and tufaceous limestone capping Creechbarrow Hill (193 m O.D.), which is mainly composed of Reading Beds, London Clay and Bagshot Beds dipping at 10° northwards on the limb of the Purbeck anticline. Arkell makes a correlation between the Creechbarrow limestone and the Bembridge Limestone of the Isle of Wight (early-Oligocene?). He argues that associated flints indicate contemporaneous erosion of nearby chalk, on the grounds that they are so large (up to 14 kg) that travel over a short distance must be inferred. "The conclusion seems inescapable that the Upper Chalk was undergoing erosion in the immediate vicinity of Creechbarrow before the deposition of the Bembridge Limestone, namely in early Oligocene or ... late Eocene times. Presumably this erosion can only have been caused by the beginning of upheaval of the Purbeck fold". There is even some evidence that might be taken to mean that upheaval was already far advanced, namely the presence 350 m south-east of Creechbarrow of pipes in the chalk summits, developed in vertical *mucronata* chalk, containing quartz grains and flints of the Creechbarrow type, indicating a former 'pre-Oligocene' surface as shown in Fig. 1. The recent correlation of the Creechbarrow Beds by Hooker (1977) with the Bournemouth Marine Beds (late-Auversian?) does not invalidate Arkell's argument of early Tertiary development of the Purbeck axis.

To such direct evidence of 'pre-Alpine' fold development and allied erosion must be added the inferences to be drawn from existing sediments within the Tertiary Basins. Wooldridge and Linton refer to the spasmodic growth of the London Basin, with "the periods of deepening ushering in the several marine transgressions (of the Eocene)". In the Hampshire Basin the sedimentary record is more protracted, as evidenced by the early Oligocene formations of the southern New Forest and northern Isle of Wight (Chatwin, 1948). The total thickness of early Tertiary rocks in southern Hampshire approaches 600 m; and since these comprise sequences of

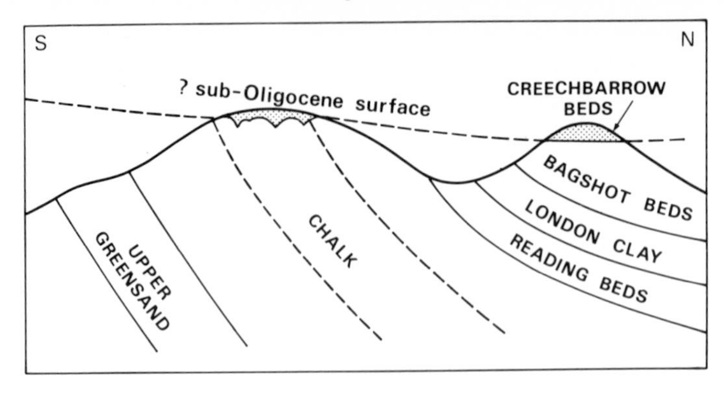


*Figure 1.* Diagrammatic section of Creechbarrow Hill, Dorset.

marine, estuarine and fluvial sediments it seems likely that the 'surface' of the Hampshire Basin oscillated about a mean position approximating to sea-level. The considerable overall thickness of the sediments, and the fact that individual formations attain greatest development along the basin axis (the Bracklesham Beds are 60 m thick in the northern New Forest and 180 m thick in the central Isle of Wight) can best be explained in terms of a spasmodic lowering of the Basin floor which was more or less offset by contemporaneous sedimentation. The question then arises as to whether synclinal deepening was allied to complementary upfolding, not only of major anticlinal axes such as those of the Weald, the Vale of Pewsey, the Vale of Wardour, and Weymouth-Purbeck-Isle of Wight (of which some evidence has been recounted), but also along less important axes (such as that of Meon-Winchester-Dean Hill-Bower Chalke).

Another problem is posed by the apparent termination of the Hampshire sedimentary record with the deposition of the Hamstead Beds (early-Oligocene). Wooldridge and Linton infer that the restricted extent of the Oligocene deposits (which they call "an almost insignificant addendum", and which probably never "extended far beyond their present limits") reflect the onset of uplift and warping premonitory of "more violent movements which followed in Oligo-Miocene times". Nevertheless, they note that the Oligocene sediments also include evidence of down-warping, which promoted two local marine transgressions; to which should be added that the uppermost 6 m of the Hamstead Beds comprise marine clays, capping freshwater deposits, indicative of a third transgression of which the full record has been lost. The simple but crucial question must be asked: did sedimentation continue in post-Hamstead times, and if so for how long? The vital evidence is largely lacking, yet pertinent observations can be made. The most telling is that the Hamstead Beds, capped by unresistant clay, still remain, some 40 million years after deposition. What has happened in the meantime? The obvious (though not the only) answer is to suggest a period of continued sedimentation (possibly extending into the Miocene period, though direct evidence of this is totally lacking), followed by a phase of erosion (largely during the Pliocene) which removed the younger deposits to lay bare the Hamstead Beds *quite fortuitously.* One implication of this interpretation is that the culminating Alpine movements were not Oligo-Miocene, but perhaps as late as "end-Miocene"—a date which has long been favoured by geologists. A problem which then arises concerns the vast amount of erosion to be assigned to the Pliocene (for example, in excess of 1500 m of Cretaceous and Tertiary strata have been worn from the crests of the Isle of Wight anticlines (White, 1912))—though this would be readily solved by the assumption of *earlier* periods of fold development, with episodes of contemporaneous erosion, in early Tertiary times.

## A model of Tertiary landform development

In order to forward the argument, it is convenient at this point to present, partly on *a priori* grounds and partly with reference to geological evidence which has been discussed, a simplified model of the possible Tertiary development of landscapes and related sediments in south-east England (Fig. 2). The essential features of this model are as follows. The major synclinal basins of the area are assumed to be affected by episodes of deepening from the onset of the Eocene to the close of the

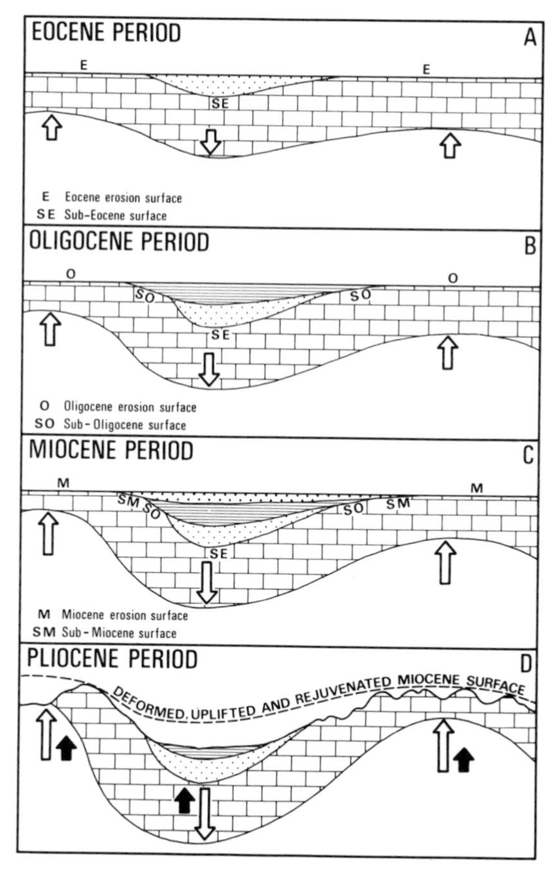


*Figure 2.* Tertiary landscape evolution of south-east England: a basic model.

Miocene. Contemporaneously, anticlinal structures were affected by complementary growth and important phases of erosion which truncated their crests in Eocene (A), Oligocene (B) and Miocene (C) times. (It is also possible to regard the 'Miocene' surface as a complex facet developed by more or less continuous denudation throughout the early Tertiary, and therefore to term it the 'early Tertiary' surface). It should be noted that, on the margins of the sedimentary basins, the surfaces would have been depressed and fossilized by overlying deposits. Outwards from the basin axes successive 'sub-Eocene', 'sub-Oligocene' and 'sub-Miocene' facets might have been preserved. It is postulated that in late Miocene/early Pliocene times a fundamental event occurred, causing interrelated erosional-sedimentary processes to be replaced by dominant erosion. This could have been the culminating folding and uplift of the 'Alpine storm', which completed existing structures and, by bodily upraising, initiated intense dissection of the 'Miocene' (early Tertiary?) surface, itself deformed by the folding. The Pliocene period has possibly encompassed some further warpings (see the discussion of Neogene tilting in the Vale of Wardour area by Green (1974)), together with incision of valley systems into the Chalk and the large-scale removal of the younger Tertiary sediments.

This model is not, of course, totally at variance with the Wooldridge and Linton interpretation, in that they recognize (a) the existence of some early Tertiary fold development (though this is seen to be of minor importance), and (b) the possibility of early Tertiary erosion away from the sedimentary basins (though they identify no elements in the present landscape inherited from it). However, in the emphasis that it lays on particular features and events the model amounts to an alternative interpretation, which can be developed further by reference to erosion surfaces and drainage development within south-east England.

## Erosion surfaces

There is no room here to recapitulate discussions of the age and origin of Tertiary erosion surfaces prior to 1939. However, it is necessary to describe in outline, with added commentary, the principal erosional phases postulated by Wooldridge and Linton.

### The sub-Eocene surface

This is regarded as largely the product of pre-Eocene erosion; "it bevels the folds produced by the pre-Tertiary movements and is evidently a peneplain locally trimmed by the waves of the Eocene sea". It now constitutes only a minor element of the landscape, and is preserved locally on the lower dip-slopes of the chalk close to the margins of the Tertiary basins and possibly, in a dissected condition, on the floors of minor chalk synclines from which infillings of Eocene strata have been removed. It is assumed by Wooldridge and Linton that elsewhere the sub-Eocene surface, upraised and distorted by the Alpine movements, has been totally destroyed by late-Tertiary erosion. This view derives from a reconstruction of the sub-Eocene surface which indicates that the surface passed well above present chalk summits (in the Wardour area the surface attained heights of 500 m or more, but the existing cuestas rise to 200-250 m).

However, at this early point in the discussion the following comments need to be made. First, there are *some* localities at which the chalk summits are evidently 'sub-Eocene', to judge from the presence of Eocene outliers (Pinchemel, 1954). Moreover, Waters (1960) has postulated the existence of early Tertiary (Eocene?) marine gravels at heights of 250-315 m in east Devon, whilst Green (1974) identifies remnants of a Palaeogene surface on residual elevations near the western margin of the Chalk outcrop in Wiltshire. Secondly, it needs to be emphasized that reconstruction of the sub-Eocene surface is necessarily a tentative procedure. On anticlinal structures *assumptions* of the amount of Chalk removed (based on mapping of fossil-zones by workers such as Brydone (1912)) provide one basis for such reconstructions. However, the effects of pre-Eocene erosion, especially of the smaller anticlinal structures, are difficult to calculate (except where evidence fortuitously remains, as on the Dean Hill anticline). Where such erosion was considerable, the height of the sub-Eocene surface is easily overestimated. Moreover, if the formation of the 'sub-Eocene' surface occurred partially within the Eocene or later periods, greater errors are likely. For instance, the reconstruction of a sub-Reading Beds surface along an anticline will give a greater 'height' than that of a sub-Bagshot Beds surface, since the latter will have resulted from erosional lowering of the former.

## The hilltop peneplain

This has been variously interpreted (for example, by Davis (1895) as an early Tertiary subaerial peneplain and by Bury (1910) as a Pliocene marine plane), but Wooldridge and Linton's hypothesis that it is subaerial in origin, lies mainly within the height range 230-260 m, and is 'post-Alpine' has commanded wide acceptance. Indeed other workers in Britain have regarded the surface as a firmly established datum. Clayton (1953) correlates the '1000-foot' (300 m) surface of the southern Pennines with "the summit surface of south-east England, the Mio-Pliocene peneplain"; whilst Brunsden (1964) states that the Lower Surface of Dartmoor (at 230-280 m) is "part of a widely developed planation surface which maintains a constant height from Cornwall to Kent". Waters (1960) has contributed an important study of the upland plain of east Devon, west Dorset and south Somerset, drawing conclusions that are in some respects in line with those of Wooldridge and Linton. It is, in fact, appropriate to focus discussion on this former surface, arguably the finest example of an upraised and dissected 'peneplain' in lowland Britain.

### *The upland plain of east Devon*

Developed discordantly across Chalk and Upper Greensand, at heights ranging from 150 m in the south to 310 m in the north, the upland plain of east Devon seemingly represents the morphological extension to the west of the Mio-Pliocene peneplain of the chalkland of Dorset and south Wiltshire. The surface is extensively capped by residual deposits, mainly of chert and flint, showing some regional contrasts either side of a line passing NW-SE close to Axminster and Lyme Regis. To the west is found 'typical' Clay-with-Flints, chert rubble in a sandy matrix, flint breccia bound by a siliceous matrix, and boulders of sarsen. Waters accepts that "a former Eocene cover may ... have furnished the sands and silica cement for the sarsens and the silica for the breccia cement", but argues that the flint and chert represent the insoluble weathering residue of the subjacent Chalk and Upper Greensand. "The weathering out of the flint and chert from the Cretaceous formations, which may have begun beneath the presumably thin Eocene cover, the cementation of the flint chips to form breccia, and the differential induration of surface layers of the sands to form sarsen boulders are all indicative of *subaerial* denudation under hot and at least seasonally humid conditions". Waters adds that these deposits are, moreover, associated with the "flattest portions" of the plateau at 230-270 m (the height of the Mio-Pliocene surface to the east).

Farther to the east the superficial deposits locally contain beach cobbles of flint, notably in the Staple Hill area (at 290-310 m), where the 'marine-trimmed surface' on which the pebbles rest is gently flexured, in conformity with the underlying geological structure resulting from the Alpine movements. Waters infers that this part of the upland plain comprises "substantial relics of an early Tertiary *marine* surface", contrasting with the more nearly plane surface to the west, with its angular drift cover.

Waters concluded that, in east Devon, there are two distinct morphological elements, a subaerial surface of late Tertiary age (equivalent to the Mio-Pliocene peneplain of Wooldridge and Linton) and a marine-trimmed early Tertiary surface, deformed by the Mid-Tertiary structural disturbances. This interpretation shows that "the morphological evolution of the area is comparable with that of south-eastern England". Nevertheless some difficult problems remain unresolved. For example, Waters's inference of an "extensive early Tertiary surface, initially base-levelled in relation to streams ... draining east towards the Hampshire Basin", which was later trimmed by "the waves of an encroaching late Eocene-early Oligocene sea" is disputable. The Blackdown Beds (see above) are equated in age with the Staple Hill gravels by Waters, yet the former are widely regarded as of 'Bagshot' age (Arkell, 1947). Another crucial question concerns the influences at work on the upland plain during the Oligocene period. The probable answer is that this was a time of predominantly subaerial denudation—as it was to the west, where erosion of the Dartmoor granite to form the Middle Surface (at above 315 m) may have furnished the sedimentary infilling of the Bovey Basin (Brunsden, 1964). The disparity in height between the surfaces in west and east Devon may be due to (a) an easterly downwarp resulting from the culminating Alpine movements, and (b) further erosional modification in east Devon during the Miocene, prior to a major rejuvenation during the Pliocene.

It seems more reasonable to regard the upland plain of east Devon as a complex, composite 'peneplain' developed by successive phases of erosion (largely subaerial, but with episodes of marine trimming to the east), constituting a major early Tertiary 'cycle'. This was by no means a period of structural stability, but must have embraced some folding and faulting. The gently flexured Marshwood Dome, for example, may have been affected by the Eocene folding of the Weymouth-Purbeck anticline (see above). Whether individual facets of the early Tertiary surface can be differentiated (as Waters has attempted) seems questionable. Further, whether regional variations in height of the upland plain derive from the 'blurring' of an erosional stairway, or result from later warpings to the east and south, is also difficult to determine.

The morphological continuity of the east Devon plain, via the Dorset Downs to the Chalk uplands west of Salisbury, is readily observable in the field, prompting the question: is the hilltop surface of the latter area, with its counterparts on the Wealden margins, in the Marlborough and Berkshire Downs, and in the Chilterns beyond the Goring Gap, also the product of a major early Tertiary cycle? Green (1974) has argued that in the Wardour area "the major part of the denudation of the Chalk outcrop appears to have taken place" (during the Palaeogene period), producing "a polycyclic land surface" which has been partially destroyed by the formation of younger surfaces of Neogene date. Elsewhere, an important line of evidence is provided by the sarsen stones, which are well distributed in south-east England but with notable concentrations in the Marlborough and Dorchester areas. Those at Marlborough have been studied in detail (Clark *et al.,* 1967), and the following main conclusions derived.

### *The evidence of sarsen stones*

First, sarsens are silicified sands and/or gravels comparable with silcretes described from Africa and Australia (Kerr, 1955). From their composition, morphology and penetration by fossil roots it can be inferred that induration occurred close to the contemporary land-surface. The view of King (1962) is that silcretes (with other duricrusts) "represent a soil mantle at the ultimate stage of pedalfer development, residual in origin and formed superficially over (wide extents) of anciently planed country". Moreover, "most of the world's duricrusts lie upon land surfaces where planation was achieved well back in the Cainozoic era". This view is much simplified; nevertheless the link between crusts and ancient surfaces is widely accepted. The occurrence of silicifications in the gravels of the upland plain of east Devon has been described by Waters (1960) and others (Woodward and Ussher, 1911). Elsewhere, as at Aston Rowant on the Chiltern summits, tabular sarsens—the apparent residue of a 'stratum'—are associated with sands and 'Clay-with-Flints' at up to 280 m O.D. In the Marlborough area, around Portesham in Dorset, and at Ashdown Park in the Berkshire Downs, the sarsens are accumulated in valley bottoms, but it is clear that they were originally located on interfluves at 220-260 m (Small *et al.,* 1970). It is, in fact, evident that sarsens were once widespread on the uppermost chalk surfaces of southern England, mainly within the altitudinal range 200-300 m, and that the occurrence of these silicifications accords with the view that the chalklands were planed, during the Tertiary, at *about* the level of the present summits.

Secondly, the age of the sarsens is difficult to determine precisely, though many views have been expressed. Woodward and Ussher (1911) report Reid's opinion that all the plateau-gravels (containing silicifications) of Devon and Dorset, as far east as Portesham, are of Lower Bagshot date. This is supported by the association of many sarsens with the Blackdown Beds. Other writers, including White (1909), Sherlock (1922) and Boswell (1927), have noted the occurrence of sarsens on or adjacent to Reading Beds (or Reading Beds remanie). A valuable observation is that of Dines and Edmunds (1929), who refer to the Fox Hills, Aldershot, where plateau gravels at 110-120 m rest on Barton Sands; beneath the gravels are many large sarsen stones. "It is possible that they are more or less *in situ* and represent an old landsurface at this level, and were formed by the action of organic acids, arising from decaying vegetation, upon the soluble silica in the sand". White (1925) produces convincing

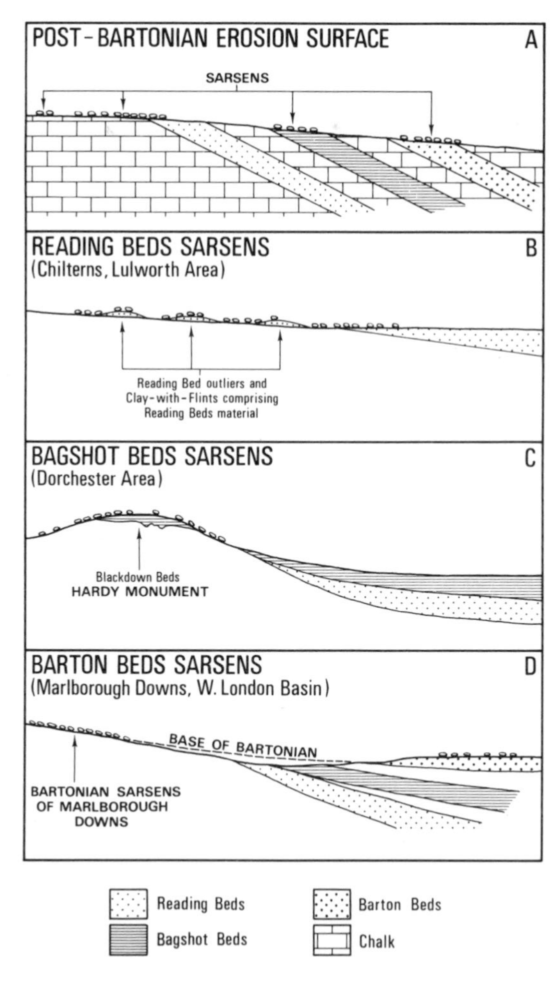


*Figure 3.* The relationships between sarsens, geological structure and erosion surfaces in south-east England. (A is an idealized 'composite' section of a regional post-Bartonian surface; B, C, D are sections of actual sites.)

arguments, on petrological grounds, for regarding the sarsens of the Marlborough Downs as of either Bartonian or post-Eocene age.

An obvious interpretation is that at least three periods of sarsen formation occurred during the Eocene (Fig. 3). However, most descriptions refer to sarsens that are not in their original positions, so that an alternative is to infer a main phase of induration at a later date, on an erosion surface truncating Eocene strata and the Chalk. In the Paris Basin the remarkable silcretes around Fontainebleau (including crusts *in situ*) are formed in fine-grained sands of Upper Oligocene age, locally overlain by Beauce marls and limestones. Cholley (1956) relates these silicifications to early stages in the development of the Miocene peneplain, specifically during the Aquitanian phase. One difficulty is that climatic and relief conditions conducive to silicification may have existed in Great Britain and Europe during several episodes of the Tertiary, even as late as the Pliocene. Numerous silicifications within the Basal Gravel Complex of South Limburg, Holland, are described by van den Broek and van der Waals (1967). These gravels rest on a late Tertiary peneplain, at below 100 m, the processes of silicification being favoured by "a quiet relief and a drainage system of shallow valleys" and a warm, seasonally arid climate. However, such a recent episode of sarsen formation in southern England seems unlikely on three scores. First, the high degree of fragmentation of the sarsen layer, if developed on a Pliocene peneplain, suggests more than the activity of Pleistocene solifluxion, though this has undoubtedly played a contributory role (Small *et al.*, 1970). Secondly the altitudinal range of sarsens approximately *in situ* (from 280 m at Aston Rowant to 110 m at Aldershot) suggests tectonic deformation, of which there is little evidence in post-Pliocene times. Thirdly, there are indications that the Pliocene was a period of valley incision in southern England, giving relief conditions unfavourable to silicification.

It is therefore concluded that the sarsens of southern England are (a) of 'early' rather than 'late' Tertiary date, (b) older than the final stages of the 'Alpine Storm', and (c) associated with erosion surfaces formed during the major early Tertiary 'cycle' postulated above.

## The Calabrian marine surface

This surface of partial marine planation, formerly regarded as of early Pliocene age, has been accepted as an important morphological element, represented by bevellings at 160-210 m (notably in the eastern North Downs). Geological evidence of the extension of the Calabrian sea into the Wessex region is so slight that many would now regard the transgression as unproven. However, for the purposes of this discussion, the transgression will be assumed, though it will be argued that even if it did occur its geomorphological impact was negligible. This was not, of course, the view of Wooldridge and Linton, and some others (Sparks, 1949), who related major features of the drainage pattern, such as the discordant character of many Wealden and Wessex streams, to the development of 'new' consequents on the emergent Calabrian surface. That this view of the role of the Calabrian transgression in the drainage evolution of south-east England cannot be sustained has been demonstrated by Jones (1974). Furthermore, evidence of the existence of a late Pliocene system of wind-gaps and valleys in the South Downs, incised 100 m below the maximum Calabrian level, has been described by Small and Fisher (1970). Revival of this valley system in the post-Calabrian period indicates the limited planational capabilities of the Calabrian sea. There is, indeed, much evidence indicating that the Calabrian transgression was itself largely controlled by the pre-existing relief, and that the resultant marine platforms were limited in extent other than in some particularly favourable localities. Initially this contention will be illustrated by two case studies.

First, in many areas where remnants of the Calabrian bench are preserved a maximum dimension of 1-2 km normal to the ancient shoreline is attained. However, some remnants may once have been rather more extensive, as in the North Downs where the surface passed onto weak Tertiary strata since removed. In the Wessex Chalklands, where the Calabrian surface is poorly preserved, the extent of the planation is inferred by Wooldridge and Linton to have been very much greater. Thus the reconstructed Calabrian shoreline of the South Downs swings northwards at the Meon Valley, before turning westwards along the chalk dip-slope south of the Pewsey Vale, and then southwards across the high ridge west of Salisbury. There is, moreover, the possibility of a former 'strait' linking the Wessex and London seas close to Basingstoke. This reconstruction is based partly on fragments of the Calabrian bench, but mainly on the presence of north-south discordant streams supposedly superimposed from the upraised sea-floor.

The considerable extent of the postulated Calabrian surface in Wessex (up to 25 km northwards *across the Chalk* from the present Hampshire Basin Tertiaries) poses a difficulty. Although marine erosion would have locally been aided where east-west synclines contained Eocene sediments, for the most part it would have been impeded by resistant Upper Chalk, both along anticlinal crests and across the broad synclinal plateau of Salisbury Plain. Some special factor must have allowed the penetration by the Calabrian sea; this can have been either a concurrent downwarping, between the major Wealden and Wardour structures, or erosive lowering prior to the transgression by Pliocene forerunners of the Avon, Test and Itchen. Evidence from the South Downs (Small and Fisher, 1970) supports the latter possibility, though a contribution by warping cannot be ruled out. A comparable penetration of the Weald by the Calabrian sea, using pre-existing river-gaps in the South Downs, is also conceivable. Such a 'Gulf and Islands' pattern is referred to by Jones (1974), who considers that "there is little evidence for such a coastline or such a sea-level".

Secondly, there are areas where the Calabrian surface seems to result not simply from marine planation, but to involve resurrection of older landscape elements (see the discussion of the South Downs dip-slope in Hodgson *et al.,* 1974). In Savernake Forest, near Marlborough, the chalk summits are bevelled at 180 m over a distance of 15 km between Martinsell Hill (289 m) and Tidcombe Down (257 m). The Calabrian sea seems, in fact, to have penetrated westwards, via this 'gap', into the

Pewsey lowland, an interpretation which is supported by spur bevellings at approximately 180 m on the scarps of the vale and isolated summits at up to 200 m (Woodborough and Picked Hills). Another gulf of the sea, penetrating westwards, is indicated by platforms to north and south of the Kennet valley near Marlborough. Thus we have evidence supporting the notion of a strongly embayed, if not 'Gulf and Islands' shoreline. However, the penetration of the Calabrian sea has been aided not only by pre-existing relief (strike-aligned lowlands related to Alpine structures), but also by the existence of the sub-Eocene surface. This is well preserved south-east of Marlborough, where it forms a 180 m 'plateau' morphologically hardly distinguishable from the Calabrian bevel. Recent exhumation of this surface, in which the Calabrian sea may have played a vital role, is indicated by the limited dissection, the presence of Eocene outliers, and a cover of 'Clay-with-Flints' sometimes containing almost pure Bagshot Sand.

The evidence increasingly supports the conclusion that the Calabrian transgression was at best of limited importance in the geomorphological evolution of south-east England. Its penetration, often hard to determine precisely, seems to have been irregular, and its planational effects insufficient to eradicate Pliocene drainage elements. The probable existence not only of an indented but a true 'Gulf and Islands' shoreline is suggested by a third case study. The chalklands marginal to the western Weald between Petersfield and Farnham are structurally complex, with numerous east-west Alpine axes (Fig. 4). The area is 'drained' by a major dry valley running northwards from West Tisted to Alton, where it leads into the Wey valley. The latter curves north-eastwards, at the base of the Chalk scarp, to Farnham, where the river formerly breached the Chalk ridge to enter the London Basin (Linton, 1930). The extended Wey valley (in the past the upper Wey clearly occupied the Tisted valley) is arguably the most discordant to structure in the whole of south-east England, and affords a good test of the hypothesis of superimposition from the Calabrian surface.

The Calabrian benches of this area have been mapped by Stevens (1959), who shows them to be composite in character and forming a pattern suggestive of a 'Gulf and Islands' shoreline. In the south the main platforms (at 200-210 m and 165-180 m) form, with intermediate benches, a morphological staircase descending westwards towards the Itchen basin. The shorelines trend north-south, along the margins of 'unsubmerged' chalk crests between Petersfield and Selborne (Fig. 4). Their configuration, in so far as they can be accurately traced, suggests a former gulf extending north-eastwards towards Alton along the line of the Tisted valley. Farther north, another group of 'unsubmerged' chalk summits occurs at Four Marks; the shorelines here suggest a westward projecting 'headland'. Another chalk 'island', separated from the Four Marks promontory by an embayment extending south-eastwards towards Alton, is indicated by the summits east of Golden Pot. To summarize, the *general* trend of the Calabrian shorelines is north-south, and the *general* gradient of the platforms is to the west. However, there is a fundamental anomaly in the drainage, with the dominant Tisted-Wey running northwards and north-eastwards. The source of the valley actually lies on the Calabrian benches;

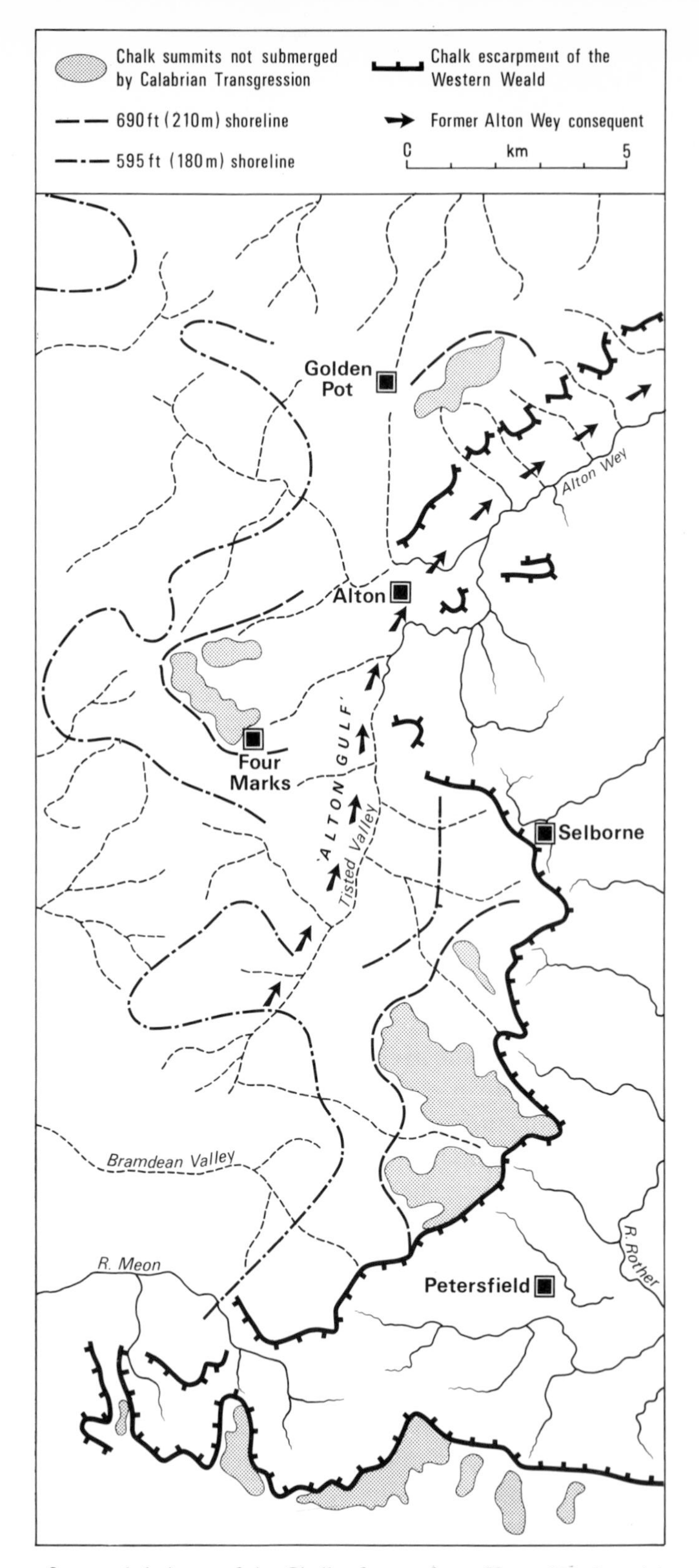


*Figure 4.* The surfaces and drainage of the Chalk of part of east Hampshire (based in part on A. J. Stevens, 1959).

subsequently it runs diagonally *up* the morphological staircase, passing through the main Calabrian shoreline by way of the 'gulf' towards Alton. The problematic nature of this drainage was recognized by Wooldridge and Linton (p. 97), who refer to it as possibly "the complex product of two cycles of erosion whose early beginnings are quite lost to us". However, in thus dismissing the Tisted-Wey valley they avoid consideration of a vital point: that in pre-Calabrian times an important *discordant* drainage element could develop, resembling in character other streams explained by them solely in terms of post-Calabrian superimposition.

The main inferences to be drawn from this third case-study are as follows. The Calabrian sea advanced from the west, utilizing the pre-existing Itchen basin. The interfluves between the Itchen and Tisted-Wey systems were partially planed, to give limited marine platforms. However, the Tisted-Wey—a Pliocene lineament—was not destroyed and, like other streams in south-east England, was revived after the Calabrian regression. A corollary must be that the Calabrian sea would have spread into the western Weald from the north-east also, along the Tisted-Wey valley from the London Basin. The 'gulf' south of Alton would in fact have been a 'strait' separating island(s) to the north from the unsubmerged chalk between Petersfield and Selborne. Detailed mapping of Calabrian shorelines within the present western Weald is impracticable, for platforms develop in the Cretaceous sands and clays here have since been destroyed.

## Drainage evolution

Wooldridge and Linton postulate that the drainage system of south-east England comprises both pre- and post-Calabrian elements. The former are represented by east-west accordant streams, mainly following anticlinal or synclinal courses, dating from the post-Alpine erosion cycle that produced the Mio-Pliocene peneplain. The post-Calabrian elements mostly follow a north-south alignment, resulting from superimposition from the emergent Calabrian sea-floor. Jones (1974) produces counter arguments, to the effect that many discordant rivers of Hampshire and Sussex predate the Calabrian transgression, "which did nothing more than cause erosional embellishments on the lower portion of the early Pliocene land surface". The discordances may be more realistically explained as due to antecedence. "For the most part it is maintained that the drainage pattern reflects that produced on the regression of the Eocene sea, and that this pattern largely maintained its unity in the face of (the Alpine) earth movements. Thus the majority of discordant rivers in southern England would appear to be best called anteconsequents".

My own investigations of Chalk geomorphology suggest conclusions similar but not identical to those of Jones. Evidence of restricted Calabrian planation, and the presence of 'Pliocene' discordant streams is given in Small and Fisher (1970) and has been discussed above. An attempt is now made to reconstruct the 'Pliocene' drainage of the Wessex chalklands and the western Weald. In undertaking this, a prime need is to discount post-Calabrian modifications. These include not only disruption of the Solent River by the Flandrian transgression (Everard, 1954), but river captures of earlier date.

Wooldridge and Linton have themselves postulated a series of captures and near captures within the basins of the Avon, Test, Itchen and Meon. These, and the former 'Candover-Hamble' stream, were interpreted as north-south Calabrian consequents, superimposed across the east-west structural grain. Synclinal axes, containing Eocene sediments, gave to the Calabrian surface the character of a belted outcrop plain. On this, outcrops of weak Eocene sands and clays led to the growth of strike elements, and thus a pattern seemingly conducive to disruption of the Calabrian consequents. At least two river captures are inferred by Wooldridge and Linton (that of the Candover-Hamble by the Alresford branch of the Itchen, and that of the Itchen by the Dever, tributary to the Test). However, in several instances imminent diversions were prevented by timely removal of the lowermost Eocenes to reveal the chalk, thus removing the erosional advantage of the strike streams. Thus the Dean Brook, tributary to the Test, failed to capture the Avon at Alderbury; and the Cheriton branch of the Itchen failed to divert the Meon at Warnford.

It is arguable that captures of this type are inherently unlikely, for geological and other reasons, and that doubt must be cast on the presumed captures. The problem is exemplified by the Dean Brook, which as described 'nearly' captured the Avon. It is evident that the Test and Avon are *broadly* comparable in terms of discharge and erosive power. As a result the height of the Test at the entry of the Dean Brook is 22 m O.D. and of the Avon near the head of the Dean Brook (the likely capture point) 35 m O.D. The gradient of the Brook is such that it rises at 75 m; thus capture is manifestly impossible at present, and unless the relative downcutting powers of the Test and Avon were *fundamentally* different in the past, equally unlikely then. It is, indeed, curious that the most *obvious* capture in the Wessex chalklands is that of the Preston-Candover valley, a strike-orientated left-bank tributary of the Test, by the north-south Candover! This leads us to a consideration of the controversial case of the Upper Meon valley.

The Meon comprises a north-west trending upper section and, south of Warnford, a north-south reach, incised across the Warnford anticlinal axis. White's suggestion (1912) that the upper Meon arose as a branch of the Itchen and was diverted southwards by the present lower Meon was discounted by Wooldridge and Linton, who argued that the Meon is a post-Calabrian consequent *nearly* captured by the synclinal Itchen (see above). However, the field evidence supports White's interpretation. The well-defined col at Wheely Farm (110 m), at the point of presumed capture, seems to be of fluvial origin rather than the exposed floor of a structural depression from which weaker rocks have been stripped. More important, the hypothesis that the Cheriton branch of the Itchen failed to develop beyond Wheely Farm for geological reasons cannot explain why another branch of the stream, developed a little to the north of the synclinal axis, was able to form the important Bramdean valley, which extends 13 km eastwards to the Wealden margins. The configuration of the Meon valley near Warnford is also suggestive of the former presence of amphitheatral embayments draining *northwards* to a *west-running* Meon. Such features are commonly developed in Wessex where minor streams have 'opened up' the weak cores of anticlinal structures (Small, 1964). The

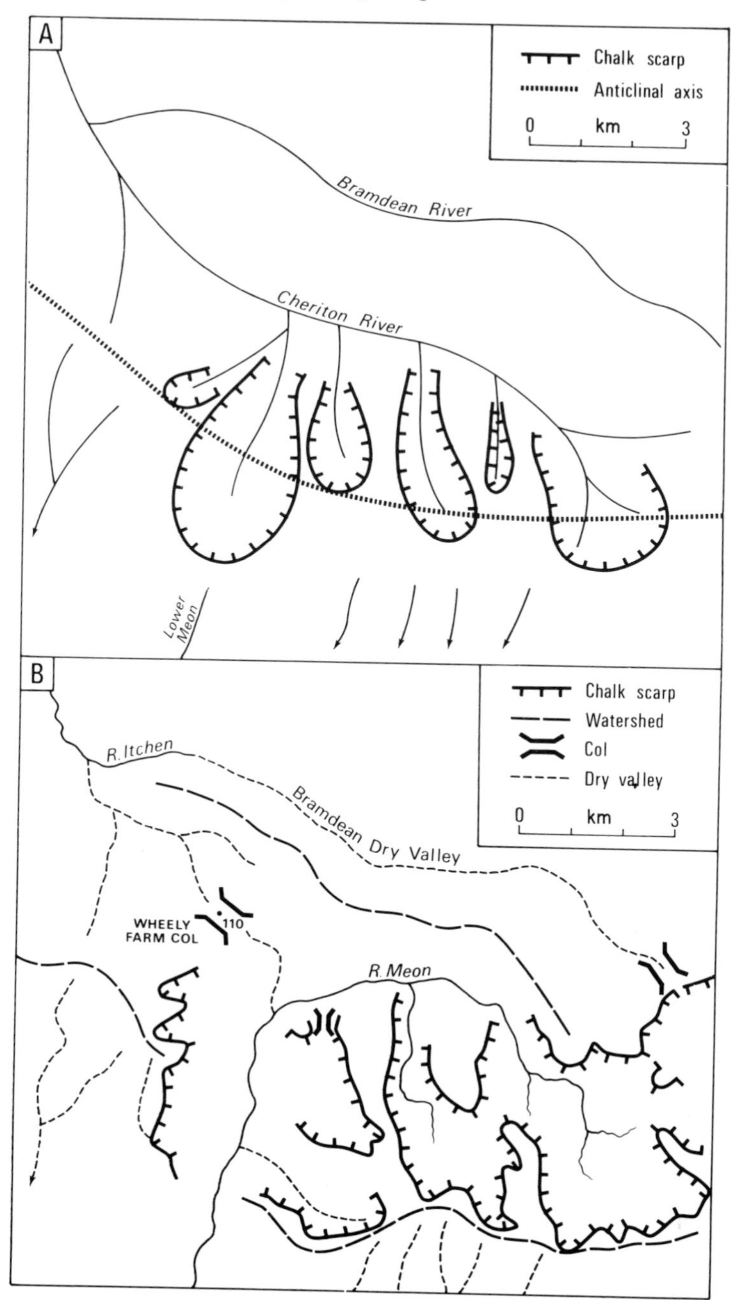


*Figure 5.* The evolution of the Upper Meon valley. (A, the initial drainage pattern; B, the present drainage pattern.)

former relief and drainage of the Meon valley are reconstructed in Fig. 5A. It is proposed that the Lower Meon, with a steeper gradient and shorter course either to the 'Pliocene' or 'post-Calabrian' shoreline, eroded headwards into the largest of the pre-existing embayments, to divert the 'Cheriton River' (as envisaged by White) but

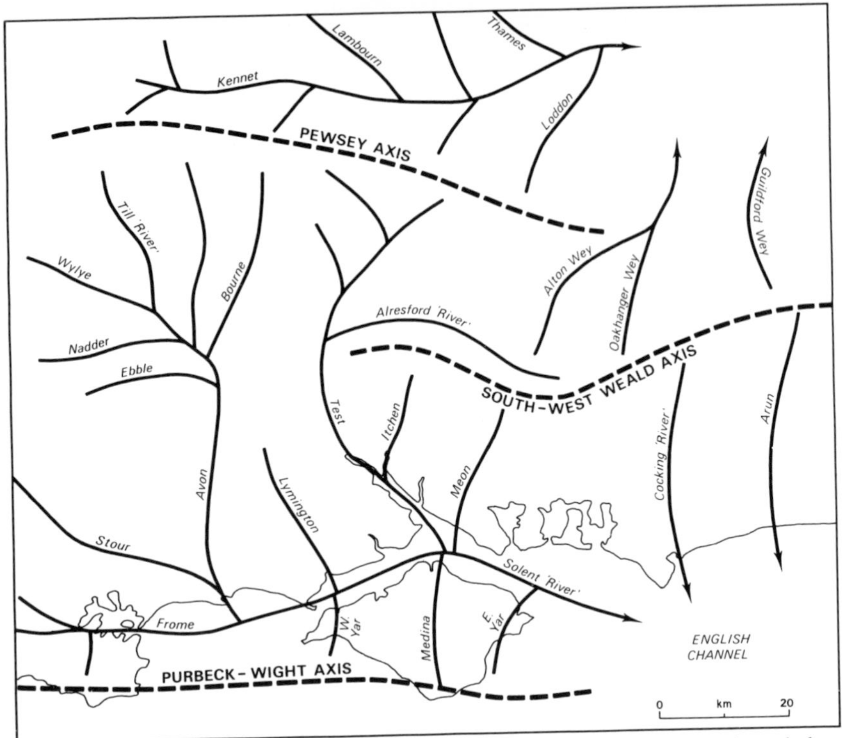


*Figure 6.* Reconstruction of the 'initial' drainage system of the Wessex chalklands and the western Weald.

to leave the 'Bramdean River' intact. The present relief and drainage are shown in Fig. 5B. Other captures of strike elements (which by definition would have had relatively gentle gradients and longer courses to contemporary shorelines) may have occurred, as at Winchester where the lower Itchen may have diverted a stream (the 'Alresford River' of Fig. 6) once flowing as far westwards as the Test.

When possible captures such as these have been identified, together with other drainage modifications associated with the breaching of Alpine folds, the 'original' drainage pattern of the Wessex chalkland can be tentatively reconstructed (Fig. 6). A similar restoration can be made for the adjacent western Weald area, where drainage diversions akin to the type envisaged by Wooldridge and Linton are more feasible. A classic case is the diversion of the western Wey system (fed by the 'Alton Wey' and 'Oakhanger Wey') into the eastern Wey ('Guildford Wey'). Another probable capture is that of the 'Cocking River' (denoted by the Cocking Gap in the South Downs) by the Rother strike-stream, tributary to the consequent Arun. Sparks (1949) argues against this diversion; but the size and form of the Cocking Gap, and its comparability in position and morphology with other wind-gaps of the South Downs (at Washington, Pyecombe and Jevington), seem best explained in terms of development by a former north-south consequent, of 'Pliocene' rather than 'post-Calabrian' age.

The major features of the 'original' drainage of the Wessex chalklands and the western Weald are as follows. Firstly, the major watersheds, oriented west-east, are coincident with major anticlinal axes (Pewsey, south-west Weald, and Purbeck-Wight), except to the west where the Wardour fold is followed by the Nadder. However, as Green (1974) shows, this may actually be a late development, in an area where the drainage was nevertheless originally west-east trending (along the Great Ridge and the Ebble). In its relationship to *major* structural features, the drainage may be categorized as highly accordant; indeed in some instances (see the Upper Test and Alresford River) the drainage is evidently adjusted to avoid crossing of the main anticlinal axes. Secondly, excepting the major synclinal consequents of the London and Hampshire Basins (the Kennet-Thames, and 'Solent River') the drainage lines are predominantly north-south or south-north, following the flanks of the main anticlinal axes. However, individual streams—usually in their lower courses—are often discordant to *lesser* Alpine structures. Thus the Avon crosses the Bower Chalke-Dean Hill fold, the Itchen and Meon the Swaythling-Portsdown anticline, the Cocking River the Dean Valley anticline, and so on. These streams seem to be the equivalent of the "low level discordances" of Jones (1974), which he attributed to "superimposition, either from later Pleistocene marine surfaces or the Tertiary cover". My own view is to regard them as mainly of 'Pliocene' age or earlier, on the grounds of evidence pointing to the existence of many prior to the Calabrian transgression (see above). Jones himself argues that the drainage is "largely derived from that of the early Tertiary", reflecting "that produced on the regression of the Eocene sea". This interpretation is in accord with the thesis of early Tertiary landscape evolution presented in this paper, though there is room for disagreement over detail. One view could be that the lineaments of the present drainage were formed during the Miocene, when the culminating Alpine movements seriously deformed the early Tertiary surface. Upwarping along the major axes may have been sufficient to create new watersheds, with consequents developing on the flanks of the folds. In their lower courses, these streams would have been powerful enough to counter upwarping along lesser axes, so that antecedence can be inferred; or, because the streams would incise courses through residual deposits of the early Tertiary surface, combined antecedence and superimposition (anteposition) can be invoked. An alternative hypothesis is that the drainage pattern is essentially early Tertiary, having developed on the early Tertiary surface as this was formed across growing fold-structures. During this long period of diastrophism and denudation, watersheds would be more likely to form along major fold-axes, and antecedence to occur where downstream sections of rivers transected lesser folds. The final Miocene movements may have done little or nothing to modify the pattern, and the late-Pliocene period in particular may have seen merely incision of the rivers well below the deformed early Tertiary surface. It is unlikely that issues such as these can now be finally resolved, since the necessary evidence has been blurred or removed by Pliocene and Pleistocene erosion.

## Conclusions

The fundamental postulate of this paper is that the most important erosion surface of south-east England is not of Mio-Pliocene age, as suggested by Wooldridge and Linton, but is an 'early Tertiary' peneplain fashioned partly during the Eocene but mainly during Oligocene and early Miocene times. The early Tertiary cycle embraced several episodes of fold growth, with contemporaneous erosion that often kept pace with uplift along anticlinal axes, and allied sedimentation in synclinal basins. The latter is thought to have terminated in the middle Oligocene, but there are reasons now to believe that deposition continued, perhaps into the Miocene period, to end with a period of diastrophism (folding *and* regional uplift) perhaps of late-Miocene age. The concept of a brief and intense 'Alpine storm', responsible for the main fold-structures of south-east England, is oversimple. Such evidence as exists points to spasmodic early growth during Eocene and Oligocene times. Moreover, Green (1974) finds some evidence of deformation, in the Wardour area, which can be "tentatively referred to the late Pliocene-early Pleistocene interval".

This reinterpretation of the Alpine folding permits a more 'Penckian' view of landform evolution in south-east England. Whereas Wooldridge and Linton firmly adopted the 'Davisian' stance of postulating a Mio-Pliocene cycle, producing the dominant surface, *after* the completion of the geological structure, it is now possible to countenance contemporaneous diastrophism and denudation. A corollary is reinterpretation of drainage development. Wooldridge and Linton infer the oldest elements in the extant system to be of post-Alpine, Mio-Pliocene date, though accepting that in many areas this original drainage had been obliterated by Calabrian marine planation. Evidence suggests that the latter, assuming it to have occurred, was a minor episode, and that the present drainage system existed in its essentials during the Pliocene. Indeed the drainage may have been inherited from the early Tertiary landsurface, on which major watersheds were initiated by the most vigorous folding. However, it is equally possible that Miocene deformation of the early Tertiary surface produced the outlines of the existing drainage.

## References


Arkell, W. J. (1947). The geology of the country around Weymouth, Swanage, Corfe and Lulworth, *Mem. Geol. Surv. U.K.*

Boswell, P. G. H. (1927). The geology of the country near Ipswich, *Mem. Geol. Surv. U.K.*

Brunsden, D. (1964). Denudation chronology of parts of south-western England, *Field Studies* **2**, 115-32.

Brydone, R. M. (1912). "The stratigraphy of the chalk of Hants". Dulau, London.

Bury, H. (1910). On the denudation of the western end of the Weald, *Q. J. Geol. Soc. Lond.* **66**, 640-92.

Chatwin, C. P. (1948). "The Hampshire Basin and adjoining areas". H.M.S.O., London.

Cholley, A. (1956). Carte Morphologique du Bassin de Paris, *Mem et Doc.* **5**, 7-104.

Clark, M. J., Lewin, J. and Small, R. J. (1967).The sarsen stones of the Marlborough Downs and their geomorphological implications, *Southampton Res. Ser. in Geog.* **4**, 3-39.

Clayton, K. M. (1953). The denudation chronology of part of the middle Trent basin, *Trans. Inst. Br. Geogr.* **19**, 25-36.

Davis, W. M. (1895). On the origin of certain English rivers, *Geog. J.* **5**, 128-46.

Dines, H. G. and Edmunds, F. H. (1929). The geology of the country around Aldershot and Guildford, *Mem. Geol. Surv. U.K.*

Everard, C. E. (1954). The Solent River: a geomorphological study, *Trans. Inst. Br. Geogr.* **20**, 41-58.

Green, C. P. (1969). An early Tertiary surface in Wiltshire, *Trans. Inst. Br. Geogr.* **47**, 61-72.

Green, C. P. (1974). The summit surface on the Wessex Chalk, *Trans. Inst. Br. Geogr. Sp. Pub.* No. 7, 127-38.

Hodgson, J. M., Rayner, J. H. and Catt, J. A. (1974). The geomorphological significance of Clay-with-flints on the South Downs, *Trans. Inst. Br. Geogr.* **61**, 119-29.

Hooker, J. J. (1977). The Creechbarrow Limestone—its biota and correlation, *Tertiary Res.* **1**, 139-45.

Jones, D. K. C. (1974). The influence of the Calabrian transgression on the drainage evolution of south-east England, *Trans. Inst. Br. Geogr. Sp. Pub.* No. 7, 139-58.

Kerr, M. H. (1955). On the origin of silcretes in southern England. *Proc. Leeds Phil. and Lit. Soc.* (*Sci. Sect.*) **6**, 328-37.

King, L. C. (1962). "The morphology of the earth". Oliver and Boyd, Edinburgh.

Linton, D. L. (1930). Notes on the development of the western part of the Wey drainage system, *Proc. Geol. Ass. Lond.* **41**, 160-74.

Pinchemel, P. (1954). "Les plaines de craie du nord-ouest du Bassin Parisien et du sud-est du Bassin de Londres et leurs bordures: Etude de geomorphologie." Colin, Paris.

Sherlock, R. L. (1922). The geology of the country around Aylesbury and Hemel Hempstead, *Mem. Geol. Surv. U.K.*

Small, R. J. (1964). Geomorphology, *in* F. J. Monkhouse (ed.) "A survey of Southampton and its region." Prepared for the meeting of the British Association, 26 August to 2 September 1964.

Small, R. J. and Fisher, G. C. (1970). The origin of the secondary escarpment of the South Downs, *Trans Inst. Br. Geogr.* **49**, 97-107.

Small, R. J., Clark, M. J. and Lewin, J. (1970). The periglacial rock stream at Clatford Bottom, Marlborough Downs, Wiltshire, *Proc. Geol. Ass.* **81**, 87-98.

Sparks, B. W. (1949). The denudation chronology of the dipslope of the South Downs, *Proc. Geol. Ass.* **60**, 165-215.

Stevens, A. J. (1959). Surfaces, soils and land use in north-east Hampshire, *Trans. Inst. Br. Geogr.* **26**, 51-66.

van den Broek, T. M. M. and van der Waals, L. (1967). The late Tertiary peneplain of south Limburg, *Soil Survey Papers* No. 3. Netherlands Soil Survey Institute, Wageningen.

Waters, R. S. (1960). The bearing of superficial deposits on the age and origin of the upland plain of east Devon, west Dorset and south Somerset, *Trans. Inst. Br. Geogr.* **28**, 89-95.

White, H. J. O. (1909). The geology of the country around Basingstoke, *Mem. Geol. Surv. U.K.*

White, H. J. O. (1912). A short account of the geology of the Isle of Wight, *Mem. Geol. Surv. U.K.*

White, H. J. O. (1925). The geology of the country around Marlborough, *Mem. Geol. Surv. U.K.*

Williams-Mitchell, E. (1956). The stratigraphy and structure of the chalk of Dean Hill anticline, Wiltshire, *Proc. Geol. Ass.* **26**, 221-7.

Woodward, R. B. and Ussher, W. A. E. (1911). The geology of the country near Sidmouth and Lyme Regis, *Mem. Geol. Surv. U.K.*

Wooldridge, S. W. and Linton, D. L. (1939). "Structure, surface and drainage in south-east England", *Trans. Inst. Br. Geogr.* **10** (revised and reprinted 1955: George Philip, London).

FOUR

# The sarsens of southern England: their palaeoenvironmental interpretation with reference to other silcretes

M. A. SUMMERFIELD AND A. S. GOUDIE
*School of Geography, University of Oxford*

Recent reassessment of the denudational scheme for south-eastern England outlined by Wooldridge and Linton (1939, 1955) has largely arisen from the realization that the Cenozoic deposits of the area have not been given sufficient consideration (Catt and Hodgson, 1976; Hodgson *et al.*, 1967, 1974). Few attempts have been made, for instance, to examine what palaeoenvironmental evidence might be provided by the Cenozoic silicified deposits (sarsens)[1]* which are so widely distributed in southern England (Kerr, 1955; Clark *et al.*, 1967; Dury, 1971; Whalley and Chartres, 1976; Dury and Habermann, 1978; Summerfield, 1979). The lack of deep burial of the Cenozoic beds, the discontinuous lenses of silicification, and the presence of rootlet holes, and even silicified roots, in many occurrences, show that these deposits have formed through surficial or penesurficial (near surface) silicification. They are, therefore, by definition, silcretes.[2] This chapter compares the petrographic, mineralogical and chemical characteristics of sarsens and silcretes in an attempt to provide provisional conclusions on the geomorphological and climatic conditions under which sarsens formed.

There is now a considerable body of literature on low latitude silcretes, particularly those from southern Africa and Australia, where they form extensive horizons up to 5 m or more in thickness. In southern Africa they occur both in association with deep weathering profiles (*weathering profile silcretes*) and with calcretes, playa (pan), aeolian and fluvial sediments and bedrock (*non-weathering profile silcretes*) (Summerfield, 1978).[3] Similar Cenozoic silicified deposits have been widely reported from France (meulières)[1] and elsewhere in north-western Europe (van den Broek and van der Waals, 1967; Dury and Habermann, 1978). Comparison with these more intensively studied occurrences, particularly those of the Paris Basin, may provide further insight into the denudational and palaeoenvironmental history

* Superscript numerals refer to Notes, which are to be found at the end of the chapter.

of southern England throughout the Cenozoic. European workers have been much concerned with the nature and role of climatic changes in the Cenozoic evolution of landforms whereas base-level changes have been the main interest of most British geomorphologists.

## Distribution

Although Cenozoic silicified deposits have been recorded in the British Isles from areas other than southern England (Montford, 1970), the great majority of occurrences are found to the south of a line from Lowestoft to the Severn estuary (Fig. 1).

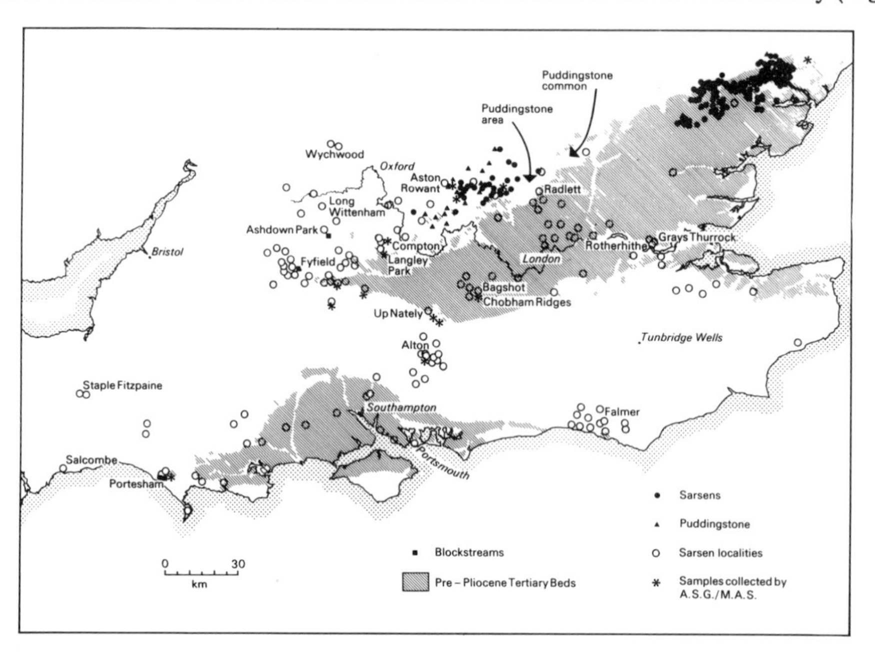


*Figure 1.* Distribution of sarsens in southern England.

This distribution is related to that of Cenozoic sediments or where these sediments were originally deposited. The present distribution of sarsen stones reflects, to a large extent, their removal by man, primarily for use as a building stone, and those remaining are only a vestige of the numbers in existence prior to man's arrival in Britain. Their present occurrence cannot, therefore, be assumed to indicate the detailed distribution of any original sarsen crust or the post-erosional remnants of such a crust.

Perhaps the greatest densities of sarsen stones are today found in Berkshire and Wiltshire, especially in the tributary valleys of the Kennet near Fyfield and Hackpen Hill (Whitaker, 1862; Adam, 1873; Carruthers, 1885; Ussher, 1906; White, 1906, 1907, 1912, 1923; Richardson, 1933) (Fig. 2). Similar valley concentrations occur in Dorset, such as in the Valley of the Stones, near Portesham (Strahan, 1898). Locally, very high densities are encountered. Small *et al.* (1970)

*Figure 2.* Sarsen valley train at Piggledene, near Fyfield, Wiltshire.

calculated that approximately 8000 to 10000 sarsen stones still lie in a 60 m by 750 m strip in Clatford Bottom, near Marlborough. These sarsen stone 'valley trains' have been produced through lateral transportation by solifluction under periglacial conditions (Clark *et al.*, 1967). Williams (1968) estimated that this process could cause movement over distances of up to 4 km over slopes as gentle as $1°30'$.

Sarsen conglomerates are common to the north-west of London (Hertfordshire Puddingstone) (Hopkinson, 1884; Green, 1890; Hopkinson and Whitaker, 1892; Hopkinson and Kidner, 1907; Sherlock, 1922; Ward, 1975) and further west towards High Wycombe (Bradenham Puddingstone) (Davies and Baines, 1953). In some areas, particularly in Hampshire and Sussex, sarsen conglomerates containing angular rather than rounded flint pebbles (flint breccias) are common (Codrington, 1870; White, 1913, 1924, 1926; Bury, 1922).

Equivalents to the sarsens of southern England are widespread in western Europe from Brittany to Bohemia and from the Garonne to Denmark. Figure 3 attempts to illustrate the main areas where 'sarsen-equivalents' occur, and gives some bibliographic guidance. In France they are generally called *meulières*, and in the Aquitaine basin and the Massif Central are a prominent component of the *sidérolithique* facies, often occurring in association with iron crusts (Millot, 1970). In parts of Belgium they are known as *rabots* (Auzel, 1930) and in Germany the blocks are given such names as *Zementquartzit*, *Braunkohlenquartzit* and *Findlings Quartzit* (Searle, 1923). Mid-latitude silcretes have also been reported from North America (Dury and Knox, 1971; Dury and Habermann, 1978).

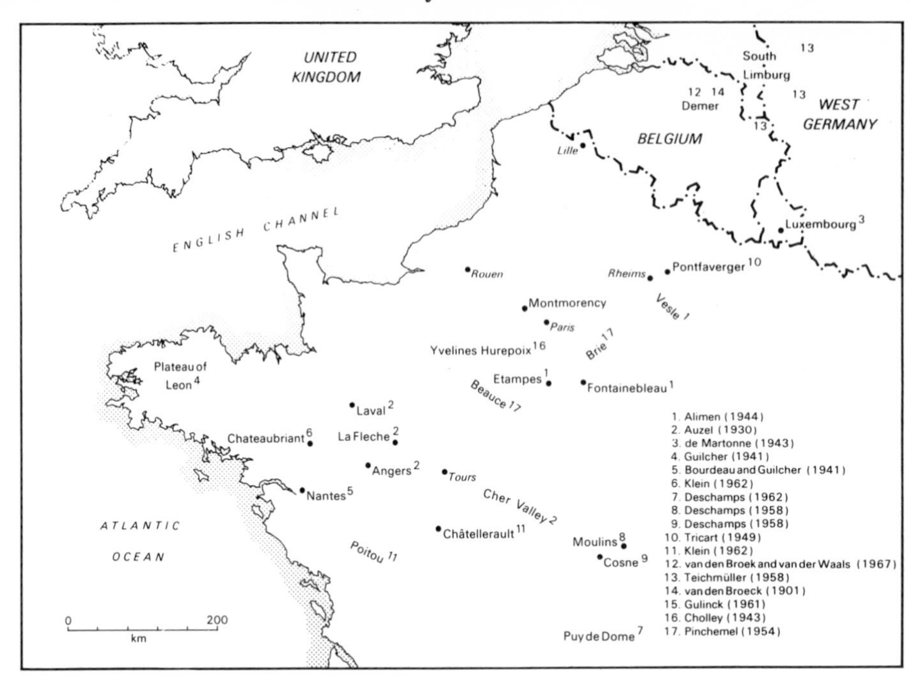


*Figure 3.* Distribution of meulière and allied silicified deposits in north-west Europe.

## Field occurrence and macromorphology

Two main types of sarsen can be distinguished on the basis of macromorphological characteristics; quartzitic varieties, which vary from those which fracture around grains to those so densely cemented by microcrystalline silica that they exhibit a sub-conchoidal to conchoidal fracture; and conglomerates, usually consisting of well-rounded flint pebbles in a quartzitic matrix (puddingstones) but in some cases composed of angular flint fragments (flint breccias). Silcretes similar to each of these types of sarsen have been recorded.

Sarsens occur predominantly as individual boulders (sarsen stones) and much more rarely as *in situ* horizons or irregular masses within various uncemented or poorly cemented beds. The *in situ* occurrences reported have almost invariably been associated with temporary artificial exposures and their *in situ* status is in doubt.

Sarsen stones vary in size from only a few centimetres across in Pleistocene gravels to boulders with long axes of 4 to 5 m (Whitaker, 1862; Sherlock, 1922; White, 1925). At Fyfield Down, near Marlborough, 77 per cent of the sarsens are between 0·3 m and 1·5 m across (Clark *et al.*, 1967). Although their three-dimensional measurements are often difficult to ascertain because of partial burial, many seem to have two axes distinctly longer than the third, giving the impression that they once formed a relatively thin layer, or layers, on, or within, the host sediment. However, their present form may be due more to jointing characteristics than to the original thickness of the silicified horizons. Boulders dislodged from silcrete horizons in southern Africa tend to be somewhat angular where columnar jointing, also a characteristic feature of many Australian occurrences, is well developed.

The hardness of sarsens, as measured by the Schmidt hammer (which gives an indirect measure of compressive strength (Day and Goudie, 1977)), is comparable to that of silcretes which, in southern Africa, commonly exhibit $R$ (rebound) values of 40 to 60. Some sarsens show evidence of case hardening on exposed surfaces, in many cases attributable to late stage surface precipitation of iron oxide. At Chobham Ridges, Surrey, a mean $R$ value of 41·2 was obtained for a fresh face on a sarsen stone compared with a mean value of 58·9 for the case hardened surface.

Many sarsen stones possess highly irregular surfaces which may be due to original variability in degree of silicification as well as to erosional modification. Channels up to 10 cm or more in depth have been found traversing the upper surface of some blocks in Piggledene, near Marlborough, Wiltshire, and some of these terminate as bowl-shaped depressions (Fig. 4). The smooth, rounded morphology of these features suggests that solution may have played a part in their formation.

Tubular cavities in sarsens, morphologically identical to those encountered in southern Africa silcretes, particularly those of the Kalahari area, have been widely observed. They range from numerous small tubular hollows up to 1 cm in diameter, noted in the sarsen stones at Avebury, to much larger features like those seen at Aston Rowant, between Oxford and High Wycombe (Fig. 5). Some of these are probably rootlet holes (Carruthers, 1885) while others may be associated with

*Figure 4.* Bowl-shaped depression on upper surface of a sarsen stone in Piggledene, near Fyfield, Wiltshire. The object for scale (a Schmidt hammer) is approximately 30 cm long. The crack visible on the right-hand side has been artificially produced.

*Figure 5.* Large tubular hollow in a sarsen stone at Chobham Ridges. Club hammer is approximately 25 cm long.

animal burrows. Silicified roots have been recorded in sarsens (Clark *et al.*, 1967) and in equivalent silicified materials in Europe (Pinchemel, 1954; van den Broek and van der Waals, 1967).

## Petrography

There are few observations in the literature on sarsen petrography. There is certainly nothing to match the classic monographs by Cayeux (1906, 1929) on the equivalent deposits in France. Brief descriptions of petrographic features are contained in many of the early accounts of sarsens and Kerr (1955) employed primarily petrographic criteria in interpreting sarsens as analogous deposits to silcretes. However, her comparative analysis was limited by a failure to appreciate the variety of petrographic features exhibited by silcretes from contrasting environments and formed in different host materials. Recent Whalley and Chartres

(1976) have presented a preliminary investigation of sarsen petrography employing both scanning electron microscopy and thin-section observations.

Petrographic studies of silcrete in Australia (Williamson, 1957; Smale, 1973; Taylor and Smith, 1975; Langford-Smith, 1978; Watts, 1978a) and southern Africa (Frankel and Kent, 1938; Mountain, 1951; Frankel, 1952; Summerfield, 1978) provide a sound basis for the comparative interpretation of sarsen petrography. Quartz constitutes both the dominant detrital and authigenic component of sarsens. Minor feldspar may be present together with a very minor heavy mineral fraction. Disseminated aggregates of sequioxides and clay minerals can be identified in some samples. This simple mineralogy is to a large extent paralleled by that of most silcretes, although the latter are somewhat variable in terms of their minor mineral component. Both anatase and iron oxide, in various states of hydration, are common in weathering profile silcretes, while finely disseminated orthoclase feldspar, and possibly glauconite, occur in the green pan silcretes of the Kalahari (Summerfield, 1978, p. 137).

A number of silcrete classifications based primarily, or solely, on petrographic criteria have been proposed (Passarge, 1904; Storz, 1926; Frankel, 1952; Hutton *et al.*, 1972; Smale, 1973). Some of these have been applied only on a regional basis to a limited range of silcrete types and there is clearly a need for a systematic classification applicable to all surficial and penesurficial silicified deposits.

Silcretes may be classified on the basis of skeletal grain content, skeletal grain relationships, matrix characteristics and major fabric differentiation features (Table I) (Summerfield, 1978, pp. 99-100, 1979). The GS-fabric varieties, in which the skeletal grains constitute a self-supporting framework, are classified according to the type of silica present between the skeletal grains. The F- and M-fabric types are subdivided on the basis of the presence or absence of glaebules (Brewer, 1964, pp. 259-60).

Silcretes exhibit a variety of specific petrographic features. Glaebules, from less than 1 mm to several centimetres across, occur both as nodules (undifferentiated internal fabric) (equivalent to Frankel's (1952) 'syngenetic nodules') and concretions (generally concentric fabric). Glaebules are locally abundant in silcretes, both in Australia and in southern Africa, where, except for inherited glaebules from host materials, they are confined to the weathering profile occurrences of the Cape coastal zone.[4] Colloform features (Frankel and Kent, 1938) occur as opaque, cream or reddish to brown concentrations of sesquioxides or titanium dioxide (as anatase) sometimes with iron oxide staining. Vughs (Brewer, 1964, p. 189), commonly equant to prolate but occasionally acicular or irregular, range in size up to 2 mm across. In Kalahari silcretes they are often lined with, and frequently completely filled by, megaquartz, microquartz, chalcedony (length-slow and length-fast) or by a combination of these (complex vugh-fills). In Cape coastal zone weathering profile silcretes vugh-fills by late-stage silicification are much less common and are usually incomplete.

Of the main silcrete fabric types (i.e. GS-, F-, M- and C-fabric) only the M-fabric variety appears to be absent in sarsens, although it does occur in the meulières of

TABLE I

*Petrographic classification of silcrete*

| Silcrete type | Characteristics |
|---|---|
| *Grain-supported fabric (GS-fabric)* | Skeletal grains (>30 μm across) constitute a self-supporting framework. |
| GS-fabric OC overgrowth silcrete | Intergranular spaces filled by optically continuous overgrowths. |
| GS-fabric chalcedonic overgrowth silcrete | Intergranular spaces filled by chalcedonic overgrowths. |
| GS-fabric microquartz silcrete | Matrix composed of microquartz, cryptocrystalline silica or opaline silica. |
| *Floating fabric (F-fabric)* | Skeletal grains (>30 μm across) 'float' in the matrix and do not form a self-supporting framework. Skeletal grain content >5%. |
| F-fabric massive silcrete | Glaebules absent. |
| F-fabric glaebular silcrete | Glaebules present. |
| *Matrix fabric (M-fabric)* | Skeletal grain (>30 μm across) content <5%. |
| M-fabric massive silcrete | Glaebules absent. |
| M-fabric glaebular silcrete | Glaebules present. |
| *Conglomeratic fabric (C-fabric)* | Detrital component includes pebbles (>4 mm across). |

France, being particularly common in silicified limestones. On the basis of samples examined in thin-section and reports in the literature, GS-fabrics are much more common in sarsens than in silcretes. The OC overgrowth variety is particularly abundant in sarsens, a microquartz matrix being less common and often localized, sometimes even in an individual thin-section. No samples with chalcedonic overgrowths have been encountered but in any case this variety is comparatively rare in silcretes.

Usually the optically continuous overgrowths in GS-fabric OC overgrowths sarsens are not readily distinguishable from the host detrital grains in thin-section. In some cases dust rings, usually of iron oxide, demarcate the original grain outline. Most rarely overgrowths are evident from occasional euhedral faces or from a lack of inclusions. Euhedral faces on overgrowths are common where void-filling is incomplete (Fig. 6).

*Figure 6.* GS-fabric OC overgrowth silcrete (sarsen) from near Avebury, Wiltshire. This is a common sarsen type, with cementation predominantly accomplished by overgrowths on large skeletal grains. Well-developed euhedral faces can be seen (arrowed bottom-centre) and separate stages of overgrowth development are visible on one grain (arrowed left) due to iron oxide coatings.

Although less common than GS-fabric varieties, F-fabric massive sarsens are petrographically indistinguishable from silcrete equivalents (Figs 7 and 8). As with the latter they vary considerably in their matrix-skeletal grain ratios, the sample from near Newbury, Berkshire illustrated in Fig. 7 having a very low skeletal grain content. Some of the skeletal grains display dissolution embayments and the matrix contains irregular zones of dull brownish material (possibly anatase) with localized iron oxide staining. This is supported by the chemical analysis of this sample (76-S-7) (see below). The matrix of F-fabric sarsens examined consists of microquartz with, in a few cases, localized zones of cryptocrystalline silica.

Glaebular fabrics have not been recorded in sarsens and could by definition only occur in association with F- and M-fabrics. Except where glaebules are inherited from the host material, as with the silicification of nodular calcrete in southern Africa, they appear to be confined to silcrete formed in association with weathering profiles.

Puddingstones can be categorized as C-fabric sarsens. They consist of flint (chert) pebbles in a GS-, or occasionally F-fabric, matrix. The flint pebbles, from 5 to 50 mm or more across, are usually well rounded, but also occur as highly angular fragments. A notable feature is that even where the matrix is predominantly of the GS-fabric OC overgrowth type, there is commonly a zone of about 2 mm around each pebble of microquartz rather than overgrowths within intergranular voids (Fig. 9). This may be due to late stage overgrowth and grain dissolution adjacent to

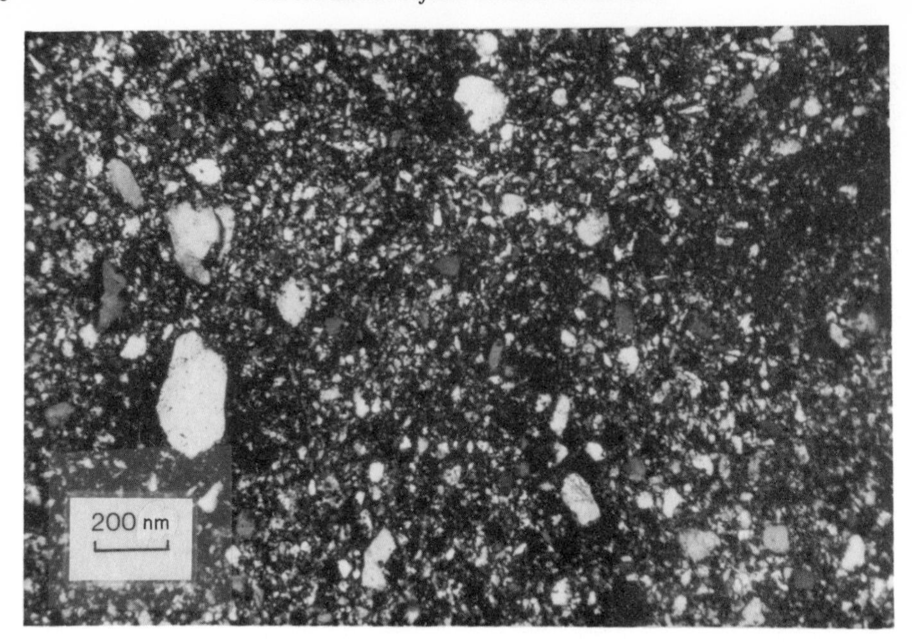


*Figure 7.* F-fabric massive silcrete (sarsen) from near Newbury, Berkshire. This sample has a high matrix content.

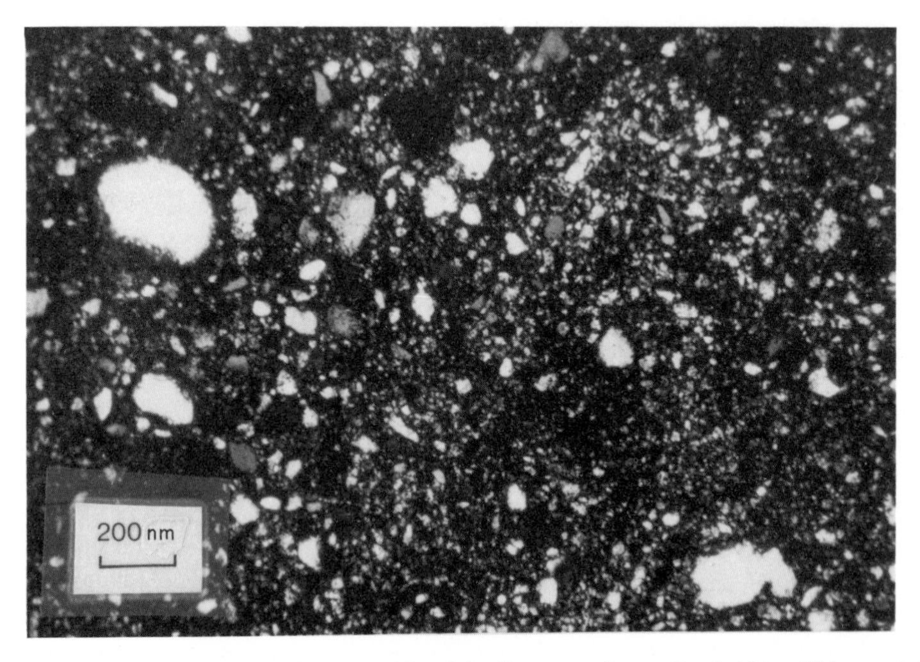


*Figure 8.* F-fabric massive silcrete from near Albertinia, Cape coastal zone, South Africa. This sample is typical of the silcretes found in association with deeply weathered Bokkeveld Shale in the Riversdale-Albertinia area in the Cape coastal zone.

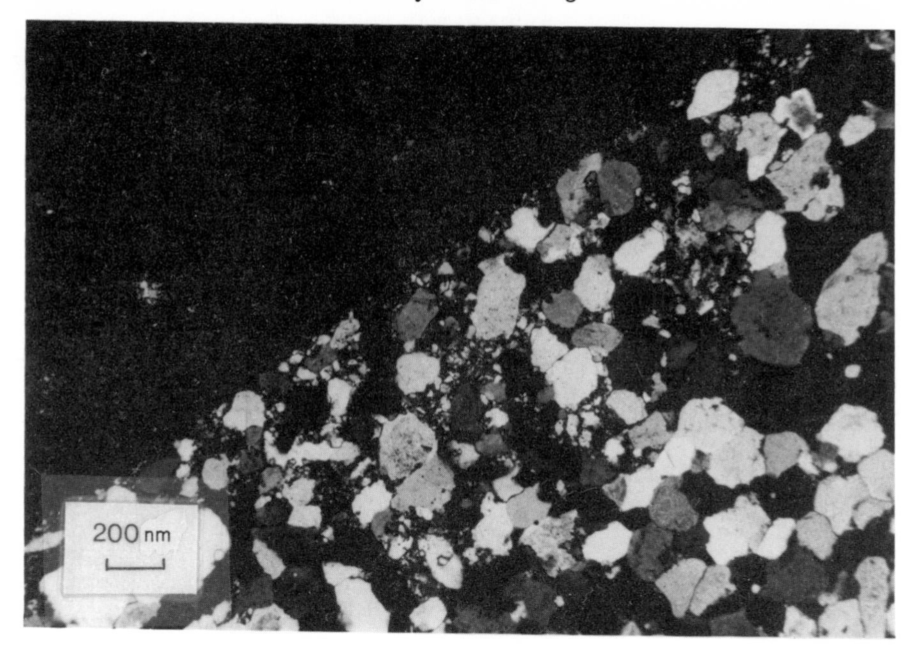


*Figure 9.* C-fabric silcrete (puddingstone) from Bradenham, Buckinghamshire. The transition from optically continuous overgrowths to microquartz can be seen as the pebble surface is approached.

the pebble-matrix interface, where migrating solutions would be concentrated, followed by later microquartz precipitation associated with decreasing permeability or changes in the characteristics of circulation solutions. Alternatively optically continuous overgrowths and microquartz may have formed contemporaneously, the concentrated solution movement in the pebble-matrix interface zone, or the presence of interfering ions, favouring microquartz precipitation rather than overgrowth development. In some puddingstones optically continuous overgrowths in the matrix are absent. Pebble-matrix contacts are usually very sharp, but in a few samples fretting on pebble surfaces has been observed. Angular flint fragments in conglomeratic sarsens have been recorded as being fractured prior to silicification (Bury, 1922). In some cases, however, such fracturing may be due to brecciation during diagenesis. Brecciation in association with pebble dissolution appears to have been responsible for the generation of an angular pebble silcrete conglomerate recorded from an exposure south of the Makgadikgadi Pans, Botswana (Summerfield, 1978, p. 370). Equant to prolate and irregular interconnected vughs are common along the matrix-pebble interface in this sample.

In addition to the comparison of general fabrics sarsens and silcretes may be compared in terms of specific petrographic features. In Table II estimates of the relative frequency of selected features recorded in thin-sections are presented for sarsens and for southern African silcretes, both from the Kalahari Beds and the Cape coastal zone (Summerfield, 1978, p. 279). The table shows that while the relative frequency of most specific petrographic features in sarsens differs from that in southern African silcretes, there are also marked differences between Kalahari

TABLE II

*Relative frequency of specific petrographic features in sarsens and southern African silcretes*

| Petrographic feature | Sarsens | Non-weathering profile (Kalahari Beds) silcretes | Weathering profile (Cape coastal zone) silcretes |
|---|---|---|---|
| Microquartz matrix | Common | Very abundant | Very abundant |
| Optically continuous overgrowths | Abundant | Absent | Absent* |
| Chalcedonic overgrowths | Absent | Uncommon | Absent |
| Glaebules | Absent | Rare | Common |
| Colloform features | Rare | Absent | Abundant |
| Vugh-fill features | Absent | Abundant | Uncommon |
| Length-slow chalcedony vugh-fills | Absent | Common | Rare |
| Length-fast chalcedony vugh-fills | Absent | Common | Absent |

* Optically continuous overgrowths have been recorded in silcretes not associated with weathering profiles in the Oliphants River area and to the west of Grahamstown.

Beds and Cape coastal zone silcretes. The significant point here is that the great variety of materials produced by surficial and penesurficial silicification under contrasting environmental conditions and in various host materials means that exact petrographic analogues of southern African silcretes are not necessarily to be expected in sarsens, at least with the same relative frequency.

The interpretation of fabric types in southern African silcretes (Summerfield, 1978, pp. 44-130) can be applied to sarsens. The common GS-fabric OC overgrowth type can be attributed to cementation through simple void-filling of a sandy quartzose host material. Overgrowth development implies a low intergranular clay content and that the circulating solutions were probably not highly charged with silica and had low concentrations of other ions. The presence of clay between the original quartz grains, or abundance of ions in the circulating solutions, would tend to give rise to microquartz or cryptocrystalline silica precipitation (Millot, 1970).

F-fabrics can be generated by skeletal grain displacement, dissolution or by replacement of an F-fabric host material such as a clayey weathering deposit containing floating quartz grains. In most, if not all, southern African silcretes, F-fabrics have been produced by replacement with, in some cases, subsidiary grain dissolution (Summerfield, 1978, pp. 123-24) and this observation would appear to be applicable to sarsens. This leads to the conclusion that F-fabric sarsens have not developed from sandy quartzose sediments but in weathering or other deposits containing floating quartz grains in a fine-grained or clayey matrix.

The fabric of C-fabric sarsens is clearly related to the host material in terms of the size and shape of the detrital components, and their matrix characteristics are determined by the same factors as outlined for the GS- and F-fabric types.

## Chemistry

Silcretes are, by definition, highly siliceous materials with only minor amounts of other elements (Table III). The $SiO_2$ content frequently exceeds 95% and exceptionally reaches 99%. The small but variable amounts of $Fe_2O_3$ and $Al_2O_3$ present appear to be related primarily to the degree of iron oxide staining and clay content. Titanium is of particular interest in that it is relatively abundant in silcretes associated with weathering profiles (commonly between 1·5 and 2% $TiO_2$) but is present in only trace amounts in non-weathering profile silcretes (usually less than 0·25% $TiO_2$). The amount of titanium present may, therefore be a useful indicator of silcrete origin in the case of non-*in situ* samples of unknown deriviation (such as sarsen stones). Only in highly acidic conditions can titanium be mobilized as a hydroxide and subsequently concentrated and precipitated in the matrix. Calcium may be a significant component in silcretes (especially in cal-silcretes). Potassium is relatively abundant in the (glauconite-rich?) green pan silcretes of the Kalahari area.

Unfortunately, apart from the puddingstone analysis reported by Kerr (1955) there are no published chemical analyses of sarsens known to the authors. Discussion is, therefore, limited to the five analyses produced by Summerfield (1978) (Table III). These suggest that sarsens are chemically identical to silcretes. The $TiO_2$ content is of particular significance. In four samples the amount of $TiO_2$ is negligible but sample 76-S-7 (the F-fabric massive silcrete referred to above) contains 2·08% $TiO_2$. As no non-weathering profile silcrete has such a high $TiO_2$ content it is very tempting to attribute this sample to weathering profile formation on this fact alone. However, further corroborative evidence for this is provided by the petrographic characteristics of this sample. Conversely, the low $TiO_2$ content of the other samples (all of which are of the GS- or C-fabric type) argues strongly against a weathering profile origin.

## Dating of sarsen formation

The dating of the period, or periods, of sarsen formation has been the subject of much debate and uncertainty. Even where *in situ* occurrences are present these are only capable of giving a maximum for silicification; that is, the age of the host material itself. For sarsens, all the probable host materials appear to be of Cenozoic age, although silcretes have been identified in pre-Cenozoic strata in Britain (Waugh, 1967, 1970a; Summerfield, 1978, p. 293). As with silcretes in Australia and southern Africa, fossil evidençe has been of little assistance in indicating the date of sarsen formation. In the case of the vast majority of occurrences which are not *in situ* the problem of dating is even more complex requiring a two-stage process of deduction of first establishing the host material, and then determining the most likely period of silicification.

In view of the lack of reliable reports of *in situ* sarsen occurrences in Britain and the relatively large number of well-documented observations of *in situ* exposures of meulières and associated deposits in France, the problem of dating sarsen formation can be usefully examined in this broader context. The proximity of these silicified

TABLE III

*Bulk chemical analyses of silcretes and sarsens*

| | $SiO_2$ | $Al_2O_3$ | $TiO_2$ | $Fe_2O_3$ | MgO | CaO | $K_2O$ | MnO | $P_2O_5$ | Ignition loss |
|---|---|---|---|---|---|---|---|---|---|---|
| Mean composition of silcrete from published and unpublished sources[1] | 95·84 | 0·66 | 1·27 | 0·77 | 0·15 | 0·12 | 0·09 | 0·01 | 0·03 | 1·06 |
| (no. of analyses in brackets) | (235) | (233) | (230) | (235) | (216) | (223) | (188) | (182) | (184) | (219) |
| Mean composition of 51 weathering profile silcretes from Cape coastal zone, South Africa[2] | 94·70 | 0·71 | 1·82 | 1·34 | 0·13 | 0·10 | 0·02 | 0·01 | 0·04 | 1·16 |
| Mean composition of 13 non-weathering profile silcretes from Kalahari Beds, southern Africa[3] | 93·92 | 1·43 | 0·13 | 0·66 | 0·55 | 0·12 | 0·68 | 0·07 | 0·03 | 2·41 |
| F-fabric massive sarsen (76-S-7)[4], Langley Park, north of Newbury | 96·12 | 0·29 | 2·08 | 0·80 | - | 0·03 | - | <0·01 | 0·02 | 0·66 |
| GS-fabric OC overgrowth sarsen (76-S-14), Aston Rowant, west of High Wycombe | 99·56 | 0·22 | 0·07 | <0·01 | - | <0·01 | - | <0·01 | 0·01 | 0·14 |

TABLE III (cont.)

| | $SiO_2$ | $Al_2O_3$ | $TiO_2$ | $Fe_2O_3$ | MgO | CaO | $K_2O$ | MnO | $P_2O_5$ | Ignition loss |
|---|---|---|---|---|---|---|---|---|---|---|
| C-fabric sarsen (76-S-17), East Tilstead, Hampshire | 98·56 | 0·20 | 0·16 | 0·07 | - | 0·02 | - | <0·01 | 0·02 | 0·96 |
| C-fabric sarsen (76-S-20), Flaunden, Buckinghamshire | 97·90 | 0·24 | 0·02 | 0·54 | 0·37 | 0·02 | - | <0·01 | 0·04 | 0·88 |
| GS-fabric OC overgrowth sarsen (76-S-27), Piggledene, Wiltshire | 98·79 | 0·15 | 0·03 | - | 0·47 | 0·02 | - | 0·01 | - | 0·54 |

Data sources:

1. Bosazza (1936, 1937, 1938, 1939); Frankel (1952); Frankel and Kent (1938); Gunn and Galloway (1978); Hutton *et al.* (1972); Joplin (1963); Netterberg (1969) in Goudie (1973, p. 28); Senior and Senior (1972); Summerfield (1978); Taylor and Smith (1975); Visser (1964); Watts (1977); Williamson (1957).
2. Summerfield (1978, Table 6.6, p. 151). All analyses by XRF and normalized to 100%.
3. Summerfield (1978, Table 6.5, p. 150). All analyses by XRF and normalized to 100%.
4. All sarsen analyses from Summerfield (1978, Table 8.3, p. 282). All analyses by XRF and normalized to 100%.

## TABLE IV

*Sarsen and meulière occurrences in the Palaeogene succession in Britain and France*

| | | HAMPSHIRE BASIN | LONDON BASIN | PARIS BASIN & AQUITAINE | | |
|---|---|---|---|---|---|---|
| MIOC | AQUITANIAN? CHATTIAN | | | Calcaire de Beauce* Calcaire de l'Orléanais (31,32) | Aquitanien | |
| OLIGOCENE | RUPELIAN | Absent ⇧ Upper Hamstead Beds | | Meulière de Montmorency* Calcaires d'Etamps (28,33) Sables et grès de Fontainebleau | Stampien | STAMPIEN |
| | LATTOFRIAN | Lower Hamstead Beds* | | Calcaire de Sannois Caillasses d'Orgement Calcaire de Brie* (29,30,32) | Sannoisien | |
| | | Bembride & Osborne Beds* Upper & Middle Headon Beds | | Marnes blanches de Pantin* Marnes bleues d'Argenteuil* Gypse et marnes intercalées* | Ludien | BARTONIAN |
| EOCENE | BARTONIAN | Lower Headon Beds* Barton Beds | Absent ⇧ | Sables de Marnes Calcaire de St. Ouen* Sable de Cresnes Palaeosols post-auversiens* | Marinésien | |
| | AUVERSIAN | Upper Bracklesham Beds | Upper Bagshot Beds (24-27) | Sables et grès de Beauchamp Sable d'Auvers Calcaire de St. Ouen* | Auversien | |
| | LUTETIAN | Middle Bracklesham Beds | Middle Bagshot Beds* (25) | Calcaire grossier Calcaire de Provins | LUTETIEN | |
| | CUSIAN | Lower Bracklesham Beds (2) | Lower Bagshot Beds* (25) | Grès de Belleu Sables à Unios Argile de Laon Sables de Cuise | Cuisien | YPRESIAN |
| | YPRESIAN | London Clay | London Clay | | | |
| PALAEOCENE | SPARNACIAN | Woolwich & Reading Beds* (1) | Woolwich & Reading Beds* (8-23) | Argile à lignite* Argile plastique* (30?) | Sparnacien | |
| | THANETIAN | Absent | Thanet Sands (3-7) | Sables de Bracheux (34) | THANETIEN | |
| | MONTIAN | Absent | Absent | Les Meulières de Vertus (30) | DANO-MONTIEN | |
| CRETACEOUS | | | | | | |

Sources for stratigraphy: Curry (1966), Pomerol (1973) and Rayner (1967).
Sources for sarsen and meulière occurrences: (1) Arkell (1947); (2) Reid (1902); (3) Dines and Edmunds (1925); (4) Hinton and Kennard (1899-1900); (5) Dines *et al.* (1954); (6) Dewey *et al.* (1924); (7) Dewey *et al.* (1924); (8) Sherlock and Pocock (1924); (9) Whitaker (1889); (10) Blundell and Monckton (1911); (11) White (1909); (12) Hopkinson and Kidner (1907); (13) Prestwich (1854); (14) Whitaker (1885); (15) Boswell (1927); (16) Hopkinson and Whitaker (1892); (17) Green (1890); (18) Hopkinson (1884); (19) Whitaker (1862); (20) Boswell (1929); (21) Sherlock (1922); (22) Sherlock and Noble (1922); (23) White (1909); (24) Jones (1901); (25) Arkell (1947); (26) Dewey and Bromehead (1915); (27) Dines and Edmunds (1929); (28) Alimen (1944); (29) Cailleux (1947); (30) Tricart, (1949); (31) Cholley (1943); (32) Cayeux (1929); (33) Kulbicki (1953); (34) Pomerol (1973).
* Unit mainly or wholly of non-marine origin.

deposits, and the probability of a broadly similar climatic and tectonic history being experienced in both southern Britain and north-western France during the Cenozoic, suggests that conditions conducive to sarsen/meulière formation probably occurred contemporaneously in the two areas.

The detailed correlation of Cenozoic strata between the Hampshire and London Basins and the Paris and Aquitaine Basins is still uncertain but Table IV attempts to present a consensus of views of both British and French workers (Curry, 1966; Rayner, 1967; Pomerol, 1973). *In situ* occurrences of sarsens and meulière reported in the literature are indicated in the appropriate stratigraphic units. Unfortunately reports of *in situ* sarsen occurrences must be treated with caution as they probably describe sarsens associated with sediments from particular units rather than formed within them (Clark *et al.*, 1967). Nevertheless, Table IV does suggest the occurrence of sarsens in association with strata ranging in age from Sparnacian to Auversian. In the more complete Palaeogene succession in the Paris and Aquitaine Basins meulière are recorded in Palaeocene, Oligocene and early Miocene sediments. Further afield, Gulinck (1961) has proposed a Miocene age for 'erratic' quartzitic boulders associated with sands and lignite in Belgium and silicification throughout the Cenozoic until the Late Pliocene has been proposed for the silicified sands of South Limburg (van den Broek and van der Waals, 1967).

Although Clark *et al.* (1967) considered the possibility of three main phases of sarsen formation (Thanetian and Sparnacian, Cuisian to Auversian, and Bartonian) they pointed out that the main period of silicification may have occurred later in conjunction with an erosion surface transecting various Eocene formations. Clark *et al.* (1967) argued that in the Marlborough area erosion during this period, which was almost certainly post-Bartonian, probably cut across the Chalk. In this case re-worked Cenozoic deposits overlying the Chalk may have given rise to a sarsen crust which after renewed erosion was broken up to produce the sarsen stones found today. Kerr (1955) maintained that the presence of sarsen pebbles in the Blackheath Beds reported by Sherlock (1947) proves that sarsen formation was pre-Sparnacian. A possibility here is that sarsen pebbles were transported laterally and incorporated into the Blackheath Beds at a later date. The fact that Sherlock (1947) describes the sarsen pebbles as being rare tends to support this explanation.

Although *in situ* occurrences are much more common in France than southern England the problem remains of whether silicification was contemporaneous with, or post-dated, deposition of the host sediment. All the major meulière occurrences are in beds of Aquitaine age or older, presumably because tectonic instability produced geomorphological conditions inimical to surficial or penesurficial silicification during and after the late Miocene. Since Miocene deposits are absent from southern England (and no late Oligocene formations are to be found in the London Basin), the presence of silicified deposits of late Oligocene and early Miocene age in the Paris Basin is of particular interest. Cholley (1943) has proposed a Miocene meulière surface to account for many of the occurrences in the Paris Basin, although he also suggested Eocene and Oligocene silicification. In view of the proximity of the two areas, any meulière formation during the early Miocene and probably late Oligocene in the Paris Basin would be expected to be accompanied by sarsen formation in southern England assuming appropriate local climatic and geomorphological conditions.

## Origin and palaeoenvironmental interpretation of sarsens

The similarities evident between southern African silcretes and sarsens (and other similar mid-latitude silicified deposits) in terms of their macromorphological, petrographic, mineralogical and chemical characteristics strongly suggest that they share common origins.

A variety of models have been proposed for silcrete formation (Goudie, 1973, pp. 121-46). These include fluvial (Stephens, 1964, 1966, 1971), sheet flood (Teakle, 1936; Litchfield and Mabbutt, 1962), pan/lacustrine (Du Toit, 1954), detrital (Haughton *et al.*, 1937; Frankel, 1952), capillary rise (Woolnough, 1927, 1928, 1930; Visser, 1937; Frankel and Kent, 1938; Williamson, 1957; Grant and Aitchison, 1970), groundwater/watertable fluctuation (Conrad, 1969, p. 333; Senior and Senior, 1972; Smale, 1973; Wopfner, 1974; Alley, 1977) and downward percolation (*per descensum*) models (Whitehouse, 1940; David and Browne, 1950; Teakle, 1950; Prescott and Pendleton, 1952; Jessup, 1961; Watts, 1978b).

Few of the numerous descriptive accounts of sarsens consider either their origin or the environmental conditions pertaining during their formation. Kerr (1955) briefly considered sarsen genesis and by a general comparison with silcretes saw the applicability of Woolnough's (1930) capillary rise model. In considering the sarsens of the Marlborough area Clark *et al.* (1967) also seemed to favour a capillary rise mechanism since they envisaged a climatic regime during formation "marked by (i) heavy seasonal rainfall, and (ii) a dry season in which evaporation exceeded precipitation".

By analogy with the chemical and petrographic characteristics of southern African silcretes, formation in weathering profiles is highly unlikely for most sarsens (i.e. GS-fabric, $TiO_2$-poor samples). On petrographic evidence most GS-fabric sarsens represent passively cemented quartz sands. F-fabric sarsens, however, may have originally formed part of a weathering profile. This seems particularly likely where the skeletal grain content is low and where $TiO_2$ is relatively abundant (e.g. sample 76-S-7, Photo 5). This distinction between 'weathering profile' and 'non-weathering profile' sarsens is important because it implies contrasting host materials, modes of formation and environments.

Silica may be precipitated by evaporation, cooling, a change in pH from above to below 9, reaction with certain cations, adsorption on to solids, reaction with organic compounds and the life processes of organisms (Siever, 1962). Evaporation is only likely to be effective in producing a very thin surface crust, as in the experimental work of Paraguassu (1972), since the very formation of this crust drastically reduces evaporation of silica solutions beneath it. The effect of cooling on silica solubility (Alexander *et al.*, 1954; Krauskopf, 1956; Siever, 1962) is insufficient to make it a significant mechanism in the temperature ranges experienced in surficial or pene-surficial environments. The marked decrease in silica solubility below pH 9 provides an important precipitation mechanism where this threshold is frequently crossed. The role of cations and the effect of adsorption on silica solubility is complex. There is general agreement about the efficacy of $Al^{3+}$ in reducing silica solubility (Iler, 1955, pp. 14-16; Okamoto *et al.*, 1957; Siever, 1972) but contradictory evidence for the effect of other ions (e.g. Greenberg and Price, 1957; Wey and Siffert, 1961; Keller and Reesman, 1963). The ability of aluminium and iron oxides to adsorb silica is well documented (Jones and Handreck, 1963, 1965; McKeague and Cline, 1963; McPhail *et al.*, 1972) but the effect is also present with other oxides (Harder, 1965, 1971; Harder and Flehmig, 1970). Although silica abstraction from undersaturated solutions by silica-secreting organisms is well-known (Jørgensen,

1953), in general the effect of organic compounds on silica solubility and precipitation is poorly understood. It appears that the presence of different organic compounds may increase or decrease silica solubility (Emery and Rittenberg, 1952; Mandl *et al.*, 1952; Siever and Scott, 1963; Evans, 1964).

The presence of optically continuous overgrowths in many GS-fabric sarsens implies low cation concentrations, and pH fluctuations may have been important in promoting silica solution and re-precipitation within the host sediment. However, as silica concentrations in surface and ground waters (Livingstone, 1963; Davis, 1964) are frequently above the equilibrium of quartz (approximately 4-11 p.p.m. below pH 9 at 25 °C (Morey *et al.*, 1962; Wanneson, 1963; MacKenzie and Gees, 1971; Siever, 1972), although well below that of amorphous silica (approximately 100-120 p.p.m. below pH 9 at 25 °C (Alexander *et al.*, 1954; Siever, 1957; Morey *et al.*, 1962)), given sufficient time quartz could precipitate from interstitial solutions without pH fluctuations of the influence of any other factor. Where microquartz has been precipitated as an interstitial cement the presence of cations may have inhibited overgrowth development.

In F-fabric sarsens, where replacement of an F-fabric host material is implied, abundant cations would have been available in the host material matrix to encourage the development of microquartz or cryptocrystalline silica rather than well-ordered macrocrystalline overgrowths. If the matrix of the host material was originally clay, replacement by silica would have been accompanied by the release of abundant $Al^{3+}$ ions, such aluminium mobilization implying highly acidic leaching conditions (Loughnan, 1969, p. 33) or the action of organic complexes.

For silcretes formed in association with weathering profiles the release of silica from silicate minerals within the profile provides the most probable silica source and this may apply to some F-fabric sarsens. However, for sarsens manifestly not originating in weathering profiles, silica could not have been provided by *in situ* silicate decomposition although it may have been introduced by surface waters from a remote site of chemical weathering, as suggested by Stephens (1964, 1966, 1971) for the silcretes of inland Australia. This silica source seems particularly applicable to silicified fluvial sediments. Dissolution and reprecipitation of quartz *in situ* provides another possible silica source, especially if the surface environment was sufficiently arid to generate 'high solubility' quartz dust (Clelland and Ritchie, 1952; Dempster and Ritchie, 1952; Bergman *et al.*, 1963) and enhanced grain surface solubilities through abrasion (Kitto and Patterson, 1942) due to aeolian action. Alternatively quartz dust may have been produced externally and transported to the sarsen forming environment. Contemporary silica fall-out rates may reach up to 5 m $Ma^{-1}$ (0·005 mm $a^{-1}$) (Summerfield, 1978, p. 211). If the presence of optically continuous overgrowths is taken to indicate the presence of solutions low in cations, this favours an atmospheric rather than a fluvial silica source for GS-fabric OC overgrowth sarsens. Surface waters would introduce a wide range of dissolved products of which silica would probably be a minor constituent, whereas aeolian transported dust from arid areas can consist of up to 75% silica (Junge, 1958).

Capillary rise models of sarsen formation involving surface evaporation present two main difficulties; the limited vertical distance over which capillary rise can occur, especially in relatively coarse materials such as sand (Mohr and van Baren, 1954, p. 80; Baver, 1956, p. 249), and the drastically reduced rate of evaporation once cementation commences at the surface. The downward percolation (*per descensum*) model is a more viable possibility and has been recently favoured in accounting for subsurface silcrete development (Summerfield, 1978; Watts, 1978b). Silica supplied from above, from atmospheric inputs, surface waters, or taken into solution by waters percolating through the host material or an overlying deposit, may be precipitated by various mechanisms below the surface. Once initiated cementation would lead to reduced porosity and permeability. This would in turn aid the silica precipitation process through the ponding of downward percolating waters above the zone of decreasing permeability. The zone of silicification, which could still be near enough to the surface to be in the zone of root activity (which in semi-arid environments may be tens of metres below the surface), may be initially determined by variations in permeability within the host material. Water-table fluctuations, produced by seasonal or longer term variations in infiltration, could also lead to a zone of silicification. Silica may also be precipitated from shallow groundwater at favourable sites, but this mode of formation would not necessarily be expected to produce laterally extensive but relatively thin zones of silicification. Due to the lack of *in situ* sarsen occurrences, and doubts about their age and the climatic and geomorphological conditions pertaining during their formation, palaeoenvironmental deductions from sarsens can best be made with reference to low latitude silcretes.

There is general agreement in the literature that silcrete formation is restricted to land surfaces with minimal relief. Some earlier workers linked the development of silcrete, and duricrusts in general, to the peneplain (Woolnough, 1927, 1930; Visser, 1937; Frankel and Kent, 1938), an unfortunate mis-use of the term in view of the palaeoclimatic inferences they also made (Langford-Smith and Dury, 1965). Although silcrete formation has been reported on slopes of up to 4·5 degrees (Mabbutt, 1965), relief must be insufficient to promote rapid drainage of the subsurface or erosion active enough to destroy the silcrete horizon as it forms. Local foci of internal drainage (pans) where silica in surface waters could be concentrated by evaporation, appear to be favoured sites for silcrete formation in the Kalahari. Sedimentation rates must also be low otherwise the landsurface would not be stable for a period of time sufficient to allow a silcrete horizon to form. Relief subdued enough for extensive periodic inundations by floodwaters is required by Stephens' (1971) model.

The evidence for the climatic conditions favouring silcrete formation is more equivocal. The majority of workers have advocated an arid or semi-arid climate but more recently this view has been questioned (Summerfield, 1978; Wopfner, 1978; Young, 1978). Although there appears to be strong evidence of silcrete formation in the Kalahari area under a relatively arid climate, and some occurrences may indeed be forming at present (MacGregor, 1931; Wayland, 1953; Smale, 1973), it is difficult to see how the silcretes of the Cape coastal zone could have formed under such a climatic regime. As argued above, petrographic and chemical evidence indicating aluminium and titanium mobilization during silcrete formation in weathering profiles points to at least periodically high acidity (probably related to high rates of organic activity); such conditions cannot easily be reconciled with semi-aridity. Similar conclusions have been reached by Watts (1977) who suggested an initial humid phase followed by a subsequent drier period during silcrete formation in northern New South Wales and southern Queensland. Mohr and van Baren (1954) and Whitehouse (1940) have described silicification within a lateritic profile under a humid tropical climate. These observations point to the generalization, for southern African occurrences at least, that silcretes formed in association with weathering profiles developed under a humid climate while those not associated with weathering profiles developed under an arid or semi-arid climate.

From these provisional conclusions the common GS- and C-fabric (non-weathering profile) sarsens can be inferred to have formed on a stable landsurface, of minimal relief and with low rates of sedimentation and erosion, under a semi-arid or arid climate. In contrast, strong circumstantial evidence of a weathering profile origin for at least some F-fabric sarsens implies similar geomorphological, but much more humid climatic, conditions during formation.

The considerable climatic changes suggested by these conclusions are supported by palaeoclimatic evidence from other Cenozoic deposits. Although bauxite, lignite and laterite associated with some of the basalt sequences of the North Atlantic early Cenozoic igneous province, together with vertebrate, plant and microfossil evidence, point to a relatively uniform warm humid climate in the early Cenozoic throughout the North Atlantic area (Nilsen and Kerr, 1978), palaeoclimatic evidence from the continental deposits of north-west Europe indicates much greater climatic variability in the mid- and late-Cenozoic. In France, the meulières are only one of many indicators of environmental conditions during this period. The climate of the Paris and Aquitaine Basins seems to have enabled the formation at different times of deeply weathered lateritic and kaolinitic horizons (Rondeau, 1975), croûte calcaire and encroûment calcaire (calcretes) (Prost, 1961; Crouzel and Meyer, 1975; Menillet, 1975), evaporitic beds, and ergs (Dewolf and Mainguet, 1976). These indicators suggest that climatic conditions ranged from arid through to humid.

## Conclusion

The previous neglect of sarsens by geomorphologists and geologists attempting to reconstruct the denudational history of southern England may perhaps be explained by the problems of interpreting such scattered and discontinuous deposits. The realization that analogous materials occur in a number of present-day tropical and sub-tropical environments provides the basis for a comparative analysis of sarsens. The evidence of a surface or near surface origin for sarsens confirms their status as silcretes. Since it is now possible to make some broad provisional deductions about environmental conditions during silcrete formation this can also be attempted, by analogy, for sarsens.

The polygenetic nature of silcretes and the differentiation of contrasting environments during formation on the basis of petrographic and chemical characteristics leads to the idea that sarsens developed both under humid and semi-arid to arid conditions. GS-fabric sarsen formation probably occurred through silicification of a sandy quartzose host sediment by percolating waters charged with silica supplied by aeolian or perhaps fluvial inputs or by *in situ* dissolution of the host sediment above the silicified zone. Water table fluctuations may have also aided silica concentration. The lack of an associated weathering profile indicates a semi-arid or arid climate during silicification. C-fabric sarsens probably formed in the same way, the only difference being the nature of the host material.

The apparently much rarer F-fabric sarsens which, in the case of at least one sample, show identical petrographic and chemical characteristics to the weathering profile silcretes of the Cape coastal zone of southern Africa, are regarded as having a distinctly different origin to the GS- and C-fabric types. Formation within a weathering profile, through the *in situ* mobilization and precipitation of silica under a relatively humid climate, is suggested.

References to the European literature on sarsen-equivalents and other weathering phenomena formed during the Cenozoic attest to a complex sequence of arid and humid phases rather than the stable warm humid climate often proposed for this period in Britain. Further comparative studies between sarsens and the equivalent deposits of north-west Europe are likely to provide a fruitful contribution to our understanding of sarsen genesis, and, more generally, the denudation chronology and palaeoenvironmental history of southern England. The problem of dating sarsen formation should be aided by reference to the more complete Cenozoic stratigraphic record preserved in north-west Europe.

## Acknowledgements

We are grateful to Drs R. J. Small and M. J. Clark who provided initial bibliographic material. Drs R. J. Small, N. L. Watts and G. Taylor kindly commented upon an early draft of this paper. This work was carried out while M.A.S. held a Natural Environment Research Council Research Studentship. The field work expenses of M.A.S. in South Africa were supplemented by a Fellowship from the Trustees of the Sir Henry Strakosch Memorial Trust. The Marico Mineral Co. (Pty.) Ltd. provided assistance in South Africa and subsequent financial support. The figures were drawn by Miss Margaret Loveless.

## Notes

(1) The terminology which has been applied to Cenozoic silicified deposits requires clarification. The following definitions are used here:

| | |
|---|---|
| Sarsens | Cenozoic siliceous deposits found in Britain, whether *in situ* or not, formed by surficial or penesurficial silicification. |
| Sarsen stones | Individual sarsen boulders (not *in situ*). |

| | |
|---|---|
| Puddingstones | Conglomeratic sarsens with rounded pebbles. |
| Flint breccia | Conglomeratic sarsens with angular pebbles. |
| Meulières | Cenozoic siliceous deposits found in France (equivalent to sarsens). |

(2) The following definition is applied here: Silcrete is an indurated product of surficial or penesurficial silicification formed by the cementation and/or replacement of bedrock, weathering deposits, soil, unconsolidated sediments or other materials through low temperature physico-chemical processes and not by metamorphic, volcanic, plutonic or moderate to deep burial diagenetic processes.

An arbitrary minimum $SiO_2$ content for silcrete of 85 per cent is employed. Application of the term 'silcrete' to sarsens is supported by the original use of the word. Lamplugh (1902) coined the term 'silcrete' to describe silicified superficial deposits in Britain and only subsequently was it applied to similar tropical occurrences (Lamplugh, 1907). To avoid the proliferation of parochial terminology it would be desirable to describe sarsens and allied materials simply as silcretes. The term sarsen is retained in this paper for the purpose of clarity in making comparisons.

(3) The distinction between *weathering profile* and *non-weathering profile* silcretes merits emphasis since it is critical to the interpretation of silcretes and sarsens. In the weathering profile type, silcrete occurs as irregular silicified zones or as a laterally extensive layer or layers either within the profile itself, or forming the uppermost unit of the profile. In southern Africa this type of silcrete appears to be restricted to the Cape coastal zone (see Note 4). Non-weathering profile silcretes include silicification of bedrock, unconsolidated and unweathered sediments, calcrete and the like. There is no associated marked chemical weathering in the sense of decomposition of bedrock or sediments and no associated deep weathering profile. In southern Africa non-weathering profile silcretes are largely confined to the Kalahari area.

(4) The term Cape coastal zone is used to describe the area in which the main weathering profile silcrete occurrences in southern Africa are found. It consists of a coastal belt up to 80 km wide stretching from the Oliphants River, some 270 km north of Cape Town, in the west, to the vicinity of East London in the east.

## References

Adams, J. (1873). On the sarsen stones of Berkshire and Wiltshire, *Geol. Mag.* **10**, 198-202.

Alexander, G. B., Heston, W. M. and Iler, R. K. (1954). The solubility of amorphous silica in water, *J. Phys. Chem., Ithaca* **58**, 453-455.

Alimen, H. (1944). Roches gréseuses à ciment calcaire du Stampien, étude pétrographie, *Bull. Soc. géol. Fr.* (Ser. 5) **14**, 307-328.

Alley, N. F. (1977). Age and origin of laterite and silcrete duricrusts and their relationship to episodic tectonism in the mid-north of South Australia, *J. Geol. Soc. Aust.* **24**, 107-116.

Arkell, W. J. (1947). The geology of the country around Weymouth, Swanage, Corfe and Lulworth, *Mem. Geol. Surv. U.K.*

Auzel, M. (1930). Premiers résultats d'une étude des meulières du Bassin de Paris, *Revue Géogr. phys. Géol. dyn.* **3**, 303-362.

Baver, L. D. (1956). "Soil physics", 3rd edn. Wiley, New York.

Bergman, I., Cartwright, J. and Caswell, C. (1963). The disturbed layer on ground quartz powders of respirable size, *Br. J. Appl. Phys.* **14**, 399-401.

Blundell, G. E. and Monckton, H. W. (1911). Excursion to Hook Nately and Basingstoke, *Proc. Geol. Ass.* **22**, 240-243.

Bosazza, V. L. (1936). Notes on South African materials for silica refractories, *Trans. Geol. Soc. S. Afr.* **39**, 465-478.

Bosazza, V. L. (1937). The physical properties of some South African raw materials for silica bricks, *J. Chem. Metall. Min. Soc. S. Afr.* **37**, 590-602.

Bosazza, V. L. (1938). Reply to the discussions on his paper entitled "Notes on South African materials for silica refractories", *Proc. Geol. Soc. S. Afr.* **41**, 46-51.

Bosazza, V. L. (1939). The silcretes and clays of the Riversdale-Mossel Bay area, *Fulcrum, Johannesburg* **2**, 17-29.

Boswell, P. G. H. (1927). The geology of the country near Ipswich, *Mem. Geol. Surv. U.K.*

Boswell, P. G. H. (1929). The geology of the country around Sudbury, *Mem. Geol. Surv. U.K.*

Bourdeau, J.-M. and Guilcher, A. (1941). Observations sur l'Éocène continental de la banlieue nantaise, *C. r. somm. Séanc. Soc. géol. Fr.* 68-70.

Brewer, R. C. (1964). "Fabric and mineral analysis of soils". Wiley, New York.

Bury, H. (1922). Some high level gravels of north-east Hampshire, *Proc. Geol. Ass.* **33**, 81-103.

Calleux, A. (1947). Concretions quartzeuses d'origine pédologique, *Bull. Soc. geol. Fr.* (Ser. 5) **17**, 475-482.

Carruthers, W. (1885). Notes on fossil roots in the sarsen stones of Wiltshire, *Geol. Mag. New Ser. Decade III* **2**, 361-362.

Catt, J. A. and Hodgson, J. M. (1976). Soils and geomorphology of the Chalk in south-east England, *Earth Surface Processes* **1**, 181-193.

Cayeux, L. (1906). "Structure et Origine des Grès du Tertiaire Parisien". Ministère des trav. publ., Paris.

Cayeux, L. (1929). "Les Roches Sédimentaires de France", Vol. 1. Ministère des trav. publ., Paris.

Cholley, A. (1943). Recherches sur les surfaces d'érosion et la morphologie de la région parisienne, *Annls. Géogr.* **52**, 1-19; 81-97; 161-189.

Clark, M. J., Lewin, J. and Small, R. J. (1967). The sarsen stones of the Marlborough Downs and their geomorphological implications, *Southampton Res. Ser. in Geogr.* **4**, 3-40.

Clelland, D. W. and Ritchie, P. D. (1952). Physico-chemical studies on dusts. II. Nature and regeneration of the high-solubility layer on siliceous dusts, *J. Appl. Chem. Lond.* **2**, 42-48.

Codrington, T. (1870). On the superficial deposits of the south of Hampshire and the Isle of Wight, *Q. J. Geol. Soc. Lond.* **26**, 528-551.

Conrad, G. (1969). "L'Evolution Continentale Post-Hercynienne du Sahara Algérien", Thèse Sci., Paris.

Crouzel, F. and Meyer, R. (1975). Encroûtements calcaires dans L'Oligo-Miocene du Bassin d'Aquitaine, *Bull. Soc. géol. Fr.* **17** (Suppl. 4) 112-114.

Curry, D. (1966). Problems of correlation in the Anglo-Paris-Belgium basin, *Proc. Geol. Ass.* **77**, 437-467.

David, T. W. E. and Browne, W. R. (1950). "The geology of the Commonwealth of Australia", Vols 1-3. London.

Davies, A. M. and Baines, A. H. J. (1953). A preliminary survey of the sarsen and pudding-stone blocks of the Chilterns, *Proc. Geol. Ass.* **64**, 1-9.

Davis, S. (1964). Silica in streams and groundwater, *Am. J. Sci.* **262**, 870-891.

Day, M. J. and Goudie, A. S. (1977). Field assessment of rock hardness using the Schmidt Test Hammer, *Br. Geomorph. Res. Grp. Tech. Bull.* **18** (Shorter Technical Methods II) 19-29.

De Martonne, E. (1943). Le relief du Luxembourg d'après G. Baeckeroot, *Annls Géogr.* **52**, 301-304.

Dempster, P. B. and Ritchie, P. D. (1952). Surface of finely-ground silicon, *Nature, Lond.* **169**, 538-539.

Deschamps, M. (1958). Analogie entre les grès et arkoses de la Brenne (Indre) et ceux de Cosne (Allier), *C. r. hebd. Séanc. Acad. Sci. Paris* **247**, 2167-2170.

Deschamps, M. (1962). L'altération des anatexites dans le sidérolithique du Lembron (Puy-de-Dôme), *C.r. hebd. Séanc. Acad. Sci. Paris* **254**, 1831-1833.

Dewey, H. and Bromehead, C. E. N. (1915). The geology of the country around Windsor and Chertsey, *Mem. Geol. Surv. U.K.*

Dewey, H., Bromehead, C. E. N., Chatwin, C. P. and Dines, H. G. (1924). The geology of the country around Dartford, *Mem. Geol. Surv. U.K.*

Dewolf, Y. and Mainguet, M. (1976). Une hypothèse éolienne et tectonique sur l'alignement et l'orientation des buttes Tertiaires du Bassin de Paris, *Revue Géogr. phys. Geol. dyn.* **18**, 415-426.

Dines, H. G. and Edmunds, F. H. (1925). The geology of the country around Romford, *Mem. Geol. Surv. U.K.*

Dines, H. G. and Edmunds, F. H. (1929). The geology of the country around Aldershot and Guildford, *Mem. Geol. Surv. U.K.*

Dines, H. G., Holmes, S. C. A. and Robbie, J.A. (1954). Geology of the country around Chatham, *Mem. Geol. Surv. U.K.*

Dury, G. H. (1971). Relict deep weathering and duricrusting in relation to the palaeo-environments of middle latitudes, *Geog. J.* **137**, 511-522.

Dury, G. H. and Habermann, G. M. (1978). Australian silcretes and northern-hemisphere correlatives, *in* T. Langford-Smith (ed.) "Silcrete in Australia", pp. 223-259. Armidale, N.S.W.

Dury, G. H. and Knox, C. J. (1971). Duricrusts and deep-weathering profiles in southwest Wisconsin, *Science, N.Y.* **174**, 291-292.

Du Toit, A. L. (1954). "The Geology of South Africa", 3rd edn. Oliver and Boyd, Edinburgh.

Emery, K. O. and Rittenberg, S. C. (1952). Early diagenesis of California basin sediments in relation to origin of oil, *Bull. Am. Ass. Petrol. Geol.* **36**, 735-806.

Evans, W. D. (1964). The organic solubilization of minerals in sediments, *Adv. org. Geochem. Earth Sci. Ser. Monogr.* **15**, 263-270.

Frankel, J. J. (1952). Silcrete near Albertinia, Cape Province, *S. Afr. J. Sci.* **49**, 173-182.

Frankel, J. J. and Kent, L. E. (1938). Grahamstown surface quartzites (silcretes), *Trans. Geol. Soc. S. Afr.* **40**, 1-42.

Goudie, A. (1973). "Duricrusts in tropical and subtropical landscapes". Clarendon, Oxford.

Grant, K. and Aitchinson, G. D. (1970). The engineering significance of silcretes and ferricretes in Australia, *Eng. Geol.* **4**, 93-120.

Green, U. (1890). Excursion to Shenley, *Proc. Geol. Ass.* **11**, 169-171.

Greenberg, S. A. and Price, E. W. (1957). The solubility of silica in solutions of electrolytes, *J. Phys. Chem., Ithaca* **61**, 1539-1541.

Guilcher, A. (1941). Sur présence de grès quartzites du type grès à sables sur le plateau du Léon (Finistère), *C. r. somm. Séanc. Soc. géol. Fr.* **XX**, 79-80.
Gulinck, M. (1961). Note sur le Boldérien d'Opgrimbie (Campine) et rémarques sur les grès «erratiques» du Limbourg, *Bull. Soc. belge Géol. Paléont. Hydrol.* **70**, 197-302.
Gunn, R. H. and Galloway, R. W. (1978). Silcretes in south-central Queensland, *in* T. Langford-Smith (ed.), "Silcrete in Australia", pp. 51-71. Armidale, N.S.W.
Harder, H. (1965). Experimente zur "Ausfallung" der Kieselsäure, *Geochim. Cosmochim. Acta.* **29**, 429-442.
Harder, H. (1971). Quartz and clay mineral formation at surface temperatures, *Spec. Pap. Miner. Soc. Japan* **1**, 106-108.
Harder, H. and Flehmig, W. (1970). Quartzsynthese bei tiefen Termperaturen, *Geochim. Cosmochim. Acta* **34**, 295-305.
Haughton, S. H., Frommurze, H. F. and Visser, D. J. L. (1937). "The geology of a portion of the coastal belt near the Gamtoos Valley, Cape Province". Geol. Surv., Pretoria.
Hinton, M. A. C. and Kennard, A. S. (1899). Contributions to the Pleistocene geology of the Thames valley. I. The Grays Thurrock area, *Essex Naturalist* **11**, 336-351.
Hodgson, J. M., Catt, J. A. and Weir, A. H. (1967). The origin and development of Clay-with-Flints and associated soil horizons on the South Downs, *J. Soil Sci.* **18**, 85-102.
Hodgson, J. M., Rayner, J. H. and Catt, J. A. (1974). The geomorphological significance of Clay-with-Flints on the South Downs, *Trans. Inst. Br. Geogr.* **18**, 17-29.
Hopkinson, J. (1884). Excursion to Radlett, *Proc. Geol. Ass.* **8**, 452-458.
Hopkinson, J. and Kidner, H. (1907). Excursion to Bushey and Croxley Green, Watford, *Proc. Geol. Soc.* **20**, 94-97.
Hopkinson, J. and Whitaker, W. (1892). Excursion to St Albans, *Proc. Geol. Ass.* **12**, 342-344.
Hutton, J. T., Twidale, C. R., Milnes, A. R. and Rosser, H. (1972). Composition and genesis of silcretes and silcrete skins from the Beda Valley, Southern Arcoona Plateau, South Australia, *J. Geol. Soc. Aust.* **19**, 31-40.
Iler, R. K. (1955). "The colloid chemistry of silica and the silicates". Cornell Univ. Press, Ithaca, N.Y.
Jessup, R. W. (1961). A Tertiary-Quaternary pedological chronology for the south-eastern portion of the Australian arid zone, *J. Soil Sci.* **12**, 199-213.
Jones, L. H. P. and Handreck, K. A. (1963). Effects of iron and aluminium oxides on silica in solution in soils, *Nature, Lond.* **198**, 852-853.
Jones, L. H. P. and Handreck, K. A. (1965). Studies of silica in the oat plant. III. Uptake of silica from soils by that plant, *Pl. Soil* **23**, 79-96.
Jones, T. R. (1901). History of the sarsens, *Geol. Mag.* (Ser. 4) **8**, 115-125.
Joplin, G. A. (1963). Chemical analyses of Australian rocks, Part 1, *Bull. Bur. Miner. Resour. Geol. Geophys. Aust.* **65**.
Jørgensen, E. G. (1953). Silicate assimilation by diatoms, *Physiologia Pl.* **6**, 301-315.
Junge, C. E. (1958). Atmospheric chemistry, *Adv. Geophys.* **4**, 1-108.
Keller, W. D. and Reesman, A. L. (1963). Dissolved products of artificially pulverized, silicate minerals and rocks. II, *J. Sedim. Petrol.* **33**, 426-437.
Kerr, M. H. (1955). On the origin of silcretes in southern England, *Proc. Leeds Phil. Lit. Soc. (Scientific Sect.)* **6**, 328-337.
Kitto, P. H. and Patterson, H. S. (1942). The rate of solution of particles of quartz and certain silicates, *J. Ind. Hyg. Toxicol.* **24**, 59-74.
Klein, C. (1962). La Brenne et ses abords: Essai d'interpretation morphologiques, *Norois* **9**, 245-263.

Krauskopf, K. B. (1956). Dissolution and precipitation of silica at low temperatures, *Geochim. Cosmochim. Acta* **10**, 1-26.

Kulbicki, G. (1953). Constitution et genèse des sédiments argileux sidérolithiques et lacustres du Nord de l'Aquitaine, *These Sci. Toulouse et Sci. Terre* **4**, 5-101.

Lamplugh, G. W. (1902). Calcrete, *Geol. Mag.* **9**, 75.

Lamplugh, G. W. (1907). The geology of the Zambesi Basin around the Batoka Gorge (Rhodesia), *Q. J. Geol. Soc. Lond.* **63**, 162-216.

Langford-Smith, T. (1978). "Silcrete in Australia". Armidale, N.S.W.

Langford-Smith, T. and Dury, G. H. (1965). Distribution, character and altitude of the duricrust in the north-west of New South Wales and the adjacent areas of Queensland, *Am. J. Sci.* **263**, 170-190.

Litchfield, W. H. and Mabbutt, J.A. (1962). Hardpan in soils of semi-arid Western Australia, *J. Soil Sci.* **13**, 148-159.

Livingstone, D. (1963). Chemical composition of rivers and lakes, *Prof. Pap. U.S. Geol. Surv.*, 440-G.

Loughnan, F. C. (1969). "Chemical weathering of the silicate minerals". Elsevier, New York.

Mabbutt, J. A. (1965). The weathered land surface of Central Australia, *Z. Geomorph.* **9**, 82-114.

MacGregor, A. M. (1931). Geological notes on a circuit of the Great Makarikari Salt Pan, Bechuanaland Protectorate, *Trans. Geol. Soc. S. Afr.* **33**, 89-102.

Mackenzie, F. T. and Gees, R. (1971). Quartz: Synthesis at earth-surface conditions, *Science, N. Y.* **173**, 533-535.

Mandl, I. Grauer, A. and Neuberg, C. (1952). Solubilization of insoluble matter in nature. I. The part played by salts of adenosinetriphosphate, *Biochim. Biophys. Acta* **8**, 654-663.

McKeague, J. A. and Cline, M. G. (1963. Silica in soil solution (II). The adsorption of monosilicic acid by soil and other substances, *Can. J. Soil Sci.* **43**, 83-96.

McPhail, M., Price, A. L. and Bingham, F. T. (1972). Adsorption interactions of monosilicic and boric acid on hydrous oxides of iron and aluminium, *Proc. Soil Sci. Soc. Am.* **36,** 510-514.

Menillet, F. (1975). Niveaux calcaires finement rubonés en milieu continental, hydromorphe et confiné à paléogéographic simple: l'example des Calcaires de Beauce (stampien sup.-Aquitanien du Bassin de Paris), *in* T. Vogt (ed.), "Types de Croûtes et leur Repartition Regionale", pp. 18-21. Colloque, Univ. Louis Pasteur, Strasbourg.

Millot, G. (1970). "Geology of Clays". Springer-Verlag, New York.

Mohr, E. C. J. and van Baren, F. A. (1954). "Tropical Soils". N.V. Uitgeverij, W. van Hoeve, The Hague and Bandung.

Montford, H. M. (1970). The terrestrial environment during Upper Cretaceous and Tertiary times, *Proc. Geol. Ass.* **81**, 181-204.

Morey, G. W., Fournier, R. O. and Rowe, J. J. (1962). Solubility of quartz in water in the temperature interval from 25° to 300°, *Geochim. Cosmochim. Acta* **26**, 1029-1043.

Mountain, E. D. (1951). The origin of silcrete, *S. Afr. J. Sci.* **48**, 201-204.

Nilsen, T. H. and Kerr, D. R. (1978). Palaeoclimatic and palaeogeographic implications of a lower Tertiary laterite (latosol) on the Iceland-Faeroe Ridge, North Atlantic region, *Geol Mag.* **115**, 153-182.

Okamoto, G., Okura, T. and Goto, K. (1957). Properties of silica in water, *Geochim. Cosmochim. Acta* **12**, 123-132.

Paraguassu, A. B. (1972). Experimental silicification of sandstone *Bull. Geol. Soc. Am.* **83**, 2853-2858.

Passarge, S. (1904). "Die Kalahari". Reimer, Berlin.

Pinchemel, P. (1954). "Les Plaines de Craie". Armand Colin, Paris.

Pomerol, C. (1973). "Stratigraphie et Paléogéographie. Ere Cénozoique (Tertiaire et Quaternaire)". Masson, Paris.

Prescott, J. A. and Pendleton, R. L. (1952). Laterite and lateritic soils, *Tech. Commun. Commonw. Bur. Soil Sci.* **47**.

Prestwich, J. (1854). Druid sandstones, *Q. J. Geol. Soc. Lond.* **10**, 123-135.

Prost, A. (1961). Nouvelles doneés sur le marno-calcaire de Brie et sur l'origine de la meulièrisation de cette formation, *C. r. hebd. Séanc. Acad. Sci. Paris* **253**, 1977-1979.

Rayner, D. H. (1967). "The stratigraphy of the British Isles". Cambridge Univ. Press.

Reid, C. (1902). The geology of the country around Ringwood, *Mem. Geol. Surv. U.K.*

Richardson, L. (1933). The geology of the country around Cirencester, *Mem. Geol. Surv. U.K.*

Rondeau, A. (1975). Cuirasses de fer et cuirasses de silice en France occidentale, *Bull. Ass. Géogr. Fr.* **424-425**, 161-164.

Searle, A. B. (1923). "Sands and crushed rocks". Henry Froude and Hodder and Stoughton, London.

Senior, B. R. and Senior, D. A. (1972). Silcrete in southwest Queensland, *Bull. Bur. Miner. Resour. Geol. Geophys. Aust.* **125**, 23-28.

Sherlock, R. L. (1922). The geology of the country around Aylesbury and Hemel Hempstead, *Mem. Geol. Surv. U.K.*

Sherlock, R. L. (1947). "London and Thames Valley". Br. Reg. Geology. Geol. Surv. U.K.

Sherlock, R. L. and Noble, A. H. (1922). The geology of the country around Beaconsfield, *Mem. Geol. Surv. U.K.*

Sherlock, R. L. and Pocock, R. W. (1924). The geology of the country around Hertford, *Mem. Geol. Surv. U.K.*

Siever, R. (1957). The silica budget in the sedimentary cycle, *Am. Miner.* **42**, 821-841.

Siever, R. (1962). Silica solubility, 0 °C - 200 °C, and the diagenesis of siliceous sediments, *J. Geol.* **70**, 127-150.

Siever, R. (1972). The low temperature geochemistry of silicon, *in* K. H. Wedepohl and K. Turekian (eds), "Handbook of geochemistry", Vol. II-14. Springer-Verlag. Berlin.

Siever, R. and Scott, R. A. (1963). Organic geochemistry of silica, *in* I. A. Berger (ed.), "Organic geochemistry", pp. 579-595. Pergamon, Oxford.

Smale, D. (1973). Silcretes and associated silica diagenesis in southern Africa and Australia, *J. Sedim. Petrol.* **43**, 1077-1089.

Small, R. J., Clark, J. J. and Lewin, J. (1970). The periglacial rockstream at Clatford Bottom, Marlborough Downs, Wiltshire, *Proc. Geol. Ass.* **81**, 87-98.

Stephens, C. G. (1964). Silcretes of central Australia, *Nature, Lond.* **203**, 1407.

Stephens, C. G. (1966). Origin of silcretes of Central Australia, *Nature, Lond.* **209**, 497.

Stephens, C. G. (1971). Laterite and silcrete in Australia: a study of the genetic relationships of laterite and silcrete and their companion materials, and their collective significance in the weathered mantle, soils, relief and drainage of the Australian continent, *Geoderma* **5**, 5-52.

Storz, M. (1926). *In* E. Kaiser (ed.), "Die Diamantenwüste Sudwest-Afrikas", Vol. 2, Ch. 25. Reimer, Berlin.

Strahan, A. (1898). Geology of the Isle of Purbeck and Weymouth, *Mem. Geol. Surv. U.K.*

Summerfield, M. A. (1978). The nature and origin of silcrete with particular reference to southern Africa, unpublished D. Phil. Thesis, Univ. of Oxford.

Summerfield, M. A. (1979). Origin and palaeoenvironmental interpretation of sarsens *Nature* **281**, 137-139.

Taylor, G. and Smith, I. E. (1975). The genesis of sub-basaltic silcretes from the Monaro, New South Wales, *J. Geol. Soc. Aust.* **22**, 377-385.

Teakle, L. J. H. (1936). Red and brown hardpan soils of the acacia semi-desert scrub of Western Australia, *J. Agric. West Aust.* **13**, 480-493.

Teakle, L. J. H. (1950). The red and brown hardpan soils of Western Australia, *J. Aust. Inst. Agric. Sci.* **16**, 15-17.

Teichmüller, R. (1958). Die Niederheinische Braunkohlenformation, *Fortschr. Geol. Rheinld Westf.* **2**, 721-750.

Tricart, J. L. F. (1949). "La Partie Orientale du Bassin de Paris: Etude Morphologique". Sedes, Paris.

Ussher, W. A. E. (1906). The geology of the country between Wellington and Chard, *Mem. Geol. Surv. U.K.*

van den Broeck, E. (1901). Grès erratiques du Sud du Démer et dans la région de la Herck, *Bull. Soc. belge Géol. Paléont. Hydrol.* **15**, 627-631.

van den Broek, J. M. M. and van der Waals, L. (1967). The late Tertiary peneplain of South Limburg (The Netherlands): silicifications and fossil soils; a geological and pedological investigation, *Geologie Mijnb.* **45**, 318-332.

Visser, D. J. L. (1937). The ochre deposits of the Riversdale district, Cape Province, *Bull. No. 9, Geol. Surv. Pretoria.*

Visser, J. N. J. (1964). Analyses of rocks, minerals and ores, *Rep. S. Afr. Geol. Surv. Handbook,* No. 5.

Wanneson, J. (1963). Essais sur les propriétés géochimiques de la silice, *C. r. hebd. Séanc. Acad. Sci. Paris* **256**, 2888-2890.

Ward, D. J. (1975). Report of field meetings to Radlett, Herts., *Tertiary Times* **2**, 154-158.

Watts, S. H. (1977). Major element geochemistry of silcrete from a portion of inland Australia, *Geochim. Cosmochim. Acta* **41**, 1164-1167.

Watts, S. H. (1978a). A petrographic study of silcrete from inland Australia, *J. Sedim. Petrol.* **48**, 987-994.

Watts, S. H. (1978b). The nature and occurrence of silcrete in the Tibooburra area of north-western New South Wales, *in* T. Langford-Smith (ed.), "Silcrete in Australia", pp. 167-185. Armidale, N.S.W.

Waugh, B. (1967). Environmental and diagenetic studies of the permo-Triassic and carbonate sediments of the Vale of Eden, unpublished PhD. Thesis, Univ. of Newcastle-upon-Tyne.

Waugh, B. (1970a). Formation of quartz overgrowth in the Penrith Sandstone (Lower Permian) of north-west England as revealed by scanning electron microscopy, *Sedimentology* **14**, 309-320.

Waugh, B. (1970b). Petrology, provenance and silica diagenesis of the Penrith Sandstone (Lower Permian) of north-west England, *J. Sedim. Petrol.* **40**, 1226-1240.

Wayland, E. J. (1953). More about the Kalahara, *Geog. J.* **119**, 49-56.

Wey, R. and Siffert, B. (1961). Réactions de la silice monomoléculaire en solution avec les ions $Al^{3+}$ et $Mg^{2+}$, *Colloques int. natn. Rech. scient.* **105**, 11-23.

Whalley, W. B. and Chartres, C. J. (1976). Preliminary observations on the origin and sedimentological nature of sarsen stones, *Geologie Mijnb.* **55**, 68-72.

Whitaker, W. (1862). On the greywethers of Wiltshire, *Q. J. Geol. Soc. Lond.* **18**, 271-274.

Whitaker, W. (1885). The geology of the country around Ipswich, Hadleigh and Felixstowe, *Mem. Geol. Surv. U.K.*

Whitaker, W. (1889). The geology of London and part of the Thames Valley, *Mem. Geol. Surv. U.K.*

White, H. J. O. (1906). On the occurrence of quartzose gravel in the Reading Beds at Lane End, Bucks, *Proc. Geol. Ass.* **19**, 371-377.

White, H. J. O. (1907). The geology of the country around Hungerford and Newbury, *Mem. Geol. Surv. U.K.*

White, H. J. O. (1909). The geology of the country around Basingstoke, *Mem. Geol. Surv. U.K.*

White, H. J. O. (1912). The geology of the country around Winchester and Stockbridge, *Mem. Geol. Surv. U.K.*

White, H. J. O. (1913). The geology of the country near Fareham and Havant, *Mem. Geol. Surv. U.K.*

White, H. J. O. (1923). The geology of the country south and west of Shaftesbury, *Mem. Geol. Surv. U.K.*

White, H. J. O. (1924). Geology of the country around Brighton and Worthing, *Mem. Geol. Surv. U.K.*

White, H. J. O. (1925). The geology of the country around Marlborough, *Mem. Geol. Surv. U.K.*

White, H. J. O. (1926). The geology of the country near Lewes, *Mem. Geol. Surv. U.K.*

Whitehouse, F. W. (1940). Studies in the Late Geological History of Queensland, *Publs Univ. Sydney Geol. Dep.* **2**, No. 1.

Williams, R. B. G. (1968). Some estimates of periglacial erosion in southern and eastern England, *Biul. peryglac.* **17**, 311-335.

Williamson, W. O. (1957). Silicified sedimentary rocks in Australia, *Am. J. Sci.* **255**, 23-42.

Wooldridge, S. W. and Linton, D. L. (1939). Structure, surface and drainage in south-east England, *Trans. Inst. Br. Geogr.* **10**, 1-124.

Wooldridge, S. W. and Linton, D. L. (1955). "Structure, surface and drainage in south-east England", 2nd edn. George Philip, London.

Woolnough, W. G. (1927). The duricrust of Australia, *J. Proc. Roy. Soc. N.S.W.* **61**, 24-53.

Woolnough, W. G. (1928). Origin of white clays and bauxite and chemical criteria for peneplanation, *Econ. Geol.* **23**, 887-894.

Woolnough, W. G. (1930). The influence of climate and topography in the formation and distribution of products of weathering, *Geol. Mag.* **67**, 123-132.

Wopfner, H. (1974). Post-Eocene history and stratigraphy of north-eastern South Australia, *Trans. Roy. Soc. S. Aust.* **98**, 1-12.

Wopfner, H. (1978). Silcretes of northern South Australia and adjacent regions, *in* T. Langford-Smith (ed.), "Silcrete in Australia", pp. 93-141. Armidale, N.S.W.

Young, R. W. (1978). Silcrete in a humid landscape: the Shoalhaven Valley and adjacent coastal plains of southern New South Wales, *in* T. Langford-Smith (ed.), "Silcrete in Australia", pp. 195-207. Armidale, N.S.W.

FIVE

# The soils and superficial deposits on the North Downs of Surrey

D. T. JOHN

*School of Geography, Kingston Polytechnic, Kingston upon Thames*

Current theories on the landform evolution of the North Downs of Surrey make varying assumptions about the soils and superficial deposits. This chapter summarizes the findings of a study (John, 1974) made to ascertain (1) which, if any, of these assumptions are valid and (2) the geomorphological implications of such new interpretations as may be necessary. The part of the North Downs surveyed lies between the meridians of Guildford and Gatton (Fig. 1), excluding the valley of the Mole. It is an area of great importance for the denudational history of the Chalk cuestas of south-east England as a whole, being the only one where the key landforms and deposits are all well represented.

Northwards the dipslope passes beneath Eocene sediments at elevations ranging from 30 to 90 m O.D., and comprises three distinct topo-lithographic sectors, delimited by the 150- and 200-m contours. That below 150 m is relatively steeply inclined, scored by subparallel dip-valleys frequently less than 15 m deep, and is shown on the New Series geological maps (sheets 285 and 286, drift) as being virtually drift-free. Between 150 and 200 m the dipslope flattens into a marked platform, mantled by Eocene residuals, patches of sand and gravel of presumed early Quaternary marine origin*, and dissected sheets of flinty, clayey drift. In the west of the area under review, the latter two formations, along with occurrences of Lower Greensand detritus, were mapped together by the Geological Survey as 'Netley Heath Deposits' because of their apparently confused interrelation. The platform either bevels the crest of the escarpment or, at heights of about 200 m, passes into isolated summit-ridges that rise to over 215 m and which are also covered by flinty clay. The consequent valleys on the two highest sectors of the dipslope rarely connect with those described above, although their alignments are often strikingly accordant. Instead, from their sources near the escarpment, they

* Shown on sheet 285 as doubtfully Pliocene. However, the marine fossils found in these deposits on Netley Heath date from the Red Crag, which is now correlated with the early Pleistocene Calabrian stage of the Italian marine succession.

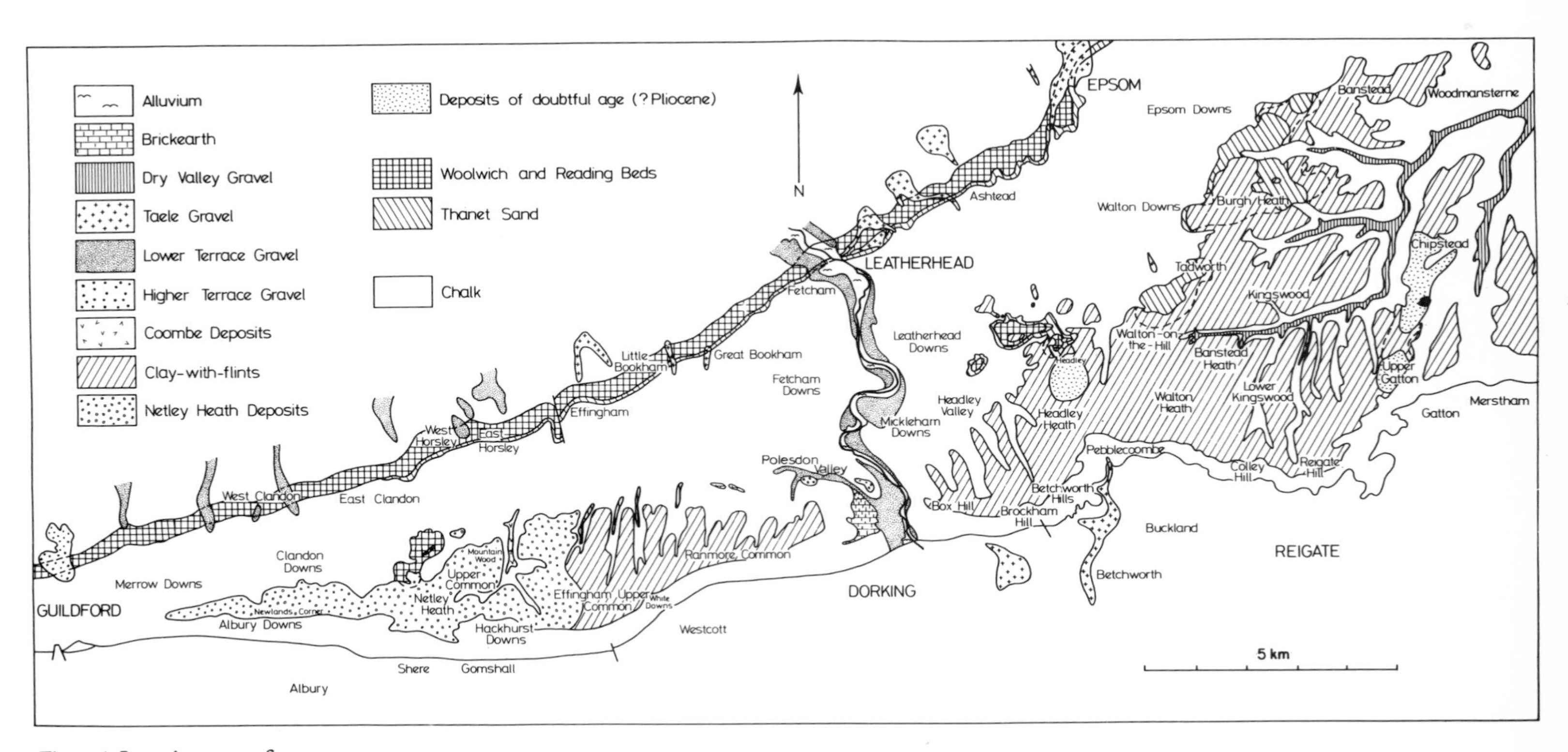


*Figure 1.* Location map of area.

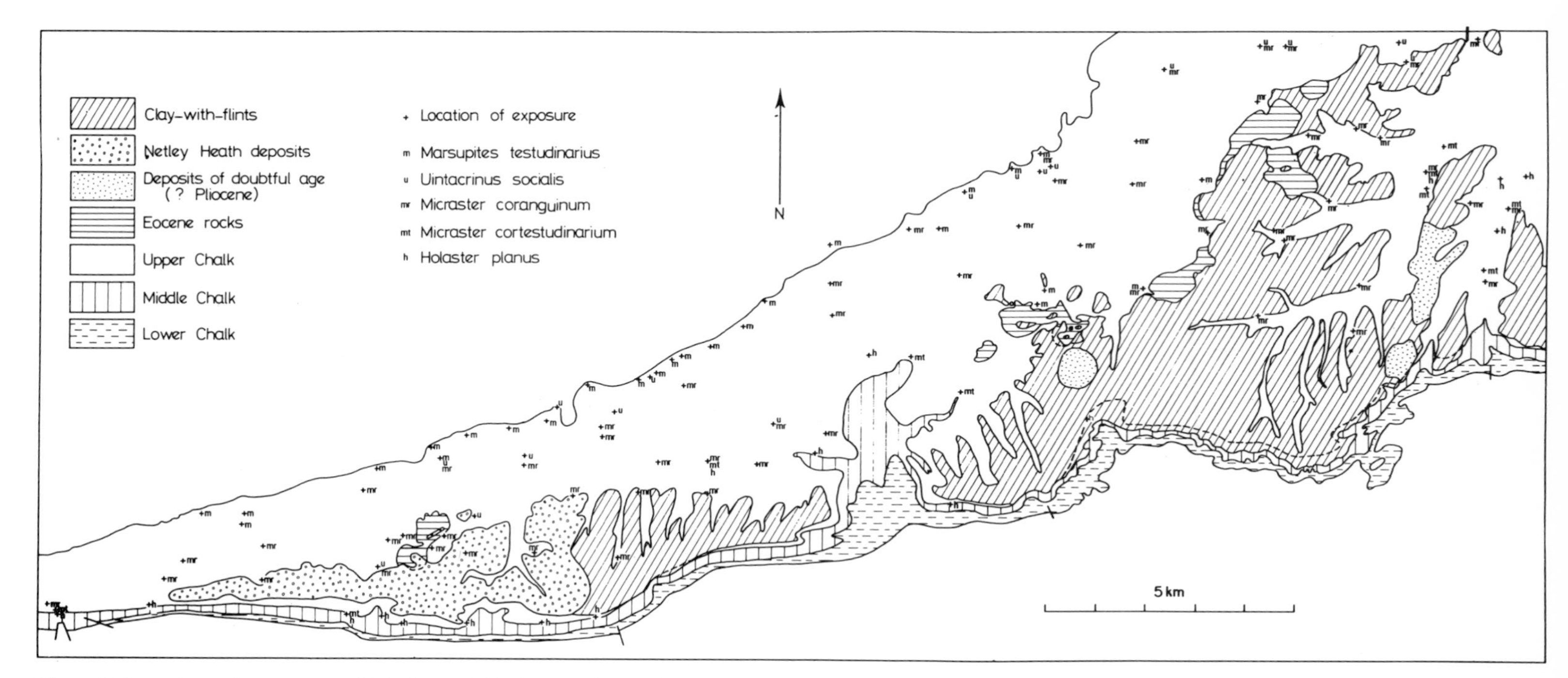


*Figure 2.* Location of exposures which have yielded zonal fossils for the Upper Chalk. This map is based upon the following published data: Davies, G. M. (1918); Davis, A. G. (1926); Dewey, H. D. and C. E. N. Bromehead (1921); Dines, H. G. and F. H. Edmunds (1929, 1933); Gray, D. A. (1965); Jukes-Browne, A. J. (1904, 1906); Kirkaldy, J. F. (1958); Whitaker, W. (1912); Young, G. W. (1905, 1906, 1907-8, 1915a, 1915b, 1917).

typically develop into steep-sided channels incised up to 50 m below the platform, before uniting with very much larger strike-valleys. These strike-valleys effectively separate the platform from the steeper northern part of the dipslope. Just two valleys, located north of Gomshall and Betchworth respectively, traverse the entire dipslope. They are broader and deeper than the other consequent valleys of the area and appear not to bear a simple relationship to the dip of the underlying Chalk. The valley which notches the escarpment at Gatton has analogous characteristics, but lies mainly east of the study area. The escarpment itself overlooks a Gault vale—the Vale of Holmsdale—onto the Lower Greensand cuesta farther south.

Surprisingly little is known of the structure of the Chalk in this part of Surrey, though an analysis of published data on the zones of the Upper Chalk yields much relevant information. The *Marsupites* and *Uintacrinus* zones outcrop in successive narrow bands adjoining the Eocene outcrop (Fig. 2), followed by a more extensive tract of older *Micraster coranguinum* Chalk. As the dipslope platform is approached, however, the two younger zones are again encountered, in a remarkably continuous chain of outliers. West of the Mole they would appear to be of *Uintacrinus* beds only, whereas to the east *Marsupites* Chalk is also preserved. Interestingly most of the outliers also support Eocene residuals and are terminated southwards by the large, equally persistent strike-valleys which practically bisect the dry-valley network. This repeated juxtaposition of outliers with strike-valleys is hardly fortuitous, and must signify a line of folding, to the south of which the dip of the Chalk is greatly reduced. The fold axis seems to be of regional significance, moreover, as is evident from Bury's (1910) comment that "Prestwich's sections (across the North Downs) appear at first sight to show a great difference of angle between the plateau and the pre-Eocene Chalk surface; but the presence ... of thin Eocene outliers upon the former indicates that this difference is to some extent due to a flexure which has affected both formations, especially along a line somewhat to the north of the plateau" (p. 652).

Dip measurements confirm that the Chalk is less steeply inclined south of the flexure. Thus between Gatton and Woodmansterne—an area where the dipslope platform is at its widest and flattest—records of numerous exposures (Jukes-Browne, 1904; Dines and Edmunds, 1933) show that the Chalk is nearly horizontal. All along the escarpment, however, the dip increases appreciably once more, under the influence of an *en echelon* series of monoclines which affect both the Chalk and the Lower Greensand (Dines and Edmunds, 1929, 1933). West of Albury the main fold axis is in the Lower and Middle Chalk, where dips of 18° to 20° occur. Beyond Albury it passes into the Lower Greensand and the dip of the Chalk at the escarpment lessens to about 5° on average. A similar pronounced upturning of the edge of the Chalk was traced eastwards of Gatton as far as the Darent by Brown (1924), and this has since been substantiated by Dines *et al.* (1969).

These observations clearly offer a simple structural explanation for the familiar tripartite character of the North Downs dipslope, at least in Surrey. They suggest that the changes in declivity at 150 and 200 m O.D. coincide with regional-scale disturbances, one located along the northern boundary of the platform, the other at or near to the escarpment. Between these disturbances the gradients of both the Chalk strata and the dipslope are comparatively small.

Despite the evidence presented, Wooldridge and Linton (1955) saw the three dipslope facets as erosion surfaces of different ages and origins. The lowest is demonstrably an extension of the sub-Eocene plane, exhumed in the Pleistocene. The platform they regarded as an early Quaternary wave-cut feature, which accords with the supposed early Quaternary marine sediments found upon it. In contrast the summit-ridges are interpreted as vestiges of a mid-Tertiary peneplain, retaining a regolith of "true Clay-with-Flints" (p. 55). The two highest surfaces, it was argued, were eroded out of a former projection of the sub-Eocene plane and are not significantly warped.

Much of the appeal of this model lay in its comprehensive application to south-east England. It has subsequently been challenged on several grounds, quite apart from the lack of any detailed assessment of the role of geological structure in determining Chalkland relief. In so far as the issues involved may have a bearing on the present area, a short review is warranted.

Pinchemel's (1954) study attached more importance to superficial deposits and concluded that the major landscape elements on the Chalk of southern Britain are inherited from an early Tertiary polycyclic surface. Green (1969) accepted the presence of such a surface on the north-west of Salisbury Plain, but nevertheless re-affirmed the widespread existence of a mid-Tertiary summit peneplain cut at the expense of the earlier surface. On the other hand, immediately to the north, on Marlborough Downs, Clark *et al.* (1967) had previously described the sub-Eocene surface as being very extensive, and inferred that the highest parts of the Chalk belong to an Eogene (essentially Oligocene and early Miocene) surface, which they believed was the most basic component in the landscape of south-east England. In their estimation, moreover, the early Quaternary or Calabrian transgression achieved little bevelling of the Chalk; a conclusion since endorsed by Jones (1974) from a re-appraisal of accepted ideas on drainage evolution in south-east England. No radically different ideas to those of Wooldridge and Linton have emerged for the Chilterns, yet it is evident that drift, rich in Eocene waste, actually extends right across the conjectured post-Eocene surfaces of the dipslope to the crest of the escarpment (Avery *et al.*, 1959; Loveday, 1962). Around the Weald only the South Downs of West Sussex have been re-examined systematically. Here Hodgson *et al.* (1967, 1974) found that the dipslope conforms closely to the sub-Eocene plane, as the Clay-with-Flints developed upon it, even on the highest ground, is substantially derived from Eocene material more or less in place. Within the Weald, Worssam (1973) established that intense erosion took place in the later part of the Pleistocene, and hence enquired why there is still so much relief if part of the Weald had already been reduced to elevations of about 200 m at the start of the Pleistocene? In effect, only if the concept of a 200-m platform is discarded can the full erosive potential of the Pleistocene be accommodated. More generally, Kellaway *et al.* (1975) have asserted that the flinty clay on the Chalk of southern England is really a 'Chalky-Tertiary' till of early Pleistocene age. Developing this line of argument, Shephard-

Thorn (1975) suggested for consideration the view that the presumed Coralline and Red Crag deposits along the North Downs might have been emplaced by ice from the North Sea. In this context it can be added that Letzer (1973), on the basis of grain-size data, has also claimed that the so-called Red Crag or Calabrian sand situated within the study area, at Headley Heath, may in fact be a coversand of northerly provenance.

## The nature, distribution and derivation of the soils and superficial deposits

The soils and associated superficial deposits were mapped as series (Robinson, 1949), using criteria and procedures employed by the Soil Survey of England and Wales (Soil Survey Staff, 1960). Comparisons of texture, colour, particle morphology and heavy mineral content were made with every potential parent sediment in the Eocene beds that outcrop within and near to the mapping area, to help determine the provenance of the soil parent materials (where this was not obvious) and to learn whether their derivation was direct or not. Wet and dry sieving and pipette analyses were performed according to British Standard 1377 (1967). The apertures of the sieves (in mm) were:

cobbles 101·6, 88·9, 76·2, 63·5
gravel 50·8, 38·1, 25·4, 19·0, 12·7, 6·35, 3·18
sand 2·0, 1·0, 0·6, 0·5, 0·335, 0·25, 0·21, 0·18, 0·15, 0·125, 0·14, 0·09, 0·075, 0·063, 0·053

The grain-size data are given as curves of 'percentage passing' and, where appropriate, in terms of standard statistical parameters (Folk and Ward, 1957), whilst colour descriptions are based largely on Munsell notation. Three morphological attributes are defined for the gravel and cobble fractions: roundness in relation to Krumbein's (1941) pebble silhouettes; sphericity with reference to Sneed and Folk's (1958) maximum projection sphericity; and form by means of their ten sphericity-independent shape classes. The heavy mineral determinations apply to 0·075-0·09 mm grains, unless stated otherwise.

The field and laboratory investigations revealed two basic kinds of superficial deposit, namely interfluve deposits and slope deposits. The former are unstratified and remarkably uniform over a minimum east-west distance of 23 kilometres*, notwithstanding massive dissection by valley systems dominated by north-south elements. The slope deposits are so designated because they occur wholly or partly within valleys, as opposed to being confined exclusively to the intervening tracts. Apart from location they differ also in that generally they are crudely stratified and texturally highly variable. Emphasis is given to the interfluve deposits, since they rest on the surface(s) whose origin is in dispute.

There are three generic groups of interfluve deposits. On the lowest dipslope sector is a flinty, sandy drift (Fig. 3), rarely more than 1 m thick and brown in

* How far any or all of these interfluve deposits extend outside the present area remains to be seen.

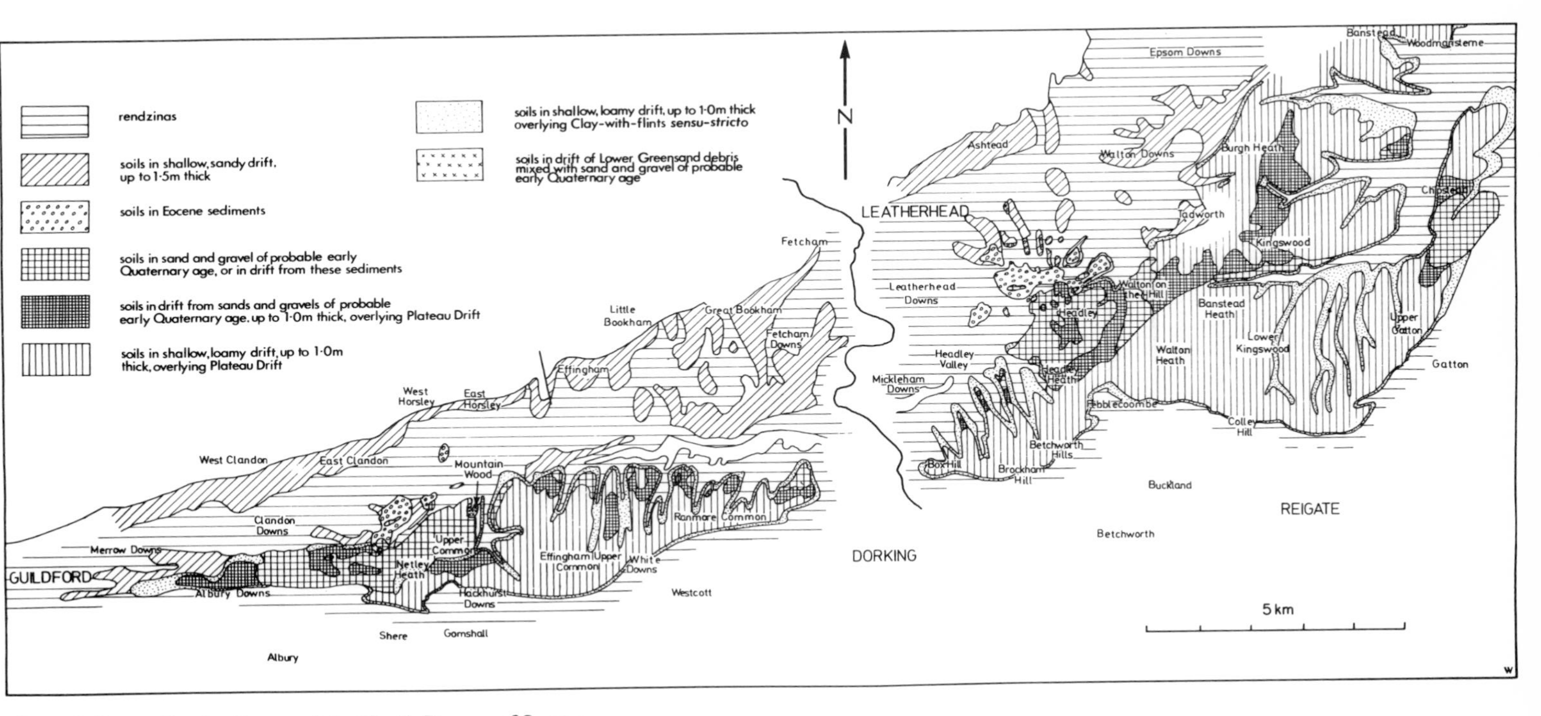


*Figure 3.* Generalized soil map of the North Downs of Surrey.

colour, although where clay-enrichment due to *lessivage* has taken place reddish-brown hues may also be apparent. The drift rests on weathered Chalk, and the interface is normally distorted by patterned-ground phenomena. West of Effingham the drift is characterized by medium and fine sand, and to the east by find sand alone. Textural and heavy mineral similarities (Fig.4 and Table I)* indicate that the parent material of the coarser-grained variant is sand from the Bottom Bed of the Woolwich and Reading Series, and that its more extensive finer-grained counterpart is derived from the Thanet Sand. The correspondence is such that the original sediments cannot have been significantly reworked, merely thinned out and degraded. Both drift variants have close distributional ties with the respective parent formations, and along the Eocene outcrop and around Eocene outliers pass laterally into them.

TABLE I

*The relative frequencies of the 0·075-0·09 mm non-opaque† heavy mineral grains in: the Thanet Sand; the finer-grained variant of the sandy drift on the northern part of the dipslope; sand from the Bottom Bed; the coarser-grained variant of the sandy drift.*

| Heavy minerals | Thanet Sand | | Fine sandy drift | | Bottom Bed sand | | Coarser-grained drift | |
|---|---|---|---|---|---|---|---|---|
| | 193564†† | 189545 | 223586 | 145528 | 092566 | 044510 | 038497 | 020491 |
| Zircon | 32·8 | 37·4 | 32·7 | 30·7 | 53·0 | 53·2 | 47·0 | 54·5 |
| Tourmaline | 23·4 | 25·6 | 22·1 | 23·2 | 18·1 | 15·4 | 13·3 | 18·3 |
| Staurolite | 16·0 | 14·1 | 18·0 | 12·9 | 4.7 | 3·9 | 2·4 | 3·2 |
| Rutile | 14·3 | 11·8 | 11·8 | 17·4 | 12·1 | 18·4 | 25·3 | 17·0 |
| Kyanite | 9·8 | 9·8 | 9·4 | 9·5 | 2·7 | 4·8 | 4·8 | 3·5 |
| Garnet | 2·9 | 1·3 | 4·5 | 3·3 | 7·4 | 2·4 | 1·8 | 2·6 |
| Muscovite | 0·8 | | 0·4 | 2·5 | | 1·2 | 2·4 | |
| Sillimanite | | | 0·9 | | 1·3 | 0·6 | 0·6 | 1·0 |
| Sphene | | | | 0·4 | | | | |
| Spinel | | | 0·2 | | 0·7 | | | |
| Amphibole | | | | | | | 1·2 | |
| Anatase | | | | | | | 1·2 | |
| Per cent opaques | 86·7 | 81·4 | 86·0 | 88·1 | 81·7 | 82·3 | 82·5 | 81·9 |
| Total count | 1760 | 2140 | 3217 | 2019 | 812 | 1903 | 948 | 1722 |

† Only the non-opaque minerals are differentiated, as they are of primary diagnostic value.
††All the sample locations in Tables I-IV lie within the 100-km square TQ.

The overwhelming majority of the platform and all of the summit-ridges are mantled by a flinty, clayey deposit approximating to Loveday's (1962) description of the Chiltern Plateau Drift. Only the field relationships seem appreciably different. Brickearth, Eocene sediments, presumed early Quaternary deposits and, more prevalently, Clay-with-Flints *sensu stricto* underlie the Plateau Drift on the

* Locations of samples referred to in the figures and text are listed in the Appendix.

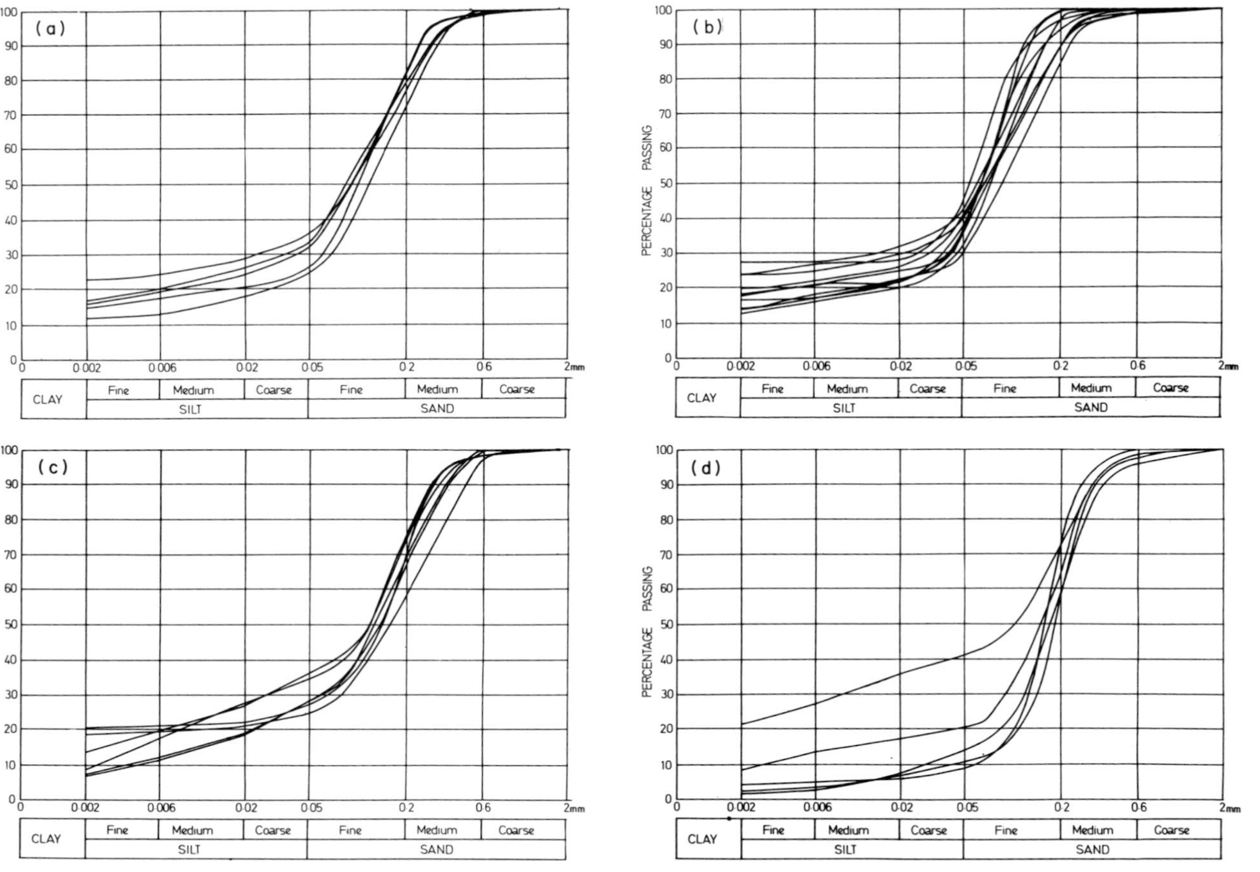


*Figure 4.* (a) Grain-size distribution curves for the fine sandy drift on the northernmost sector of the dipslope. (b) Grain-size distribution curves for weathered Thanet Sand. (c) Grain-size distribution curves for the medium sandy drift on the northernmost sector of the dipslope. (d) Grain-size distribution curves for weathered Bottom Bed sand.

Chilterns. During the study reported here, the deposit was found to rest mainly on Chalk which had been affected by cryogenic activity. Undisturbed Thanet Sand is present beneath the deposit on Burgh Heath and Walton Heath (Dines and Edmunds, 1933), whilst in other parts of these same localities there is a widespread black, flinty clay between the deposit and the Chalk (Groves, 1928). A basal layer of Clay-with-Flints *sensu stricto*, however, was met with only once, at the extremity of the Plateau Drift sheet, on the crest of the escarpment above Pebblecoombe (TQ 212530). Indeed, elsewhere along the crest of the escarpment, on Walton Heath (TQ 228527) for example, and on the dipslope at Lower Kingswood (TQ 252532), Plateau Drift was observed changing laterally into Clay-with-Flints *sensu stricto*, as the slope exceeded 2° to 3°. A single incidence of Plateau Drift resting on presumed Calabrian sediments was detected, as a detached mass about 5 $m^2$ in area, embedded in a layered sequence of sand and gravel infilling a small trench-like valley on the north-western flank of Headley Heath (TQ 202536).

The Plateau Drift in the area mapped is very much thicker than the sandy waste found at lower elevations on the dipslope, ranging from 1 to 15 m and averaging roughly 3 m. By far the most extensive facies is a mainly reddish or reddish-brown, sometimes sandy clay, commonly with greyish streaks and mottles in the top metre or so, imparted by surface-water gleying. Its stone content is extremely variable and largely comprises angular fragments of flint nodules with generally lesser quantities of dark flint pebbles of Eocene aspect. On Netley Heath much Lower Greensand debris is incorporated with the flints and north of Reigate Hill pieces of ironstone conglomerate from the Blackheath Beds are plentiful as well. Unlike the variability of the stone content, the fine-earth fraction ($<2$ mm) displays an obvious textural unity throughout (Fig. 5a). At numerous sites on the dipslope platform and on the summit-ridges this stony Plateau Drift was seen to merge at depths of 1 to 2 m with structureless inclusions of stoneless clays. These were vividly variegated in reds, purples and greys for the most part, but eastwards of the longitude of Epsom more subdued olives, greens, browns and buffs progressively increased in importance. The merging transition between these two Plateau Drift types evidently suggests that the one has weathered out of the other, with the stones having been introduced from diverse other sources. A representative sample of grain-size curves for the stoneless clays (Fig. 5b) reveals a marked textural affinity, which supports such an inference.

The variegated colours of the stoneless clays are actually identical to those of the non-marine strata of the Woolwich and Reading Beds, and exhibit the same changes as this formation displays going from west to east. These changes reflect, in the case of the Eocene deposits, the gradual transition which sets in between Leatherhead and Epsom, from the brightly mottled fluviatile sediments of 'Reading' type to the more drab-coloured estuarine equivalents of 'Woolwich' type. Since the correlation extends equally as convincingly to particle-size distribution (Fig. 5c), there can be little doubt that the stoneless clays of the Plateau Drift and their weathered derivatives are from these Lower Eocene strata. As before, the agreement is so close as to preclude any notable reworking of the original sediments.

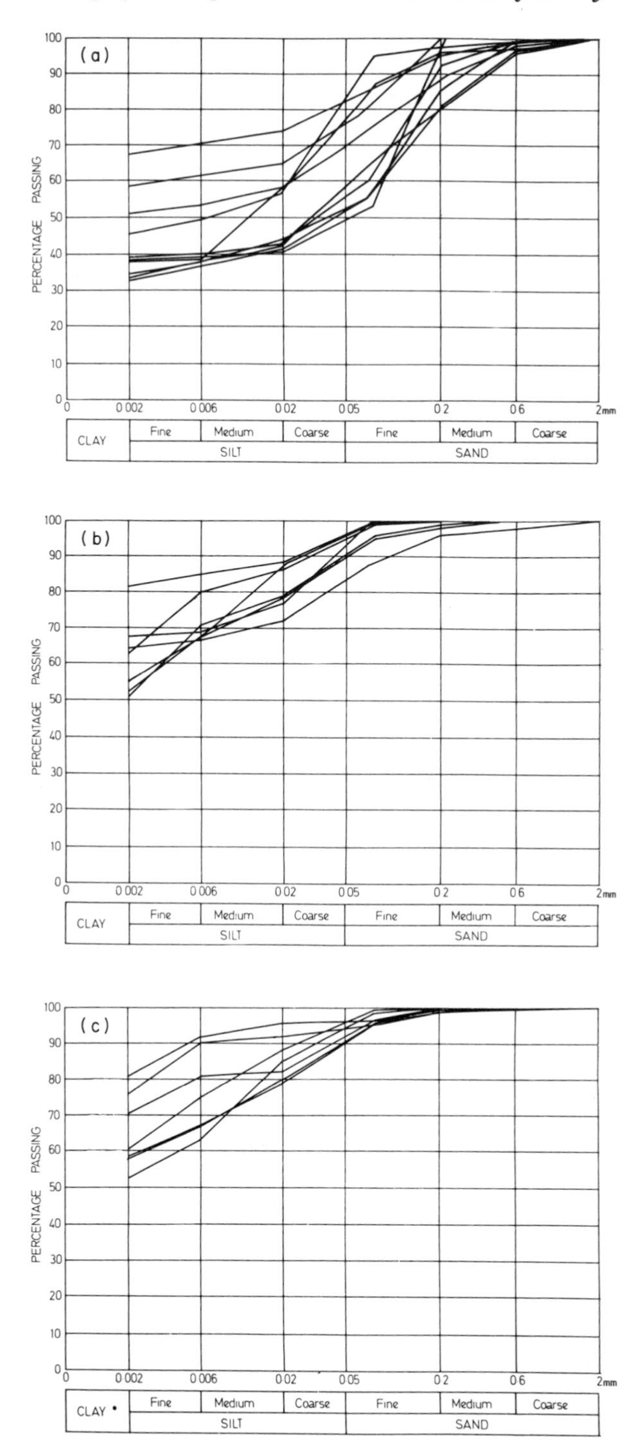


*Figure 5.* (a) Grain-size distribution curves for the Plateau Drift. (b) Grain-size distribution curves for stoneless clays in the Plateau Drift. (c) Grain-size distribution curves for Reading Beds clay.

TABLE II

*The relative frequencies of the 0·075-0·09 mm non-opaque heavy mineral grains in sediments in the Plateau Drift: red, clayey sand; yellow medium-fine sand; and pale brown loam*

| Heavy minerals | Red sand | | Yellow sand | Brown loam |
|---|---|---|---|---|
| | 196514 | 230528 | 247529 | 269537 |
| Tourmaline | 34·5 | 33·3 | 42·5 | 39·4 |
| Staurolite | 16·2 | 17·9 | 16·0 | 21·4 |
| Kyanite | 14·8 | 14·6 | 10·7 | 6·7 |
| Zircon | 13·6 | 12·6 | 11·3 | 12·4 |
| Muscovite | 12·8 | 12·2 | 0·7 | 1·1 |
| Rutile | 6·7 | 6·5 | 12·0 | 16·9 |
| Garnet | 1·2 | 1·2 | 2·0 | 1·1 |
| Sphene | 0·3 | 1·2 | 0·7 | |
| Epidote | | 0·4 | | |
| Sillimanite | | | 4·0 | 1·1 |
| Per cent opaques | 80·3 | 79·4 | 82·5 | 88·4 |
| Total count | 1749 | 1191 | 855 | 764 |

Occasionally, clayey, red sand of fine grade, yellow medium-fine sand, and pale brown loams were associated in a random, disorganized manner with the stoneless clays. The heavy minerals separated from these sediments are recorded in Table II. They match essentially those obtained by Davies (1915-16) from the Lower Eocene rocks in this part of the London Basin, particularly the Woolwich and Reading Beds. Accordingly they strengthen the argument already forwarded from considerations of texture and colour on the provenance of the Plateau Drift. The two samples of clayey, red sand listed in the table, incidentally, came from above and below Wooldridge and Linton's degraded cliff-line at about 200 m. In both cases, untarnished pyrites accounted for nearly 3 per cent of the total 'heavy' residues in the size range examined, indicating that since its deposition the sand has not been subjected to chemical weathering.

In view of a common derivation from substantially *in situ* Lower Eocene sediments, it is to be expected that the two drifts discussed so far should share certain basic distributional characteristics. Appropriately enough, both behave as would dissected, emaciated solid formations with dips close to that of the underlying Chalk. The sandy drift is thickest where dissection and slope are least, and in these situations *sols lessivés* are developed. This is easily seen south of Fetcham and again north of Headley. Significantly in this last locality the drift is still intimately associated with Thanet Sand residuals in all stages of demolition. With increased relief the drift becomes shallower and gives rise to calcimorphic brown earths—a relationship which is particularly extensive on Epsom Downs. The final stages of the drift's destruction are in progress on the closely-fretted dipslope south of

Effingham, where the calcimorphic brown earths have been all but replaced by a fine sandy phase of the silt-rich rendzinas which occur on the completely drift-free Upper Chalk.

Turning to the Plateau Drift, reference is made to two localities in which the structural attitude of the Chalk is different. The first lies between Gatton and Woodmansterne where, it will be recalled, the inclination of the Chalk is negligible. Now if a solid formation rested almost conformably on this part of the Chalk, its basal plane would be practically horizontal, so that the boundary of valley-incision through it would both mark the limits of the formation and occur everywhere at roughly the same height. As the edge of the Plateau Drift in the same locality faithfully expresses the dendritic pattern of the dry-valley network and at the same time rarely deviates from the line of the 150-m contour, it clearly has much in keeping with this notional solid formation. A similarly predictable distribution pattern exists where the Chalk has a more perceptible gradient, as is the case south of the large Polesden and Headley strike-valleys that are nearly in line with each other either side of the Mole. Once more it is helpful to imagine a solid formation resting with little discordance on the Chalk of this locality, although now the interface would be shelving in a northerly direction. The interface, and hence the formation too, would be most widely preserved next to the narrowest or upper reaches of the strike-valleys and vice versa adjacent to their distal sections near the Mole. Thus if the northernmost perimeter of this hypothetical solid formation coincided with, say, the 150-m contour, then just as this contour swings towards the escarpment as the Mole is approached, so too would the perimeter of the formation. However, the closer to the escarpment the contour is, the farther to the south of it would the boundary actually lie, because the formation would rest on a proportionately higher part of its basal plane. This is precisely the height-boundary relationship displayed by the Plateau Drift in the locality concerned.

Supposed Calabrian sediments make up the third group of interfluve deposits, and are confined to the dipslope platform. They show impressive stratigraphic and petrographic constancy, from Netley Heath in the west to Chipstead in the east. At their base is a very poorly sorted mix of flints and some Lower Greensand clasts, with interstitial fines (Fig. 6a). Ignoring the fines the gravel is moderately sorted, whereas the proportion of Lower Greensand material declines steadily eastwards, from about 20 per cent south of East Horsley to about 3 per cent north-west of Chipstead. The great bulk of the flints are dark, smoothly rounded and medium-to-coarse gravel size; the remainder buff, more crudely rounded and cobble-size. Without exception, they are smothered in cresentic percussion scars or 'chatter' marks. Published data (Dines and Edmunds, 1929, 1933), together with information from exploratory boreholes for the M25 motorway indicate that where the gravel is overlain by sand it is no more than 2 m thick, but that where the sand no longer exists as much as 7 m may be present.

The sand itself is yellow when unweathered, stoneless, massive and almost invariably of medium-fine grade (Fig. 6b). Judging from the sources cited above and from drillings made by the writer on Headley Heath (TQ 205539), thicknesses of up

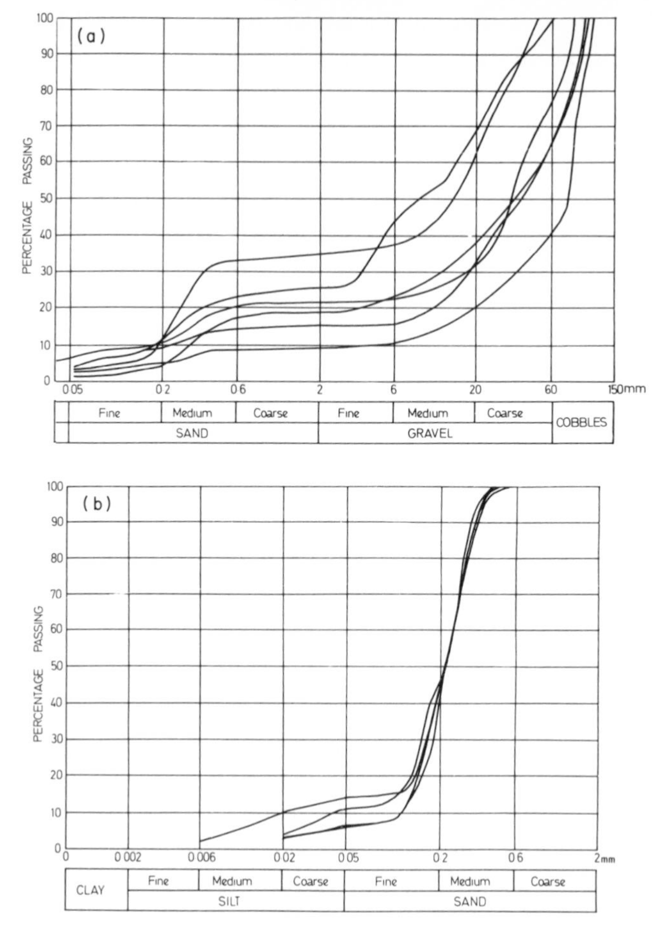


*Figure 6.* (a) Grain-size distribution curves for presumed Calabrian gravel. (b) Grain-size distribution curves for presumed Calabrian sand.

to 11 m occur. An analysis of eleven samples from scattered locations reveals that with respect to grain-size parameters the sand is nearly always moderately sorted and positively skewed. Such consistency applies also to the heavy minerals obtained from the sand and from the interstitial fines in the basal gravel (Table IIIa), as is known from much earlier work (see, for instance, Davies, 1915-16). Andalusite and monazite grains are diagnostic of the heavy mineral suite of the sand, though are omitted from Table IIIa as they are respectively coarser and finer than the grain sizes examined for comparative purposes. Complete heavy mineral separates for three samples are given in Table IIIb.

Numerous marine fossils have been found near the base of the sand on Netley Heath. Their mode of occurrence is unclear: Chatwin (1927) stated that they came from isolated boulders of shelly conglomerate, whilst Green, in the discussion of a paper by Sherlock (1929) was emphatic that the conglomerate formed ferruginous horizons within the sand. Wooldridge (1927) too referred to the fossiliferous

TABLE IIIa

*The relative frequencies of the 0·075-0·09 mm non-opaque heavy minerals grains in the presumed Calabrian sand*

| Heavy minerals | Sample location | | | | |
|---|---|---|---|---|---|
| | 205539 | 205537 | 092508 | 084499 | 275554 |
| Zircon | 55·6 | 55·9 | 50·4 | 49·6 | 49·3 |
| Rutile | 23·3 | 16·8 | 22·8 | 23·6 | 26·1 |
| Tourmaline | 13·5 | 13·3 | 11·9 | 11·6 | 13·4 |
| Staurolite | 2·9 | 7·5 | 8·7 | 6·5 | 4·9 |
| Kyanite | 3·7 | 4·8 | 5·8 | 7·3 | 3·5 |
| Sillimanite | 0·2 | 0·5 | 0·1 | 1·1 | 1·4 |
| Anatase | 0·3 | | 0·3 | 0·4 | 0·7 |
| Muscovite | | 0·8 | | | 0·7 |
| Corundum | 0·5 | 0·3 | | | |
| Spinel | | 0·3 | | | |
| Per cent opaques | 68·2 | 75·0 | 73·0 | 74·6 | 68·4 |
| Total count | 1959 | 1593 | 2757 | 1088 | 450 |

TABLE IIIb

*The relative frequencies of the non-opaque heavy mineral grains in the presumed Calabrian sand*

| Heavy minerals | Sample location | | |
|---|---|---|---|
| | 205539 | 084499 | 275554 |
| Zircon | 47·2 | 62·4 | 68·3 |
| Rutile | 15·9 | 11·9 | 11·8 |
| Tourmaline | 15·6 | 10·2 | 10·7 |
| Staurolite | 10·1 | 5·8 | 4·4 |
| Kyanite | 6·2 | 4·4 | 3·1 |
| Andalusite | 3·0 | 4·4 | 1·3 |
| Sillimanite | 1·3 | 0·6 | 0·3 |
| Monazite | 0·3 | 0·3 | 0·1 |
| Muscovite | 0·1 | | |
| Spinel | 0·1 | | |
| Corundum | | | 0·1 |
| Epidote | 0·1 | | |
| Per cent opaques | 69·2 | 72·4 | 67·2 |
| Total count | 2517 | 2045 | 2424 |

material as slabby ironstones formed within the sand. Either way, the heavy mineral assemblage cemented to the fossils is quite unlike that which typifies the sand overall (Boswell, in Dines and Edmunds, 1929), especially the high incidence of garnet.

The sand is generally weathered brown to a depth of 1 m and in places this gives way to a reddened or rubified zone which persists for another metre or more. Vertical fissures rendered grey by surface-water gleying and having a polygonal pattern in plan are developed in this rubified zone. In a large excavation on Headley Heath (TQ 205539), geliturbate and ice-wedge structures containing flinty, buff loam were observed to cut deeply into the rubified zone, truncating the associated fissures.

The stratigraphic and petrographic conformity of the gravel and the overlying sand, in conjunction with their strictly interfluvial distribution, are legitimate grounds for recognizing a new solid formation. An appropriate name would be the Headley Formation, the basal member being the Ranmore Gravel and the succeeding member the Headley Sand, after the type localities. Wooldridge (1927) has remarked on the closeness of this sedimentary sequence to that at Littleheath in the Chilterns. Lastly in this context, it is misleading to retain the name 'Netley Heath Deposits', for apart from a very restricted drift of mainly Lower Greensand material referred to below, the included deposits belong to either the Plateau Drift or the Headley Formation, and have been separately mapped as such (see Fig. 3).

The conspicuous Lower Greensand detritus in the Ranmore Gravel distinguishes it from any Eocene gravel in this part of the London Basin. The most similar deposit—the Blackheath Beds—is in any case deficient in cobbles and has an interstitial fill of fine sand (Fig. 7). The heavy mineral suite of this fine sand is also very different to that of the coarser sand of the Ranmore Gravel (Table IV). Even so, most of the gravel-size flints in the latter are derived directly from the Blackheath Beds; only the cobbles have come straight from the Chalk. To elaborate, the gravel fractions in the two deposits have similar prominence, are concentrated in the

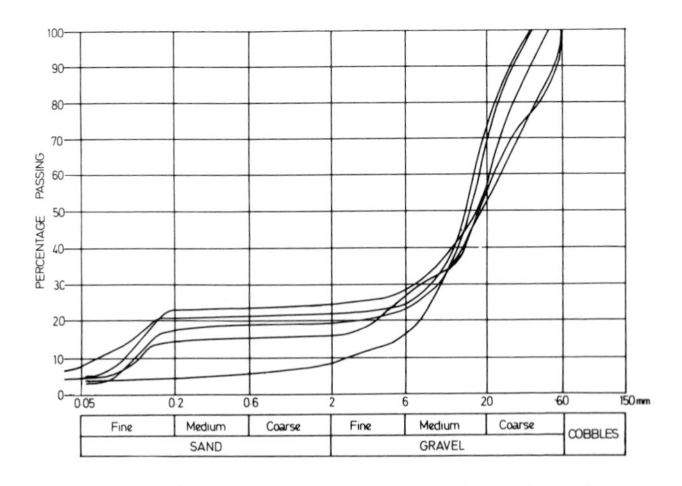


*Figure 7.* Grain-size distribution curves for disturbed Blackheath Beds gravel.

TABLE IV

*The relative frequencies of the 0·075-0·09 mm heavy mineral grains in the Blackheath Beds and in the Ranmore Gravel*

| Heavy minerals | Blackheath Beds | | Ranmore Gravel | |
|---|---|---|---|---|
| | 382578 | (2 samples) | 184524 | 156513 |
| Staurolite | 30·9 | 21·9 | 5·9 | 3·0 |
| Tourmaline | 19·9 | 32·5 | 9·6 | 20·6 |
| Rutile | 17·5 | 12·6 | 24·1 | 18·4 |
| Kyanite | 15·9 | 12·5 | 5·4 | 4·5 |
| Zircon | 14·1 | 19·8 | 54·2 | 52·1 |
| Garnet | 0·4 | 0·4 | | |
| Sillimanite | 0·4 | | 0·4 | 0·8 |
| Muscovite | 0·2 | 0·1 | 0·4 | |
| Sphene | 0·2 | | | 0·8 |
| Anatase | 0·4 | | | |
| Epidote | 0·4 | | | |
| Brookite | | 0·2 | | |
| Andalusite | | 0·1 | | |
| Per cent opaques | 76·2 | 71·3 | 72·3 | 74·0 |
| Total count | 2382 | 3469 | 1724 | 1028 |

medium to coarse grades (compare Figs 6a and 7), and are difficult to differentiate morphologically. Morphological data are set out in Tables Va, Vb and Figs 8a, b, and are based on six widely separated samples of Ranmore Gravel and five from various Blackheath Beds outliers*. With every sample, 50 flints were measured in each of the 6·35-12·7 mm, 12·7-19·0 mm and 19·0-25·4 mm sizes, and a further 25 or more from the 25·4-38·1 mm separate. The 38·1-50·8 mm, 50·8-63·5 mm and cobble-size flints were pooled with those of like diameter from other samples of the same deposit, in order to categorize a minimum of 50 individuals for each of these three grades. Reaching this total in some instances involved returning to the field to gather single specimens of the required dimensions. All the Blackheath Beds cobbles were obtained in this way, from sites on Worms Heath.

Regarding roundness, the data for the 6·35-25·4 mm flints do not allow statistical differentiation on a size-for-size basis (Table Va). They show, however, that the Eocene flints are always more highly rounded and better sorted. It is argued that this is no coincidence, rather that it reflects the slightly larger number of fractured, formerly well-rounded flints in the other deposit. The same relationship holds with the coarser gravel sizes, although the roundness values do diverge enough for a statistical separation of the 25·4-38·1 mm flints. Of the cobbles, the Ranmore Gravel examples are more angular, but the reason now is unlikely to be breakage for,

* The samples were those for which grain-size data are presented in Figs 6a and 7.

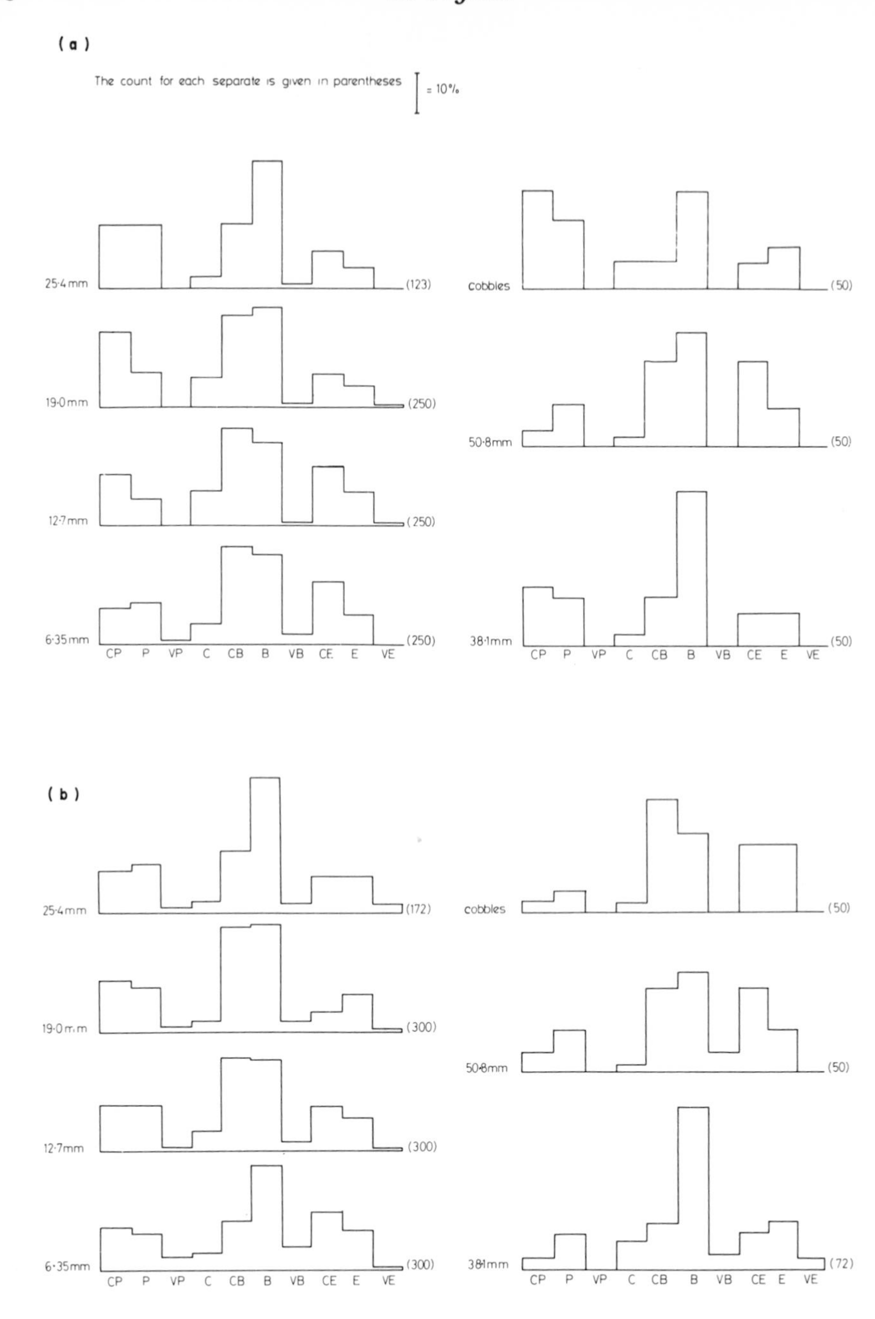


*Figure 8.* (a) Percentage form frequencies for gravel and cobbles from the disturbed Blackheath Beds. (b) Percentage form frequencies for the Ranmore Gravel.

atypically, the related sorting value is less than that of their Eocene counterparts. Admittedly the data represents just 100 clasts, yet it makes sense in that it agrees with the textural improbability of the Blackheath Beds having supplied the cobbles in the Ranmore Gravel. The sphericity values (Table Vb) repeat this pattern, of small discrepancies between the gravel grades which are not always statistically verifiable, but which are consistent in that it is generally the Eocene flints which have the highest sphericities and lowest standard deviations—or, as is contended here, the fewest breakages. With the cobbles the correlation breaks down once more, as the Ranmore Gravel specimens are more spherical and uniformly so.

The gravel-size flints in the two deposits also have a notably parity of form. Again the differences tend to be statistically insignificant when the data for only one separate are considered, but assume meaning when it is realized that the differences relate to each separate in exactly the same way. More specifically, when the chi-squared test is applied to the shape frequencies of the sieve-fractions in the 6·35-38·1 mm range, the null hypothesis being that they are the same, the following results are achieved*:

6·35-12·7 mm, $p = 0{\cdot}05$ (reject null hypothesis)
12·7 -19·0 mm, $p = 0{\cdot}26$ (accept null hypothesis)
19·0 -25·4 mm, $p = 0{\cdot}15$ (accept null hypothesis)
25·4 -38·1 mm, $p > 0{\cdot}50$ (accept null hypothesis)

Evidently only the smallest flints are discriminated with any certainty, though on consulting the relevant histograms (Figs 8a, b) it is found that each Ranmore Gravel separate has more very platy and very bladed flints and usually more very elongated shapes in addition. This cannot in every instance be due to chance, especially since the variance is anticipated by the equally small and equally persistent differences of roundness and sphericity. Furthermore if the one gravel has in part formed at the expense of the other, the creation of new shapes is to be expected. The cobbles predictably show far less resemblance of form, the most obvious distinction being the comparative lack of platy and compact platy individuals in the Ranmore Gravel.

The provenance of the Headley Sand may be treated more concisely. Detailed comparisons of grain-size parameters and heavy mineral suites failed to detect a potential source in the Eocene strata up to and including the Bagshot Beds. This is not a new observation of course, it merely extends in a more systematic way the earlier work based on heavy minerals alone. The true parent sediments lie to the south, in the Weald, for as Wooldridge (1927) established, the heavy minerals in the presumed Plio-Pleistocene marine deposits on the North Downs occur in roughly matching amounts in the Lower Greensand, particularly the Folkestone Beds. This is corroborated for Surrey by Wood's (1956) mineralogical account of the Lower Greensand in the western Weald. Rogers and Richardson's (1947) survey of the

* The test was not employed on the coarser flints as the number of 'cell' combinations entailed (measurements were made on minimum samples of just 50 in the size ranges in question) would make it insensitive.

TABLE Va

*Roundness data for flints from the Blackheath Beds (prefix B) and from the Ranmore Gravel (prefix R)*

| Size (mm) | Mean absolute roundness for deposit * | Roundness standard deviation for deposit | Mann-Whitney *U* Test of comparison of sample means for 6·35-38·1 mm flints |
|---|---|---|---|
| 6·35 -12·7 | B 0·475±0·027<br>R 0·470±0·028 | 0·220<br>0·257 | $p>0{\cdot}10$ difference not significant |
| 12·7 -19·0 | B 0·636±0·029<br>R 0·635±0·030 | 0·233<br>0·266 | $p>0{\cdot}10$ difference not significant |
| 19·0 -25·4 | B 0·742±0·025<br>R 0·695±0·027 | 0·182<br>0·218 | $p>0{\cdot}10$ difference not significant |
| 25·4 -38·1 | B 0·743±0·037<br>R 0·594±0·048 | 0·181<br>0·285 | $p = 0{\cdot}002$ difference highly significant |
| 38·1 -50·8 | B 0·78 mean for 50 flints<br>R 0·57 mean for 72 flints | 0·18<br>0·23 | |
| 50·8 -63·5 | B 0·65 mean for 50 flints<br>R 0·50 mean for 50 flints | 0·15<br>0·20 | |
| Cobbles | B 0·64 mean for 50 flints<br>R 0·61 mean for 50 flints | 0·27<br>0·10 | |

* 95 per cent confidence limits are given for flints < 38·1 mm equivalent spherical diameter.

mechanical composition of this formation in north-west Surrey further demonstrates that it is chiefly constituted of sand of the appropriate grade.

Whilst drift from the Headley Formation is fairly widespread over the dipslope platform and in the valleys cut into it, the main undisturbed residuals have a restricted, not to say curious distribution. One lies immediately west of the Mole, the others immediately west of the large dry valleys located to the north of Gomshall, Betchworth and Gatton. As stated earlier, these are the only valleys which cross the entire dipslope. The outliers, totalling four in all, provide the only instances within the area where the Headley Formation seems ever to have rested directly on the Chalk. Proof that it does so on Headley Heath has come from boreholes for the M25 motorway. The considerable thicknesses of sediments known to exist in the other outliers (Dines and Edmunds, 1929, 1933), allied with their interfluvial positions, suggest that they are also in contact, or very nearly so, with the Chalk.

The second basic category of deposits, i.e. the slope deposits, includes a largely degraded coverloam, Coombe Rock, Clay-with-Flints *sensu stricto*, sand and gravel from the Headley Formation, and a very limited development of Lower Greensand chert and sandstone with reworked Headley Sand. The coverloam has a rather

TABLE Vb

*Sphericity data for flints from the Blackheath Beds (prefix B) and from the Ranmore Gravel (prefix R)*

| Size (mm) | Mean sphericity for deposit in $\psi p$ units* | Sphericity standard deviation for deposit | Mann-Whitney $U$ Test of comparison of sample means for 6·35-38·1 mm flints |
|---|---|---|---|
| 6·35-12·7 | B 0·712±0·013<br>R 0·678±0·018 | 0·106<br>0·147 | $p$ = 0·008 difference highly significant |
| 12·7 -19·0 | B 0·728±0·010<br>R 0·693±0·011 | 0·082<br>0·096 | $p$ = 0·002 difference highly significant |
| 19·0 -25·4 | B 0·718±0·011<br>R 0·680±0·011 | 0·082<br>0·091 | $p$ = 0·005 difference highly significant |
| 25·4 -38·1 | B 0·692±0·016<br>R 0·669±0·015 | 0·079<br>0·094 | $p$ = 0·061 difference dubious |
| 38·1 -50·8 | B 0·66 mean for 50 flints<br>R 0·65 mean for 72 flints | 0·04<br>0·08 | |
| 50·8 -63·5 | B 0·67 mean for 50 flints<br>C 0·67 mean for 50 flints | 0·10<br>0·10 | |
| Cobbles | B 0·69 mean for 50 flints<br>B 0·70 mean for 50 flints | 0·09<br>0·07 | |

* 95 per cent confidence limits are given for flints < 38·1 mm equivalent spherical diameter.

uniform fine-earth component, characterized by coarse and medium silt. It extends over high and low ground, and is best preserved as a buff, usually flint-rich veneer, normally less than 1 m thick, covering large expanses of Plateau Drift. Silt is an equally important ingredient in the parent material of the rendzinas on the moderate to steep slopes of the Upper Chalk.

The rest of the deposits are layered in varying degrees, have comparatively heterogeneous textures and, save for the Clay-with-Flints *sensu stricto*, rapidly thicken downslope. An idea of the thicknesses attained on the dipslope can be gained from three boreholes for the M25 motorway that were sunk in the upper part of the Headley Valley (TQ 197543). All three penetrated coarsely stratified gravelly sand. One borehole was on the lower valley-side just below the large outlier of the Headley Formation, and entered Chalk at a depth of 7·6 m. The other two were on and near to the valley-bottom and reached Chalk at 21·2 and 21·6 m respectively. On the escarpment the thicknesses range up to about 5 m, and this much bedded, coarse- to fine-gravel-size Coombe Rock is visible a little above the foot of the escarpment in the quarry north of Betchworth station (TQ 207515). In road works next to the station and in a nearby gas-pipeline trench, finely laminated Coombe

Rock comprising mainly silt- and sand-size particles was seen to bury Upper Greensand to depths of between 1 and 3 m.

The slope deposits show every sign of having been affected by intensive patterned-ground formation, and the Coombe Rock encloses palaeosols like those described elsewhere on the North Downs by Kerney (1963). The deposits are derived from the Chalk, from the interfluve deposits—including one now all but destroyed—and merge upslope with them. The lateral transition from Clay-with-Flints *sensu stricto* into Plateau Drift that was witnessed at a number of sites is believed to be the normal relationship between these two deposits in Surrey. This would help explain why Clay-with-Flints *sensu stricto* was met with only on the shoulders of interfluves, bordering Plateau Drift, and why the latter lacks a feather-edge which thins to less than 1 m. The fine-earth of the Clay-with-Flints *sensu stricto* contains 65 to 95 per cent clay-size particles, with notable increments of medium to fine sand in the vicinity of the Headley Formation or drift derived therefrom. The colour varies from strong brown next to the Chalk to yellowish-brown where more than 50 cm thick.

Coombe Rock is ubiquitous on the slopes below the Clay-with-Flints *sensu stricto* and below the thin spreads of flinty brown sand on the lowest sector of the dipslope. Lenses and layers of Clay-with-Flints *sensu stricto* are frequently embedded in the Coombe Rock. They are of interest pedogenically in that they are more or less indistinguishable from the illuvial horizons of the contemporary *sols lessivés* developed in the former deposit. The Chalk beneath the Coombe Rock is generally shattered to depths of as much as 4 m, but near the crest of the escarpment on Merrow Downs (TQ 026491), in excavations for a reservoir, the deposit was found to overlie sediments let into the Chalk. These were wedge-shaped bodies of cross-bedded, medium-fine sand containing tabular blocks of fresh flint. Individual fissures were 1 to 2 m across at their widest, tapering downwards to depths of 4 to 7 m. The Chalk encasing the sand was broken into a loose, rubbly condition.

The remaining deposit in this category is the mass of Lower Greensand chert and sandstone with reworked Headley Sand which is confined to Mountain Wood, on the northern slopes of Netley Heath. About 2 m of this deposit trenching into undisturbed Headley Sand is exposed in a cutting for a Forestry Commission road (TQ 092508). The stratification of the upper part of the deposit appears horizontal; that of the lower part is inclined at an angle of 30° north-north-east. A sprinkling of the former gravel which contributed the chert and sandstone is still very much in evidence on the surface of the Headley Sand in the higher parts of Mountain Wood.

## The formation and age of the soils and superficial deposits

The petrographic monotony, lack of layering and distribution of the interfluve deposits indicate that they are the basically *in situ* waste of solid formations which covered the dipslope of the North Downs in Surrey before most of the present valleys were cut. Conversely, the overall variability, frequent crude stratification and position of the slope deposits testify to periglaciation.

Clearly the waste of brown to reddish-brown flinty sand and the higher-level Plateau Drift are greatly attenuated and weathered Lower Eocene strata, with variable admixtures of flints from diverse sources and some Lower Greensand material. Probably much stripping of the main Eocene cover was accomplished in Tertiary times. Nevertheless, the secondary (as distinct from the inherited) reddening in these drifts, as well as the incorporation of flints and other stones, can readily be attributed to Pleistocene events. No less pronounced secondary reddening or rubification affects the Headley Sand. Accordingly it cannot in this instance pre-date the Pleistocene, since the Red Crag fossils on Netley Heath occur near the base of what is now seen to be an undisturbed residual of this deposit, well below the horizons discoloured by soil formation and weathering. Similarly it cannot post-date the Pleistocene, for as described earlier, the rubified zone and associated grey fissures are truncated on Headley Heath by cryogenic structures. An interglacial date for the reddening is therefore most likely. The widespread periglacial distortion of the bases of the two drifts makes it reasonable to infer that the great preponderance of their flints were injected by frost-heaving, from the drift-Chalk interface. Any significant concentration of the flints beforehand by solution is discounted, however, as this would be difficult to reconcile with the extremely variable stoniness of the drifts, from scarcely any to densely crammed accumulations. The worn flints and Lower Greensand detritus in the Plateau Drift have been included in the same way, but from gravels which once overlaid the deposit or its antecedents. In short, the remnant Eocene formations under discussion are thought to have been altered by advanced oxidative weathering in the temperate stages of the Pleistocene and more radically by geliturbation in the colder ones. The virtual absence of the Thanet Sand and Bottom Bed beneath the Plateau Drift must result from the fluviatile and estuarine strata of the Woolwich and Reading Beds having originally overstepped these marine formations, roughly along a line coincident with the northern limits of the Plateau Drift.

These conclusions on the formation of the Plateau Drift differ fundamentally from those of earlier workers, who invoked glacial and/or periglacial redistribution of the parent sediments. Had the latter indeed been subjected to such reorganization, it is fair to surmise that they would have been more intimately blended, and their petrographic integrity either destroyed or made far less distinct. Moreover, were the Plateau Drift really a till it might be expected to contain occasional blocks and frequent lumps of chalk, for ice traversing such a soft, well-jointed rock would surely have acquired material from it. Decalcification seems an unlikely explanation for its complete absence, as fresh chalk is still abundant even in the top 0·3 m of the Gipping Till (Bristow and Cox, 1973), which is at least as old as the Wolstonian. The lack of Plateau Drift on slopes exceeding 2°-3° and its predictable height-boundary relationship are equally incompatible with glacial deposition, unless it happened before the incision of the valleys which now dissect the deposit, when the dipslope had negligible relief. The only time when the Chalk surface is known definitely to have been like this was at the opening of the Eocene, although exponents of the view that the sea which laid down the Headley Formation widely bevelled the Chalk would reason that it was in part returned to this condition at the start of the Pleistocene. Even if this were fact and not assumption, for the Plateau Drift to be an early Pleistocene till as Kellaway *et al.* (1975) suggest would still leave unanswered why the Plateau Drift is exclusive to interfluves when thick drift from the Headley Formation drapes valley-sides and valley-bottoms to depths of over 20 m. The constant association of Plateau Drift with gentle slopes in the same way precludes its having been derived by solifluction from once extensive residuals of Woolwich and Reading Beds. Had this in fact been so, Plateau Drift should be present around the large Reading Beds eminences at Hook Wood (TQ 075505), Cherkeley Wood (TQ 188543) and Nower Wood (TQ 195548), which it is not.

The sediments in the Headley Formation were regarded as a Calabrian marine assemblage by most previous investigators. This interpretation is accepted and so only new information will be examined. Firstly, a very high proportion of the flints in the basal gravel are from the Blackheath Beds, whose outcrop extends from Kent into Surrey, but only as far as the Croydon-Merstham monocline. Derived Blackheath Beds flints occurring as far west as Netley Heath thus indicate migration along the shore of a westward-transgressing sea, as a large river flowing in this direction contradicts the known Cainozoic history of the London Basin. Further, if the sea which left the gravel had genuinely planed the Chalk, as Wooldridge and Linton contended, a very different assortment of flints would have arisen, and the Eocene pebbles would have been more altered than they are in the prolonged erosional phase implied.

The textural uniformity, lack of bedding, moderate sorting and positive skewness of the Headley Sand probably reflect near-shore rather than littoral environments (see John, 1974, for discussion). Apart from a few intertidal forms, the same is true of the associated fossils from Netley Heath listed by Chatwin (1927). Even so, positively skewed sands do occur on beaches near river mouths (Friedman, 1961, 1967), so that it is perhaps significant that the principal residuals of Headley Sand all adjoin the few dry valleys that extend from the crest of the North Downs to the Eocene outcrop and beyond. In other words, should these valleys mark earlier lines of Wealden drainage, then the positive skewness of the Headley Sand need not necessarily prevent its acceptance as a beach deposit. Such a reconstruction would also offer an explanation of why the Headley Sand is best preserved next to these valleys, for here the supply of sediment from the parent Lower Greensand strata would have been greatest. Similarly, it would make it feasible to propose that the Headley Formation rested directly on the Chalk only in proximity to these valleys because they were the only places where sufficient of the Eocene cover had been removed hitherto by rivers.

The possibility that Wealden drainage influenced significantly the deposition of the Headley Formation raises a problem connected with the nature of the heavy mineral suite it contains. Edelman and Doeglas (1933) subdivided the Tertiary rocks of the Netherlands younger than the Asschien into an A-group and B-group on heavy mineral criteria. The A-group were entirely marine, the B-group entirely continental. Edelman (1933) correlated the first group with the Red Crag of East

Anglia and the second with the Lenham Beds, which have the same distinctive mineralogy as the Headley Formation. As Elliot (in Worssam, 1963) noted, the correlation apparently conflicts with the presence of marine fossils in the Lenham Beds, and this applies with equal force to the younger Red Crag fossils in the Headley Formation. Another difficulty in accepting the implications of Edelman's correlation, however, is that the Headley Formation extends across the interfluves between the major consequent valleys of the North Downs irrespective of having been derived in no small part from the Lower Greensand farther south. Bury (1910, 1922) rightly realized that this is puzzling if ancient Wealden rivers were the depositional agents, but understandable if the rivers were discharging onto a sea-shore. Conceivably such rivers may have supplied sediments in quantities so large to the shores around or near which the Headley Formation accumulated that the original heavy mineral attributes were essentially retained. This being so, it is easy to see how the conflict in question may have arisen.

Since the Red Crag fossils on Netley Heath were found near the base of an undisturbed residual of Headley Sand, they cannot have been incorporated after this deposit was laid down, as might be deduced from the aberrant heavy minerals in the matrix of the fossiliferous material. Moreover in response to the specific suggestion that the fossils are perhaps glacial erratics from the southern bight of the North Sea (Shephard-Thorn, 1975), it is as well to emphasize that the enigmatic heavy mineral assemblage is also very different from that of the East Anglian Red Crag and their Dutch equivalents (Boswell, in Dines and Edmunds, 1929). No doubt the fossils could have been derived from older sediments at the time of deposition, yet the interval, if any, separating the fossils and the sand cannot have been great, for the uplift of the North Downs and the excavation of the Weald must have occupied most of Pleistocene time. As for the hypothesis that the residual of Headley Sand on Headley Heath may in fact be a coversand of northerly provenance (Letzer, 1973), there are several formidable objections which can be made to it. In particular it is difficult to envisage why all the residuals of this deposit should be associated with the same basal shingle, why all of them should be preserved only on the west of major consequent valleys*, and why all of them have identical heavy minerals of obvious Lower Greensand affinity.

Coming lastly to the slope deposits, evidence has already been adduced as to their periglacial character and little further comment on their origin is needed. Generally they can be ascribed to solifluction. The exceptions include: the coverloam, which has the hallmarks of a much-disturbed loessial formation; the finely-laminated, finely-textured Coombe Rock south of the escarpment, which owes more to niveo-fluvial action; and the cross-bedded sand containing blocks of flint, found beneath Coombe Rock in the Chalk, which appears to have been deposited by running water, possibly in cavities left by the decay of ice-wedges at a time when the Chalk

* In addition to the tentative explanation given earlier on this peculiar preservation pattern, there may well be structural considerations beyond the scope of this chapter (see John, 1974) which are responsible for the deposit being largely confined to the west of the valleys concerned.

was still partially frozen. The slope deposits probably date from the recent stages of the Quaternary, as the chances of their having survived from earlier ones in the kinds of topographic situations involved are hardly favourable. Since certain of these periglacial accumulations are derived from the interfluve deposits, the relict weathering observable in the latter is therefore unlikely to be much older. The secondary reddening, especially, is unlikely to date from an early interglacial.

## A new model of landform evolution

The essential finding of the study is that basically *in situ* remains of Eocene strata cover the vast majority of the dipslope. This agrees with Pinchemel's thesis on Chalk landscape development in so far as it stresses the ubiquity of the sub-Eocene surface. Nevertheless, one of his main supporting contentions—that the flinty clays on the highest parts of the Chalk are due ultimately to a sort of *lessivage* operating under early Tertiary tropical climates—cannot be accepted here. The bulk, if not all, of the flinty clay in the area mapped came from Eocene rocks, in the Pleistocene.

Put another way, the findings of the study invalidate Wooldridge and Linton's claim that the dipslope platform is a wave-cut bench, and that the restricted summit-ridges are Mio-Pliocene peneplain remnants. Inevitably the ideas dependent upon the reality of these surfaces have also to be refuted, notably those relating to late Tertiary tectonic stability and to the upper limit of the Calabrian transgression.

Evidently the morphological configuration of the Chalk dipslope in Surrey is structurally controlled, and earlier in this paper data was presented to show that the Chalk is flexured so as to give the illusion of 'surfaces' younger than the sub-Eocene plane. The northernmost flexure was certainly active in early Eocene times, for near Headley extensive Thanet Sand residuals rest on *Marsupites* Chalk, whilst northwards the same residuals pass onto older *Micraster* Chalk. This is of great significance, as the flexure not only coincides with the northern boundary of the dipslope bench and a line of major strike-valleys, but also with the point where, it is argued, the parent strata of the Plateau Drift overstepped older, sandier Eocene formations. These sandier formations are now represented by the brown to reddish-brown drift of flinty sand on the northern part of the dipslope.

In conclusion, the dipslope regardless of altitude is viewed as being in the final stages of exhumation from beneath its original cover of Landenian sediments. The stripping of this cover and doubtless younger superjacent sediments was accomplished by Tertiary erosion and the Calabrian transgression, whose upper limit remains to be determined. The chief role of the periglacial regimes which have intervened since then has been the dissection and very extensive removal of the Headley Formation, re-exposing to weathering and erosion the older cover of vestigial Eocene sediments. The relict reddening or rubification observed in the soil profiles of the interfluve deposits most likely dates from the more recent stages of the Quaternary. Indeed it would be surprising if this were not so, as the detailed convexo-concave form of the topography of the North Downs of Surrey is largely the result of intense periglaciation, which has left widespread slope deposits and varied patterned-ground phenomena. Little removal of the interfluve deposits has occurred since the cessation of periglacial activity.

## Acknowledgements

I wish to thank the University of London for a grant from the Central Research Fund towards the cost of travel.

## Appendix

### Location of samples

All the sampling sites lie within the 100-km square TQ. The map references for each location are listed under the relevant figure numbers. The last set of references relate to the eleven samples of presumed Calabrian Sand mentioned in the text.

*Figure 4a*
145528 (4 samples)
223586

*Figure 4b*
189545 (4 samples)
193564 (4 samples)
194557 (2 samples)

*Figure 4c*
038497 (2 samples)
020491
043495
064503
148544

*Figure 4d*
092566 (3 samples)
044510 (2 samples)

*Figure 5a*
196514 (2 samples)
041492
083493
101494
205539
225554
230527
230528
233541

*Figure 5b*
205539
230527
232539
248529
258534
266530
268536

*Figure 5c*
173574 (4 samples)
084508 (2 samples)
195551

*Figure 6a*

| | |
|---|---|
| 098511 | 184524 |
| 156510 | 197536 |
| 156513 | 266574 |

*Figure 6b*

| | |
|---|---|
| 084499 | 205539 |
| 092508 | 275554 |

*Figure 7*

377576 (2 samples)
382578 (2 samples)
331543

The eleven samples of presumed Calabrian sand mentioned in the text came from the sites specified under Fig. 6b and from the following locations:

| | |
|---|---|
| 205537 (3 samples) | 203538 |
| 091503 | 204532 |
| 198538 | |

## References


Avery, B. W., Stephen, I., Brown, G. and Yaalon, D. H., (1959). The origin and development of brown earths on Clay-with-Flints and Coombe Deposits, *J. Soil Sci.* **10**, 177-95.

Bristow, C. R. and Cox, F. C. (1973). The Gipping Till: a reappraisal of East Anglian glacial stratigraphy, *J. Geol. Soc. Lond.* **129**, 1-37.

British Standard 1377 (1967). "Methods of testing soils for civil engineering purposes". British Standards Institution, London.

Brown, H. J. W. (1924). Minor structures in the Lower Greensand of west Kent and east Surrey, *Geol. Mag.* **62**, 439-51.

Bury, H. (1910). The denudation of the western end of the Weald, *Q. J. Geol. Soc. Lond.* **66**, 640-92.

Bury, H. (1922). Some high-level gravels of north-east Hampshire, *Proc. Geol. Ass.* **33**, 81-103.

Chatwin, C. P. (1927). Fossils from the ironsands on Netley Heath (Surrey), *Mem. Geol. Surv. Summ. Prog.* 1926, 154-7.

Clark, M. J., Lewin, J. and Small, R. J. (1967). The Sarsen stones of the Marlborough Downs and their geomorphological implications, *Southampton Res. Ser. Geogr.* **4**, 3-40.

Davies, G. M. (1915-16). The rocks and minerals of the Croydon regional survey area, *Proc. Trans. Croydon Nat. Hist. Sci. Soc.* **8**, 53-96.

Davies, G. M. (1918). Excursion to Ranmore Common, *Proc. Geol. Ass.* **29**, 36-8.

Davis, A. G. (1926). Notes on some Chalk sections in north-east Surrey, *Proc. Geol. Ass.* **37**, 211-20.

Dewey, H. and Bromehead, C. E. N. (1921). The geology of south London, *Mem. Geol. Surv. U.K.*

Dines, H. G., Buchan, S., Holmes, S. C. A. and Bristow, C. R. (1969). Geology of the country around Sevenoaks and Tonbridge, *Mem. Geol. Surv. U.K.*

Dines, H. G. and Edmunds, F. H. (1929). The geology of the country around Aldershot and Guildford, *Mem. Geol. Surv. U.K.*

Dines, H. G. and Edmunds, F. H. (1933). The geology of the country around Reigate and Dorking, *Mem. Geol. Surv. U.K.*

Edelman, C. H. (1933). Petrologische provincies in het Nederlandsche Kwartair, *Meded. Geol. Inst. Univ. Amst.* **43**, 1-104.

Edelman, C. H. and Doeglas, D. J. (1933). Bijdrage tot de petrologie van het Nederlandsche Tertiair, *Verh. Geol.-Mijnb. Genoot. Ned. Kolon., Geol. Ser.* **10**, 1-38.

Folk, R. L. and Ward, W. C. (1957). Brazos River bar: a study in the significance of grain-size parameters, *J. Sedim. Petrol.* **27**, 3-26.

Friedman, G. M. (1961). Distinction between dune, beach and river sands from their textural characteristics, *J. Sedim. Petrol.* **31**, 514-29.

Friedman, G. M. (1967). Dynamic processes and statistical parameters compared for size frequency distribution of beach and river sands, *J. Sedim. Petrol.* **37**, 327-54.

Gray, D. A. (1965). The stratigraphical significance of electrical resistivity marker bands in the Cretaceous strata of the Leatherhead (Fetcham Mill) Borehole, Surrey, *Bull. Geol. Surv. G.B.* **23**, 65-90.

Green, C. P. (1969). An early Tertiary surface in Wilshire, *Trans. Inst. Br. Geogr.* **47**, 61-72.

Groves, A. W. (1928). Eocene and Pliocene outliers between Chipstead and Headley, Surrey, *Proc. Geol. Ass.* **39**, 471-85.

Hodgson, J. M., Catt, J. A. and Weir, A. H. (1967). The origin and development of Clay-with-Flints and associated soil horizons on the South Downs, *J. Soil Sci.* **18**, 85-102.

Hodgson, J. M., Rayner, J. H. and Catt, J. A. (1974). The geomorphological significance of the Clay-with-Flints on the South Downs, *Trans. Inst. Br. Geogr.* **61**, 119-29.

John, D. T. (1974). A study of the soils and superficial deposits on the North Downs of Surrey, Ph.D. Thesis, University of London (unpublished).

Jones, D. K. C. (1974). The influence of the Calabrian transgression on the drainage evolution of south-east England, *in* Brown, E. H. and R. S. Waters (eds), "Progress in Geomorphology". *Inst. Brit. Geogr. Sp. Pub.* No. 7.

Jukes-Browne, A. J. (1904). The Cretaceous rocks of Britain. 3, The Upper Chalk of England, *Mem. Geol. Surv. U.K.*

Jukes-Browne, A. J. (1906). Remarks on the Upper Chalk of Surrey, *Proc. Geol. Ass.* **19**, 286-90.

Kellaway, G. A. and Redding, J. H., Shephard-Thorn, E. R. and Destombes, J. P. (1975). The Quaternary history of the English Channel, *Phil. Trans. Roy. Soc. Lond.* **A279**, 189-218.

Kerney, M. P. (1963). Late-glacial deposits on the Chalk of south-east England, *Phil. Trans. Roy. Soc. Lond.* **B246**, 203-54.

Kirkaldy, J. F. (1958). "Geology of the Weald", Geol. Ass. Guide 29.

Krumbein, W. C. (1941). Measurement and geological significance of shape and roundness of sedimentary particles, *J. Sedim. Petrol.* **11**, 64-72.

Letzer, J. M. (1973). The nature and origin of superficial sands at Headley Heath, Surrey, *Proc. Trans. Croydon Nat. Hist. Sci. Soc.* **14**, 263-8.

Loveday, J. A. (1962). Plateau deposits of the southern Chiltern Hills, *Proc. Geol. Ass.* **73**, 83-102.

Pinchemel, P. (1954). "Les plaines de craie du Bassin Parisien et du Sud-est du Bassin de Londres et leurs bordures". Colin, Paris.

Robinson, G. W. (1949). "Soils, their origin, constitution and classification", Murby, London.

Rogers, H. S. and Richardson, J. A. (1947). Mechanical analysis of the Lower Greensand of north-west Surrey, *Proc. Geol. Ass.* **58**, 259-69.

Shephard-Thorn, E. G. (1975). The Quaternary of the Weald—a review, *Proc. Geol. Ass.* **86**, 537-547.

Sherlock, R. L. (1929). Discussion on the alleged Pliocene of Bucks. and Herts., *Proc. Geol. Ass.* **40**, 357-72.

Sneed, E. D. and Folk, R. L. (1958). Pebbles in the Lower Colorado River, Texas, *J. Geol.* **66**, 114-50.

Soil Survey Staff (1960). "Field handbook". Soil Survey of Great Britain, London.

Whitaker, W. (1912). The water supply of Surrey, *Mem. Geol. Surv. U.K.*

Wood, G. V. (1956). The heavy mineral suites of the Lower Greensand of the western Weald, *Proc. Geol. Ass.* **67**, 124-37.

Wooldridge, S. W. (1927). The Pliocene history of the London Basin, *Proc. Geol. Ass.* **38**, 49-132.

Wooldridge, S. W. and Linton, D. L. (1955). "Structure, surface and drainage in south-east England". George Philip, London.

Worssam, B. C. (1963). Geology of the country around Maidstone, *Mem. Geol. Surv. U.K.*

Worssam, B. C. (1973). A new look at river capture and at the denudation history of the Weald, *Rep. Inst. Geol. Sci.* **73/17**.

Young, G. W. (1905). The Chalk area of north-east Surrey, *Proc. Geol. Ass.* **19**, 188-99.

Young, G. W. (1906). Excursion to Headley, *Proc. Geol. Ass.* **19**, 347-9.

Young, G. W. (1907-8). The Chalk area of western Surrey, *Proc. Geol. Ass.* **20**, 422-55.

Young, G. W. (1915a). Report of an excursion to Horsley and Netley Heath, *Proc. Geol. Ass.* **26**, 110-11.

Young, G. W. (1915b). Report of an excursion to Leatherhead, Polesden Valley and the gorge of the Mole, *Proc. Geol. Ass.* **26**, 320-4.

Young, G. W. (1917). Excursion to Ashtead, Headley-on-the-Hill and the valley of the River Mole, *Proc. Geol. Ass.* **28**, 38-9.

SIX

# Glaciation of the London Basin and its influence on the drainage pattern: a review and appraisal

COLIN A. BAKER*
*Department of Geography, University of London King's College*
DAVID K. C. JONES
*Department of Geography, London School of Economics and Political Science*

## Introduction

The hypothesis of two glacially-induced southward displacements of the Lower Thames has for long been considered one of S. W. Wooldridge's outstanding achievements (Wooldridge, 1938, 1957, 1960; Wooldridge and Linton, 1939, 1955; Wooldridge and Henderson, 1955; Wooldridge and Cornwall, 1964). As a result of extensive fieldwork within the London Basin, he was able to synthesize and develop upon earlier ideas (Sherlock and Noble, 1912; Gregory, 1922; Hawkins, 1923; Sherlock, 1924) and so present a coherent evolutionary model of the development of the Lower Thames drainage system. Although this hypothesis was widely accepted, it is important to note that Wooldridge himself regarded the work as incomplete (Linton, 1969), especially as the later interpretation of the Essex glacial sequence as the product of three glacial advances (Clayton, 1957, 1960, 1964) proved difficult to reconcile with the two stage diversion of the Thames proposed by Wooldridge. Indeed Clayton was forced to remark, "Some problems of dating the various diversions remain and I am not sure that the various pieces of evidence slot together as neatly as is sometimes supposed" (Clayton, 1969; p. 10).

Recent work has done much to resolve these problems. Reinterpretation of the glacial sequence (Sparks *et al.*, 1969; Turner, 1970; Baker, 1971; Bristow and Cox, 1973; Rose, 1974), together with the results of detailed studies of fluvial and fluvio-glacial deposits (Rose *et al.*, 1976; Gibbard, 1977, 1979; Green and McGregor, 1978) have resulted in an essentially monoglacial interpretation with a single deflection of the Thames. Although this new model is based more securely on

* Present address: St Lawrence College, Ramsgate, Kent CT11 7AE.

lithostratigraphic evidence, certain areas of uncertainty still remain. The objects of this chapter, therefore, are first to briefly summarize the changing views concerning the evolution of the Lower Thames and then to extend the new model by examining three aspects that have yet to be re-evaluated. These are: evidence for the so-called Chiltern Drift Glaciation, the evolution of the Mid-Essex Depression and the development of the River Lea.

## The double drainage diversion hypothesis (following Wooldridge)

The evolutionary sequence developed by Wooldridge (1938) stemmed from the identification and mapping of glacial, fluvio-glacial and fluvial deposits in the northern and western portions of the London Basin. A number of sedimentary units were distinguished which, when correlated on the basis of altitude, combined to form an impressive geomorphological stairway leading upwards from the Thames alluvial carpet to the 183 m early Pleistocene (?) 'Calabrian' marine bench.

The spatial distribution of these deposits and their relationship to the major topographic units of the London Basin are of great significance. That part of the London Basin which lies to the north of the Thames can be subdivided into three regions on the basis of topography and sedimentary deposits. To the west of the River Colne (Fig. 1), a remarkably fine sequence of Thames terraces has been preserved on the Chiltern dipslope, especially in the vicinity of Beaconsfield (Hare, 1947; Wooldridge and Linton, 1955; Sealy and Sealy, 1956). In the eastern part of the basin, east of the River Lea, lies a second area within which glacial and fluvio-glacial deposits predominate (Clayton, 1957, 1960, 1964). Between these two lies a central region, bounded by the rivers Colne and Lea, which has evolved through the interplay of glacial, fluvio-glacial and fluvial agencies. It is here that the most complex chronological problems have been posed, for although there are numerous difficulties still to be overcome in dating both the Thames terraces in the west and the glacial events in the east, it is within this central region that the two evolutionary sequences have had to be matched.

The central region consists of six major topographic units (Fig. 1), which are aligned NE-SW in harmony with the local trend of the London Basin syncline. The sequence begins with the Chiltern backslopes which descend to the Vale of St Albans. This drift-floored lowland is overlooked by a weakly-developed north facing Tertiary 'escarpment' (Reading Beds and London Clay), which forms the northern margin of the South Hertfordshire Plateau, an arcuate upland rising to between 125 and 150 m. To the south lies a second, partially drift-floored, lowland known as the Finchley Depression, bounded in turn by the Hampstead-Highgate Ridge. This narrow linear relief feature rises to a similar elevation as the South Hertfordshire Plateau and probably represents a fragment of a former continuous synclinal ridge that extended north-eastwards to the Epping Forest Ridge (Wooldridge, 1923; Hayward, 1957). The final unit is the terrace-flanked valley of the Lower Thames.

The model of drainage evolution proposed by Wooldridge and others (Fig. 2) can be briefly described as follows. Upon the retreat of the 'Calabrian' sea, the early

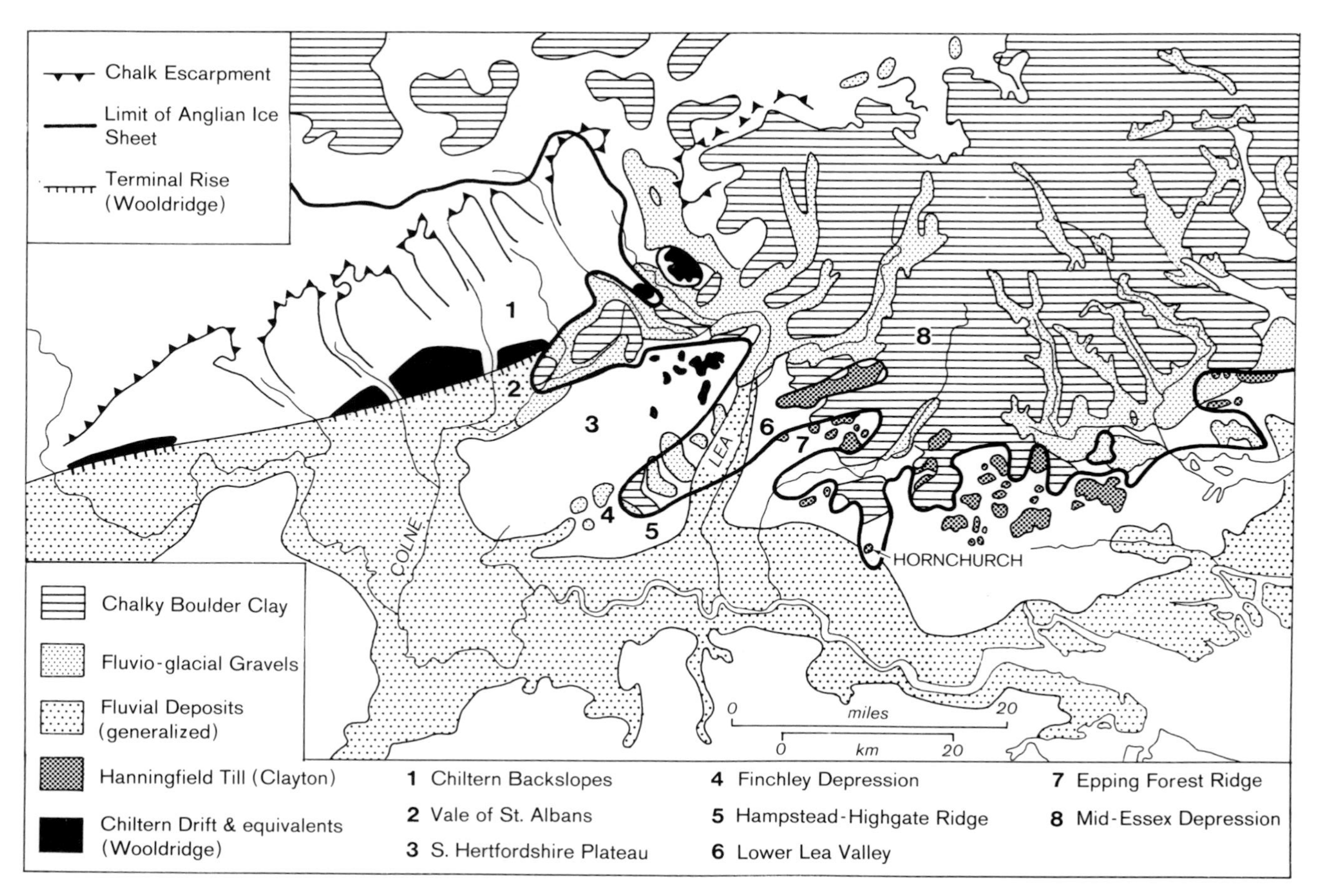


*Figure 1.* Map of the eastern London Basin showing generalized distribution of fluvial and glacial deposits referred to in the text and the main topographic units (based mainly on Wooldridge and Linton, 1955 and Clayton, 1964).

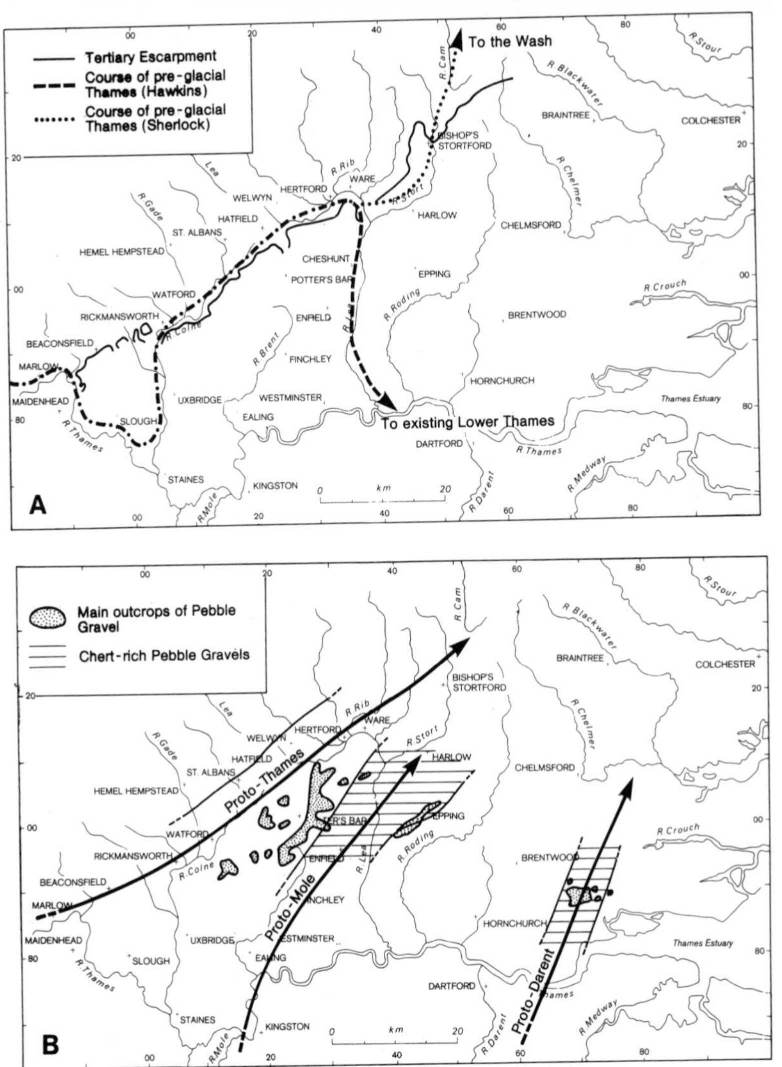


*Figure 2 A-H.* The evolution of the Thames drainage system as envisaged in the double deflection model (multiglacial interpretation) originally advanced by Wooldridge (1938) and developed by later workers. (A) Early views on the 'pre-glacial' course of the proto-Thames (Hawkins, 1923; Sherlock, 1924). (B) The Pebble Gravel course of the proto-Thames (Wooldridge, 1938).

Pleistocene proto-Thames flowed north-eastwards along the line of the Vale of St Albans (Hawkins, 1923; Sherlock, 1924; Wooldridge, 1938), and thence across central Essex to the North Sea (Fig. 2B). Three lines of evidence support such a reconstruction. First, this course represents the most logical continuation of the Middle Thames route between Reading and Beaconsfield. Second, the discontinuous Pebble Gravels that occur at about 125 m on the South Hertfordshire Plateau and elsewhere, contain 'pin-hole' chert thought to be derived from the Lower Greensand of the Weald. The concentration of this material in SSW—NNE

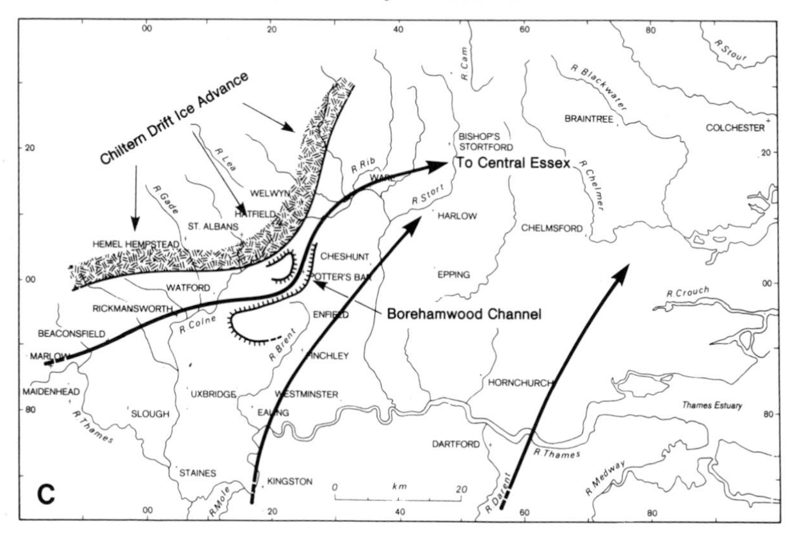


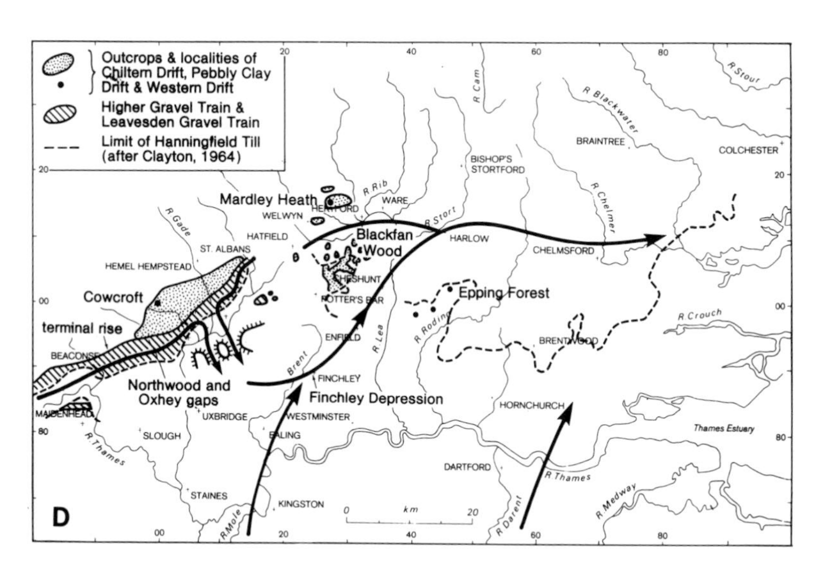


*Figure 2 (cont.).* (C) The first diversion of the proto-Thames (early stage): Chiltern Drift ice advance. (D) First diversion (late stage): Chiltern Drift, Higher Gravel Train and Leavesden Gravel Train.

oriented belts has been taken to indicate the fluvial transport of Wealden debris across the present line of the Lower Thames (Wooldridge, 1927; Wooldridge and Linton, 1955) (Fig. 2B). Third, the identification of a second and slightly lower set of Pebble Gravels (Westland Green Gravel) along the northern margin of the South Hertfordshire Plateau (Hey, 1965) confirmed the existence of north-eastward flowing drainage (Fig. 3A).

The first diversion of the proto-Thames, according to Wooldridge, was caused by a 'Chiltern Drift' glaciation, during which an ice-sheet approached the river from a

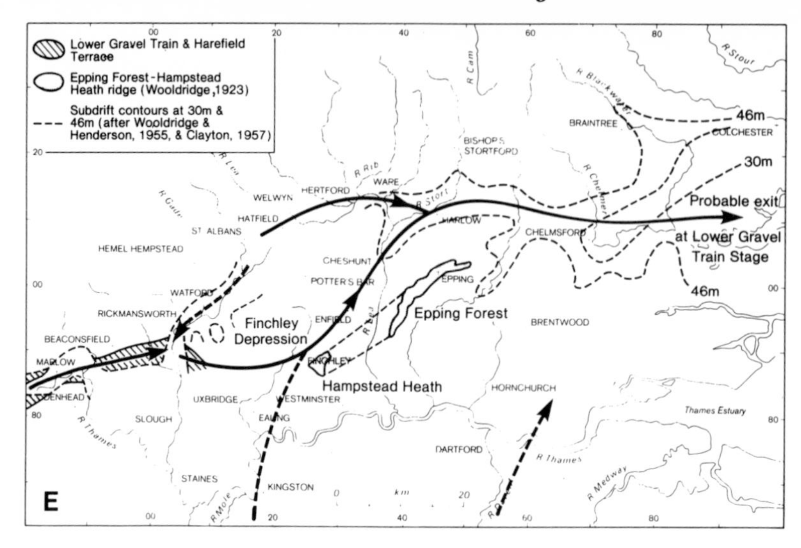


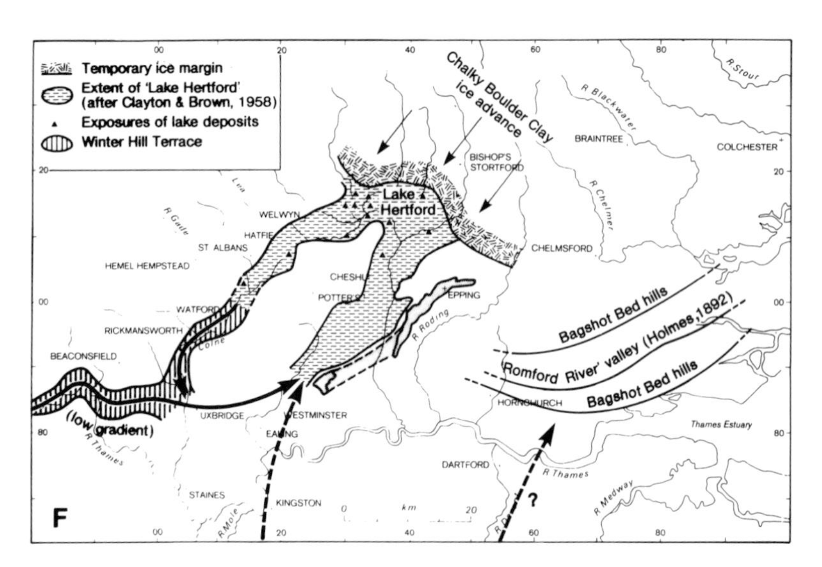


*Figure 2 (cont.).* (E) Establishment of first diversion course: Lower Gravel Train. (F) Second diversion (early stage): Chalky Boulder Clay, Lake Hertford and Winter Hill Terrace I.

north-westerly direction. Diversion occurred in two stages, first via the Borehamwood Channel (Fig. 2C), and later via two cols south of Watford to the Finchley Depression (Fig. 2D). Stratigraphic evidence for this ice advance was claimed in the 'Western Drift' or 'Chiltern Drift' (Barrow, 1919; Wooldridge, 1938; Wooldridge and Cornwall, 1964), while the presence of numerous far-travelled erratic pebbles in the Westland Green Gravels was believed to be the product of ice-rafting, probably during the early stages of the same glacial phase (Hey, 1965).

Evidence for such a shift was claimed from the terrace sequence west of the Colne.

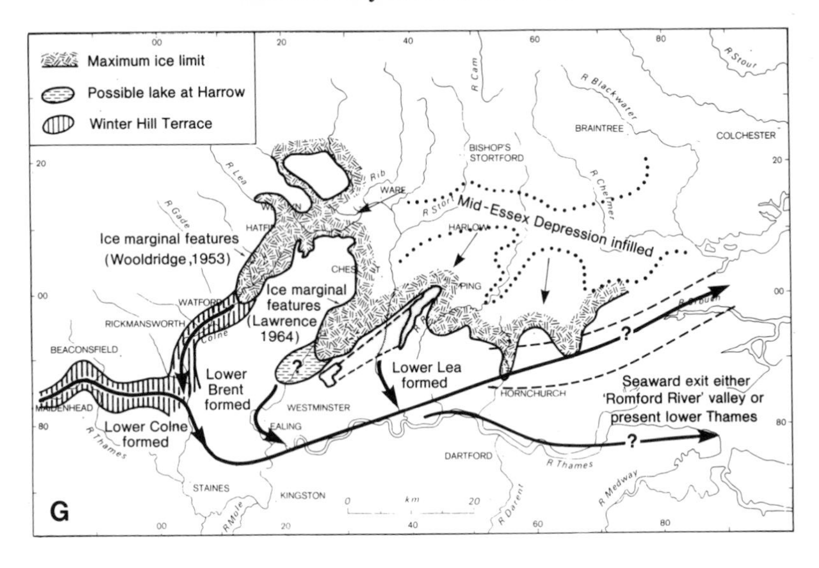


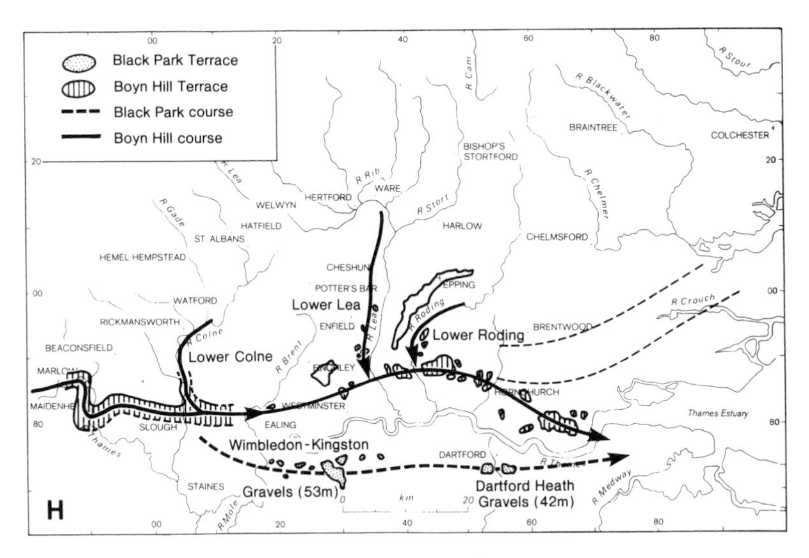


*Figure 2 (cont.).* (G) Second diversion (late stage): Chalky Boulder Clay, Eastern Drift, Winter Hill Terrace II. (H) Establishment of second diversion course: Black Park and Boyn Hill stages.

Here, the gravel spreads of the succeeding Higher Gravel Train were traced eastwards towards the Vale of St Albans until Redheath (TQ 0697), near Watford, where they appeared to be replaced by a second gravel train, the Leavesden Gravel Train, declining westwards along the northern flank of the Vale (Fig. 2D). This change in direction of slope indicated that reversal of drainage must have taken place in the western part of the proto-Vale of St Albans due to the incursion of the Chiltern Drift ice-sheet. The succeeding Lower Gravel Train (Harefield Terrace of Hare, 1947) and Winter Hill Terrace were traced eastwards across the Colne valley and south of the western extremity of the South Hertfordshire Plateau, thereby indicating that the Thames had abandoned its original course in favour of the Finchley Depression routeway (Hare, 1947) (Fig. 2E). At both the Higher and Lower Gravel Train stages, the diverted proto-Thames was believed to have rejoined its former course near Ware, and to have passed eastwards along the line of the Mid-Essex Depression (Figs. 2D and E) towards the Blackwater estuary (Wooldridge and Linton, 1955). The form of the Mid-Essex Depression was described by Wooldridge and Henderson (1955), and Clayton interpreted a subdrift gravel-covered bench at 50 m O.D. beneath Harlow New Town as part of the Lower Gravel Train terrace (Clayton, 1957). Despite this, no firm stratigraphic evidence for a proto-Thames occupation of the Finchley Depression was ever adduced.

The second major ice advance—responsible for the deposition of the Chalky Boulder Clay or 'Eastern Drift' of Wooldridge (1938)—was considered to have caused a second southward displacement of the Thames (Figs. 2F and G). This ice-sheet entered the London Basin from the north and north-east, covering most of Essex and extending two lobes south-westwards into the Vale of St Albans and the Finchley Depression. As a result the Mid-Essex Depression was buried beneath a thick mantle of drift. The original interpretation (Wooldridge, 1938; Wooldridge and Linton, 1955) envisaged a relatively simple sequence of events with a single ice advance damming the Thames and causing deflection west of Harrow, the water overflowing into a pre-existing depression excavated by the 'Romford River' (Holmes, 1892) or 'Nore River' (Wooldridge and Cornwall, 1964). Although Wooldridge makes no reference to large impounded lakes, their existence is clearly implied, with overflow routeways cutting southwards and initiating the Lower Colne and Lower Brent valleys in addition to the new course of the Thames.

Following an investigation by Clayton (1957) of the Essex glacial sequence, a more complex interpretation was put forward involving two phases of drainage development, each associated with the Chalky Boulder Clay ice advance. In the first (Fig. 2F), ice was believed to have blocked the proto-Thames near Harlow, impounding proglacial water to form 'Lake Hertford' (Clayton and Brown, 1958). The unusually low gradient of the Winter Hill Terrace west of Uxbridge (Hare, 1947), and its merging with the outwash train emanating from the western end of the Vale of St Albans, were both taken to indicate deltaic sedimentation in the upstream portion of this extensive lake. At this stage water was considered to have spilled into the 'Romford River' and caused the deflection of the Thames. The second phase (Fig. 2G) was associated with the re-advance of the ice-sheet to its maximum extension, the blocking of the Vale of St Albans and the Finchley Depression as far as Aldenham and Finchley respectively, and initiation of the Lower Brent.

Implicit in both interpretations is the assumption that the former Hampstead Heath-Epping Forest Ridge formed a continuous and effective barrier, which in the case of the 'Lake Hertford' reconstruction (Fig. 2F) must have risen to at least 68 m O.D. (Clayton and Brown, 1958). Overflow may have occurred across the lowest col in this barrier (Dury, 1958), with meltwater draining southwards into the 'Romford River'. However, initiation of the discordant Lower Lea is generally considered to have postdated the deflection of the Thames, taking place during or following the retreat of the Finchley lobe, either as a consequence of overflow (Dury, 1958) or due to headward erosion of a tributary to the new Lower Thames (Clayton and Brown, 1958).

That the Thames continued to utilize the spillway route, thereby becoming established in its present asymmetrical position within the London Basin (Fig. 2H), is based on the evidence of the Black Park Terrace (Hare, 1947) and its probable equivalents in the Wimbledon-Kingston gravels (55 m O.D.) and Dartford Heath gravels (40 m O.D.) (Wooldridge and Linton, 1955). However, the succeeding Boyn Hill Terrace (Fig. 2H) denotes a more stable position, graded as it is to the interglacial Holsteinian (Hoxnian) sea-level (Zeuner, 1959). At Hornchurch, a narrow tongue of Chalky Boulder Clay is significantly overlain at 24 m O.D. by 6 m of terrace gravels referable to the Boyn Hill stage (Holmes, 1894; Baden-Powell, 1951; Wooldridge, 1957) and provides the crucial datum within the London Basin.

## A reappraisal

Four major problems were encountered with this interpretation of drainage development. First, the age and origin of the Chiltern Drift remained obscure, as did its relationship with the Higher and Leavesden Gravel Trains. Second, the suggested occupation of the Finchley Depression by the Thames lacked stratigraphic support and was insecurely based on three lines of reasoning: (1) the linking of the Finchley moraine with the low-gradient Winter Hill Terrace some 20 km to the west by the poorly-substantiated 'Lake Hertford'; (2) the fact that both the Lower Gravel Train and Winter Hill Terrace were traced toward the Finchley Depression (Hare, 1947), and (3) that neither of these terraces had been identified within the present Lower Thames valley.

The third problem concerned the Mid-Essex Depression. Although this appeared to be the logical continuation of the Vale of St Albans routeway (Fig. 2), neither the dates of its initial occupation nor eventual abandonment were stratigraphically fixed. In reality the existence of a large bedrock depression represented the main argument for the Thames ever having used this route, the correlation of the Harlow sub-drift 'terrace' with the Harefield Terrace (Clayton, 1957) across 50 km of 'dead ground' being patently weak. Fourth, there remained the problem of the age and subdivision of the Chalky Boulder Clays of the London Basin. There was much controversy as to whether these tills were deposited by a single glacial stage (Wooldridge, 1938), or two advances equatable with the Gipping (Wolstonian) and Lowestoft (Anglian) Tills of East Anglia (Baden-Powell, 1948; West and Donner, 1956), or were the product of three distinct glacial episodes as proposed by Clayton (1957, 1960, 1964). Uncertainty also existed as to whether any part of the Chalky Boulder Clay could be correlated with the Chiltern Drift. These four problems are examined in the following sections.

## The Chalky Boulder Clay and its relationship to the second deflection of the Thames

The Anglian glacial stage (Mitchell *et al.*, 1973) is marked by the development of widespread calcareous tills in East Anglia, collectively referred to as Chalky Boulder Clay. Nineteenth century writings on this formation adopted a broadly monoglacial approach, and it is only in this century that there has emerged a tendency towards internal subdivision, both chronologically and genetically. The multiglacial view appears to have had its origins in the palaeontological work of Warren (1912) and Oakley (in Bull, 1942). The Thames double deflection model of Wooldridge seems to have provided an additional impetus; indeed, so much so that Wooldridge wrote, "It is indefensible and misleading to speak of *the* Chalky Boulder Clay glaciation, ignoring its evidently composite character" (Wooldridge and Cornwall, 1964). Apparently, Wooldridge came to believe that both the Chiltern Drift and Eastern Drift of the Vale of St Albans belonged to the Chalky Boulder Clay, even though the former was in a highly weathered condition.

The now familiar Lowestoft Till-Gipping Till dichotomy was advocated first by Baden-Powell (1948) and West and Donner (1956), but recent work (Turner, 1970, 1973; Bristow and Cox, 1973; Perrin *et al.*, 1973; West, 1973; Baker, 1977) has shown that there is no evidence for Gipping (Wolstonian) Till south of Ipswich and Cambridgeshire, and that the bulk of the Chalky Boulder Clay in East Anglia is of Lowestoftian (Anglian) age. Within the London Basin, an Anglian age is proved where Hoxnian interglacial deposits overlie the till (Sparks *et al.*, 1969; Turner, 1970; Gibbard and Aalto, 1977), thus confirming the important stratigraphic relationship at Hornchurch (Holmes, 1892).

The subdivision of the Chalky Boulder Clay, independently proposed by Clayton (1957, 1960, 1964) for Essex, involved the identification of three tills based on morphology, depth of decalcification and localized stratigraphic relationships. Investigations at Chelmsford and Harlow suggested the following evolutionary sequence (Clayton, 1957):

| | | |
|---|---|---|
| 6 | Third glaciation—Springfield Till | 'tripartite sequence' |
| 5 | Deposition of Chelmsford Gravels | |
| 4 | Second glaciation—Maldon Till | |
| 3 | Period of erosion | |
| 2 | First glaciation—Hanningfield Till | |
| 1 | Deposition of Pebble Gravels | |

Attempts to correlate this proposed glacial sequence with the East Anglian and North-west European sequences proved equivocal and inconsistent (Table I). Attempts to link it with the Thames diversion model of Wooldridge met with similar difficulties, and much confusion exists in the literature as a result (Table II).

The original scheme (Clayton, 1957) was correct in assigning the tripartite sequence (Springfield Till-Chelmsford Gravels-Maldon Till) to the Eastern Drift of the Vale of St Albans, but the regional correlation (Table I, column 3) was seriously time-transgressive in that it correlated the tripartite sequence with the

TABLE I

*Proposed correlations of the Essex glacial sequence (multiglacial interpretation) with those of East Anglia and NW Europe*

| NW Europe* | East Anglia* | Essex and Hertfordshire | | |
|---|---|---|---|---|
| | | Clayton (1957) | Wooldridge (1957) Clayton (1960, 1964) | Clayton and Brown (1958)† |
| Saale | Gipping (Wolstonian) glaciation | Springfield Till Chelmsford Gravels Maldon Till | Springfield Till | |
| Holstein (Great Interglacial) | Hoxnian Interglacial | *Erosion* (Boyn Hill Terrace) | Chelmsford Gravels (Boyn Hill Terrace) | (Boyn Hill Terrace) |
| Elster II | Lowestoft (Anglian) glaciation | Hanningfield Till | Maldon Till | Springfield Till Chelmsford Gravels Maldon Till |
| *Interstadial* | Corton Beds interstadial | | *Erosion* | *Erosion* |
| Elster I | North Sea Drift | | Hanningfield Till | Hanningfield Till |

* Following West (1968).

† There is no explicit regional correlation given by Clayton and Brown (1958), but it is clear from the text that this correlation is intended. The Springfield Till - Chelmsford Gravel - Maldon Till assemblage is grouped together in one complex glacial phase (p. 113), which represents the last glaciation to reach the London Basin. This ice advance was responsible for the formation of Lake Hertford. The Winter Hill Terrace was graded into Lake Hertford. Since the Winter Hill Terrace must pre-date the Boyn Hill Terrace, the glaciation must be pre-Hoxnian (i.e. Lowestoft). See also remarks in Sparks *et al.* (1969; p. 266).

Gipping (Wolstonian) stage, a stage which must *post-date* the Boyn Hill Terrace (Hoxnian). The Vale of St Albans glacial drift and Winter Hill Terrace, however, must pre-date the Boyn Hill Terrace (Figs 2G and 2H). In order to resolve this dilemma, Wooldridge (1957) argued that the Hornchurch boulder clay was of Maldon Till-type (Table II) and thus of pre-Boyn Hill age (Table I). This correlation was subsequently adopted by Clayton (1960, 1964), who went further in suggesting that the Eastern Drift of Finchley and St Albans was of Springfield Till-type and Saalian in age, and that the intermediate Chelmsford Gravels were of Hoxnian (or 'Great') Interglacial age (Tables I and II). Not only did this effectively split the tripartite glacial sequence, it created an embarrassing dichotomy as to which ice advance was responsible for the 'second diversion of the Thames

TABLE II

*Proposed correlations of the Essex glacial sequence (multiglacial interpretation) with the Thames double deflection model*

| Essex glacial sequence | Middle and Lower Thames sequence | | |
|---|---|---|---|
| | Clayton (1957) | Wooldridge (1957) Clayton (1960, 1964) | Clayton and Brown (1958) |
| Springfield Till | Chalky Boulder Clay of Vale of St Albans (diversion of Lower Thames to present valley) | Drift of Finchley and St Albans | Formation of Lake Hertford (into which Winter Hill Terrace graded) and diversion of Lower Thames to present valley |
| Chelmsford Gravels | | Boyn Hill Terrace | Brief withdrawal of ice sheet |
| Maldon Till | | Hornchurch drift ice diverting Lower Thames to present valley | Early advance of ice causing impounding of proto-Thames. Consequences unknown |
| Interval of erosion | Higher and Lower Gravel Trains: proto-Thames course along lines of Finchley and Mid-Essex Depressions | | |
| Hanningfield Till | Chiltern Drift: diversion of proto-Thames from original route along the line of Vale of St Albans. | | |

(Fig. 2G); if the Maldon Till ice sheet was responsible for the diversion, the Thames must have been diverted to its present course prior to the Springfield Till advance, so that the critical Finchley and St Albans tills were merely emplaced in vacant hollows. This negated the evidence upon which the original diversion hypothesis (Wooldridge, 1938) was so firmly based.

The most satisfactory correlation was alluded to in Clayton and Brown (1958) (Tables I and II). In this, the tripartite glacial sequence was retained as a continuous depositional sequence and placed chronologically before the Boyn Hill Terrace. This accords with more recent palaeobotanical work. However, the authors appeared uncertain as to which glacial advance was actually responsible for the diversion of the Thames.

A further problem is evident in Table II; namely, the underlying acceptance that the Chiltern Drift was contemporaneous with the Hanningfield Till. Although neither deposit is now accepted as older glacial till *sensu stricto* (see later), it is interesting to note that their assumed equivalence occasioned no comment in the literature concerning the extent and effectiveness of the first glaciation (Fig. 1). If the ice advance had been as widespread as this correlation demands, then the proto-Thames, at the Higher Gravel Train stage, would have been blocked by ice within the Mid-Essex Depression (Fig. 2D), thereby bringing into question the validity of the Finchley Depression routeway.

The unsatisfactory nature of these various correlations points to fundamental errors of interpretation in both the drainage diversion model of Wooldridge and the three-fold subdivision of Chalky Boulder Clay proposed by Clayton. Although the latter found some initial support (Wooldridge, 1957; Wooldridge and Cornwall, 1964), it was subsequently criticized on methodological and stratigraphical grounds (Thomasson, 1961; Turner, 1970; Baker, 1971; Bristow and Cox, 1973) and has been revised in the light of more recent evidence. The main till plateau of Essex is now generally accepted to be of Springfield Till age; the Maldon Till has been confirmed as a localized and impersistent lower till, and the Hanningford Till is either synonymous with the Springfield Till or represents decalcified and soliflucted derivatives of other formations.

A monoglacial interpretation of the Chalky Boulder Clay in southern East Anglia now enjoys wide support (Turner, 1970; Baker, 1971; West, 1973; Bristow and Cox, 1973; Rose *et al.*, 1976; Rose and Allen, 1977) although there is evidence for two minor ice fluctuations (Bristow and Cox, 1973; Gibbard, 1977; Baker, 1977). While this conclusion appeared to substantiate the Thames deflection model of Wooldridge (1938), with the second diversion being achieved as envisaged by Clayton and Brown (1958), further investigation of the fluvial and fluvio-glacial sequences has radically changed the situation. These new hypotheses are examined below.

## Monoglaciation and the development of the Thames

Recent ideas concerning the former courses and diversions of the proto-Thames are illustrated in Figs 3A-H. These are based on detailed stratigraphic investigations by Hey (1965), Rose *et al.* (1976), Gibbard (1977, 1979) and Green and McGregor (1978).

The Pebble Gravel course of the proto-Thames is partially confirmed in the distribution and elevation of Westland Green Gravels (Fig. 3A). Hey (1965) and Green and McGregor (1978) have shown that these gravels decline north-east through the area and contain a large proportion of far-travelled material. This is taken to indicate contemporary glaciation in the upper catchment of the Thames (possibly of Baventian age). That this proposed glaciation could have been responsible for the Chiltern Drift is as yet unconfirmed; the possibility cannot be discounted.

At lower elevations, the Higher and Lower Gravel Trains provide evidence of the next discernible stage (Fig. 3B). Significant differences of interpretation, compared with that of Wooldridge, are apparent here. Palaeocurrent analyses and pebble counts (Gibbard, 1977; Green and McGregor, 1978) show conclusively that the Thames continued to flow north-eastwards along the Vale of St Albans throughout Gravel Train times, with no evidence of drainage reversal. The Leavesden Gravel

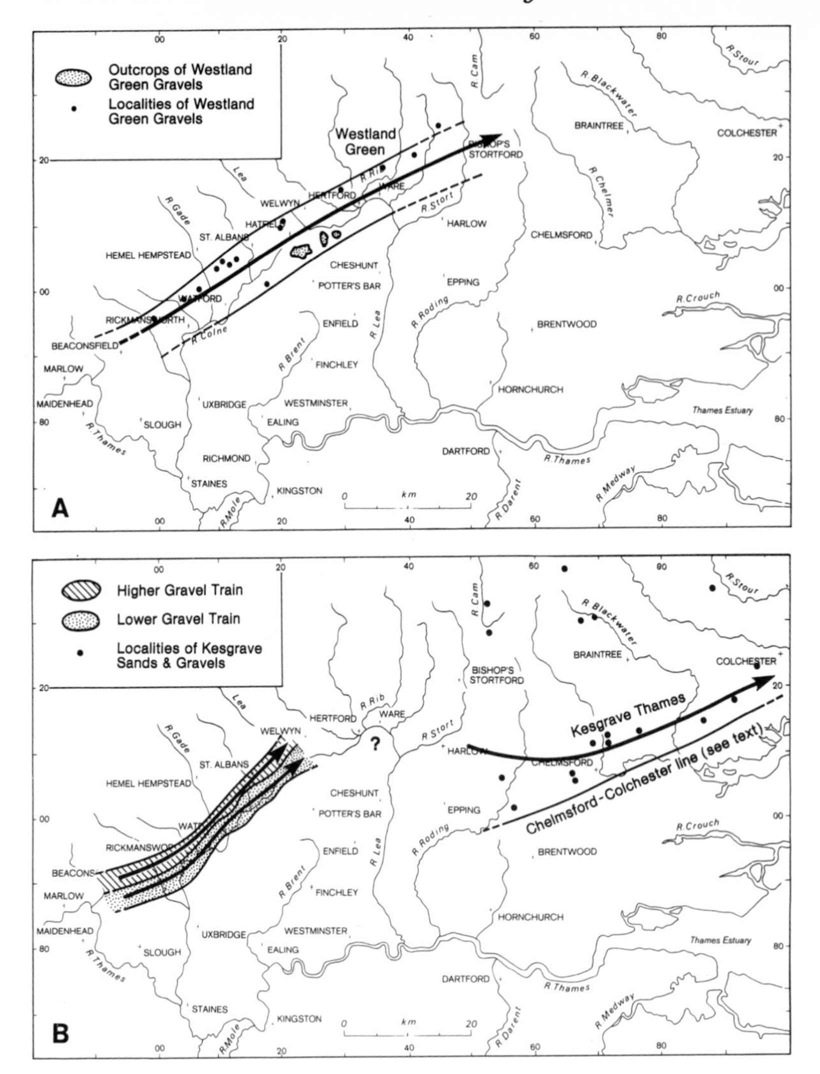


*Figure 3 A-H.* The contemporary view as to the evolution of the Thames drainage system (monoglacial interpretation). (A) Westland Green Gravel stage (Baventian?) (Hey, 1965; Green and McGregor, 1978). (B) Higher and Lower Gravel Trains and Kesgrave Sands and Gravels stages (Beestonian to early Anglian) (Rose *et al.*, 1976; Green and McGregor, 1978).

Train has no separate existence, the gravels representing a continuation of the Higher and Lower Gravel Trains (Green and McGregor, 1978). Consequently, there is no stratigraphic evidence to substantiate the first diversion of Wooldridge.

The continuous north-eastward passage through the Vale of St Albans persisted until the Westmill Lower Gravel (Winter Hill Terrace) stage (Gibbard, 1977) (Fig. 3C). At the same time (early Anglian ?), Gibbard (1979) has further shown that the Finchley Depression was almost certainly occupied by the proto-Mole-Wey, since the Dollis Hill Gravels within this lowland are rich in Lower Greensand chert. Hence it appears that the Thames never occupied the Finchley Depression, for the

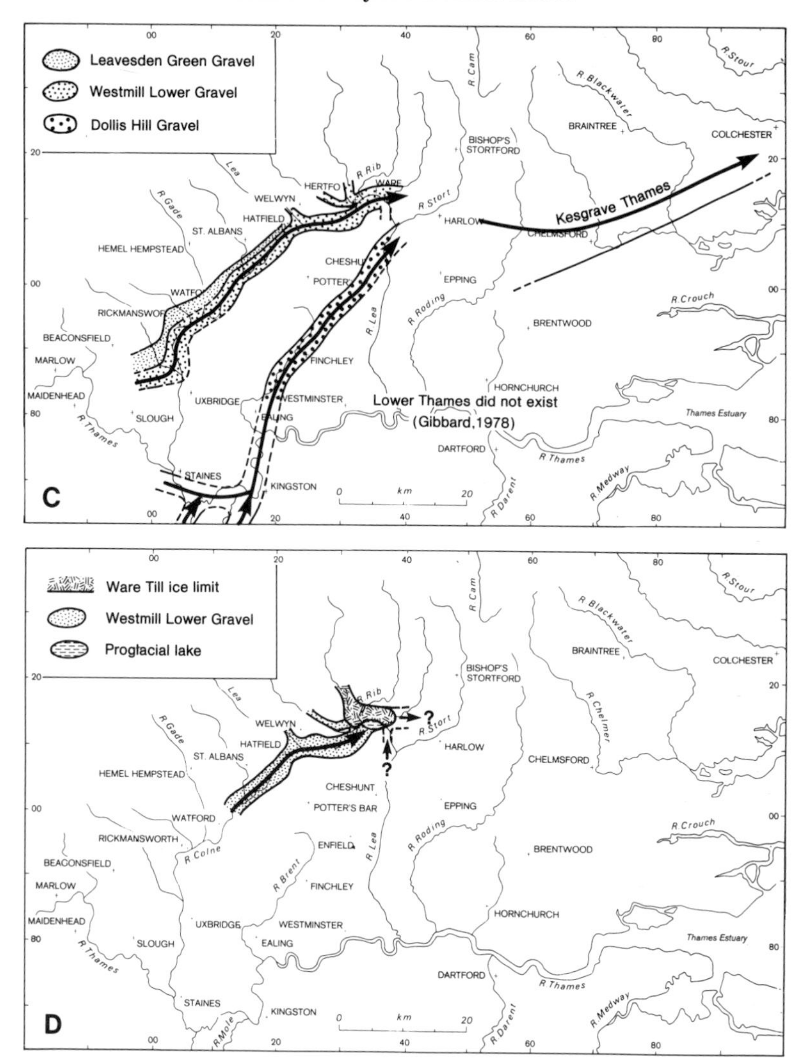


*Figure 3 (cont.).* (C) Leavesden Green Gravel, Westmill Gravel, Mole-Wey St George's Gravel and Dollis Hill Gravel (Early Anglian) (Gibbard, 1977, 1979). (D) First ice advance: Ware Till; Watton Road Silts (Anglian) (Gibbard, 1977).

post-diversion course can only be traced along the present lower course in the Black Park Terrace (Fig. 3H).

Further to the east, in Essex, the course of the pre-Anglian proto-Thames has been examined by Rose *et al.* (1976) (Fig. 3B). An extensive body of sand and gravel (Kesgrave Sands and Gravels) occurs beneath the Chalky Boulder Clay occupying much of Essex and south-east Suffolk, and infilling, in part, the Mid-Essex Depression. The height range of the formation and large-scale cross-set structures indicates a north-eastward flowing major river, which laid down a series of terraces under periglacial conditions as it migrated south-eastwards. A distinct rubified palaeosol is developed on the Kesgrave Sands and Gravels, and has been interpreted

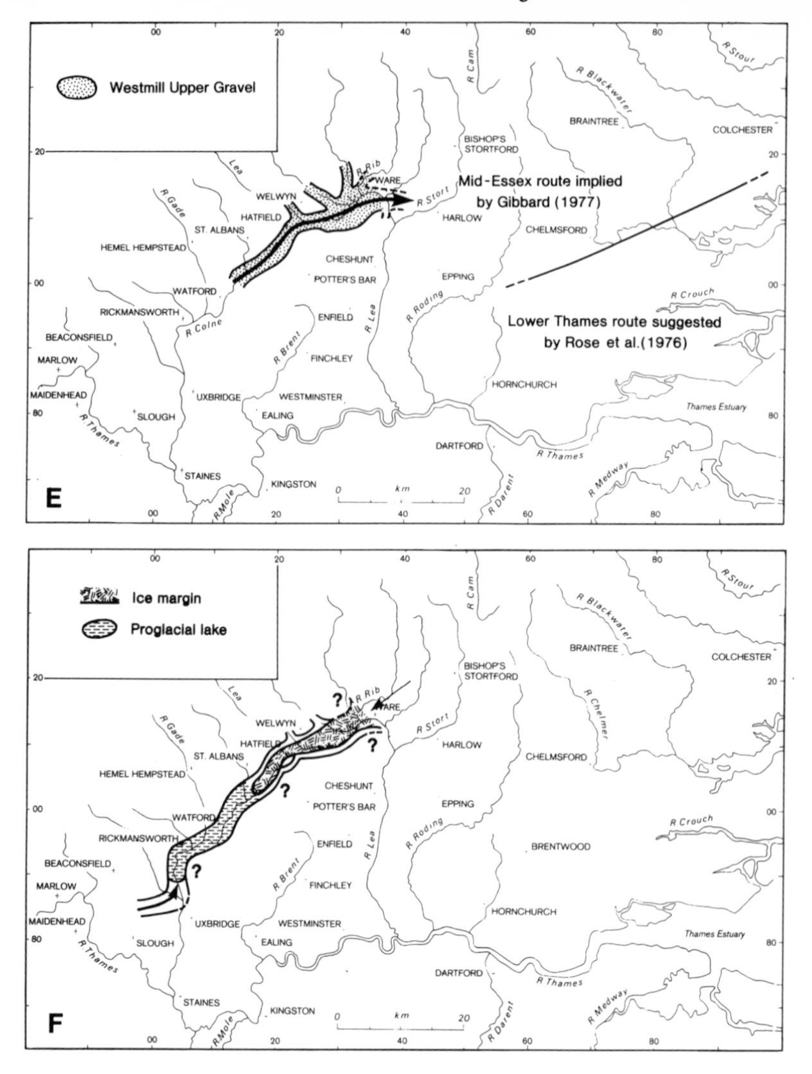


*Figure 3 (cont.).* (E) Temporary ice retreat: Westmill Upper Gravel (Anglian) (Gibbard, 1977). (F) Second ice advance (early stage): Eastend Green Till I, Moor Mill Laminated Clay, Winter Hill Terrace I (Anglian) (Gibbard, 1977).

as the product of pedogenesis in a humid, warm temperate climate, believed to be the Cromerian Interglacial stage. The Kesgrave Formation was accordingly correlated with the preceding Beestonian periglacial stage (Table III). Downcutting and migration by the proto-Thames is thus envisaged to have begun during this stage, so that by the Cromerian Interglacial, the river must have adopted a route south and east of a line from Chelmsford to Colchester (Rose *et al.*, 1976; p. 493) (Figs 3B and C). This conclusion is somewhat difficult to reconcile with Gibbard's (1979) assertion that, in the early Anglian stage (Dollis Hill Gravel), "the Lower Thames valley did not then exist". If the early Anglian proto-Thames and proto-Mole-Wey flowed north-eastwards as far as Ware and Cheshunt (Fig. 3C), and if there is no

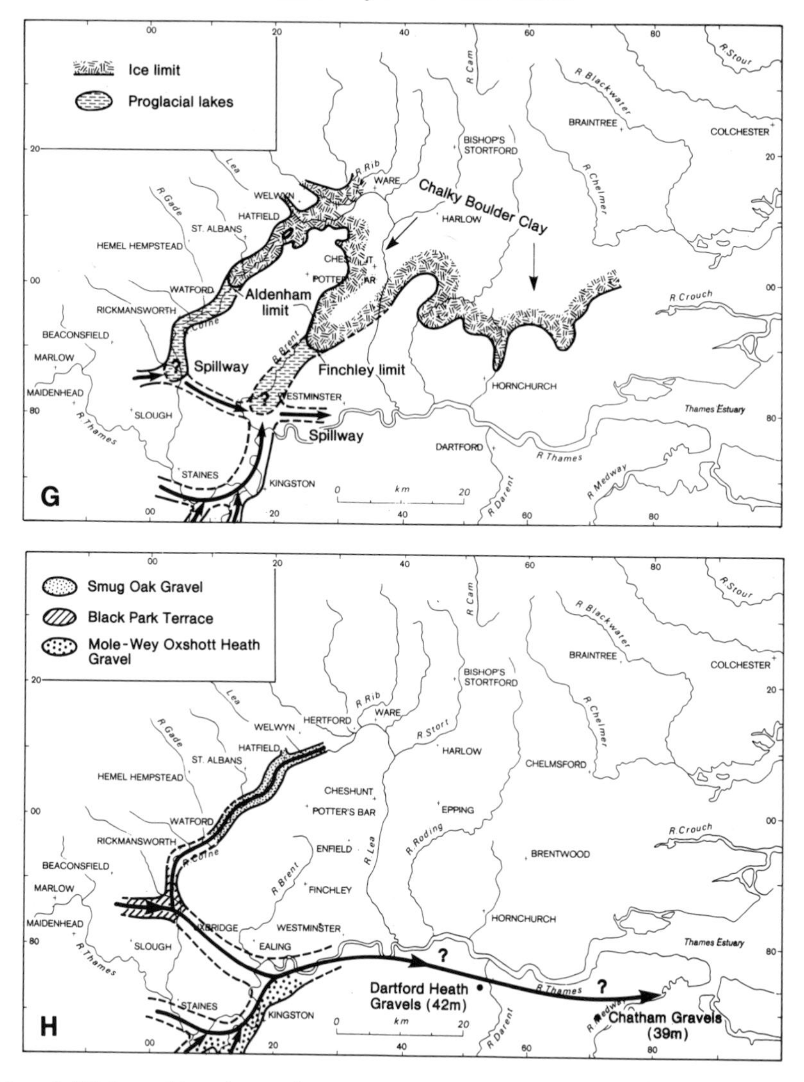


*Figure 3 (cont.).* (G) Second ice advance (late stage): Eastend Green Till II, Moor Mill Laminated Clay, Winter Hill Terrace II, Finchley Till, Coldfall Wood Laminated Clay (Anglian) (Gibbard, 1977, 1979). (H) Establishment of diversion course: Smug Oak Gravel, Black Park Terrace, Mole-Wey Oxshott Heath Gravel (Late Anglian) (Hare, 1947; Gibbard, 1977, 1979).

evidence for the river north and west of the Chelmsford-Colchester line at the Cromerian stage (Fig. 3C), where was the North Sea exit for the proto-Thames? The solution to this problem probably lies in a revised date for the Kesgrave Sands and Gravels, for some stratigraphers are of the opinion that these sediments constitute glacial outwash (partly Chelmsford Gravels) of Anglian age (Allender and Hollyer, 1973; Bristow and Cox, 1973; Baker, 1977; Rose *et al.*, 1978a, b). Indeed, the considerable height range of the Kesgrave Formation and the large volume of sediment involved suggests a fluvio-glacial origin, rather than a fluvial terrace laid down under periglacial conditions (Baker, 1977; Green in Rose *et al.*, 1978a). Much hinges on the environmental significance attached to the Valley Farm palaeosol; undoubtedly it represents a phase of subaerial exposure, vegetation colonization and pedogenesis (Baker, in discussion, Rose and Allen, 1977), but what was the status of this temperate phase? Its stratigraphic position demonstrates a pre-Chalky Boulder Clay age which could be early Anglian or pre-Anglian. An early Anglian age may apply if conditions necessary for lessivage could be obtained under interstadial, rather than interglacial, environments. Divergence of opinion on the chronology of the pre-Chalky Boulder Clay drifts in Essex is summarized in Table III. If the alternative dating for the Kesgrave Formation is accepted, the proto-Thames route south and east of the Chelmsford-Colchester line could have been adopted by the mid-Anglian stage, after initial glacial diversion (Fig. 3E).

The proto-Thames was deflected from the Vale of St Albans and mid-Essex routeway by glacial impounding and diversion within one glacial stage—the Lowestoft (Anglian) ice advance. Gibbard (1977) recognizes two phases of the local glaciation. In the first (Fig. 3D), a small tongue of ice dammed the Thames near Ware to form a localized and short-lived proglacial lake. The downstream continuation of the river either at this stage (Fig. 3D), or during temporary ice retreat (Fig. 3E), is uncertain; but the evidence of the 'Kesgrave' Thames route, on either of the two interpretations (Table III), would suggest that the mid-Essex route had been abandoned by this stage, although the Vale of St Albans route was still maintained.

The second phase of the Anglian glaciation (Figs. 3F and G) was a more extensive advance, ice attaining its maximum extension to the Aldenham and Finchley limits (Fig. 3G), and proglacial lakes are postulated by Gibbard (1977, 1979) in both the proto-Thames and proto-Mole-Wey valleys. The proto-Thames overflowed via a col near Uxbridge, draining into the second lake which itself overflowed into the

TABLE III

| Stage | Sediments and palaeosols in Essex and W. Suffolk | |
|---|---|---|
| | After Rose *et al.* (1976) | Alternative chronology |
| Anglian | Lowestoft Till (glacial)<br>Barham Sands and Gravels (fluvio-glacial)<br>Barham Loess (periglacial) | Lowestoft Till (glacial)<br>Barham Sands and Gravels (fluvio-glacial)<br>Barham Loess (periglacial)<br>Valley Farm Rubified *Sol Lessivé* (interstadial)<br>Kesgrave Sands and Gravels (fluvio-glacial) |
| Cromerian | Valley Farm Rubified *Sol Lessivé* (interglacial) | |
| Beestonian | Kesgrave Sands and Gravels (periglacial) | |

existing Lower Thames valley in the vicinity of Hammersmith (Fig. 3G). These phases of impounding were contemporaneous with the Winter Hill Terrace. It is apparent from these reconstructions that the concept of a 'Lake Hertford' (Fig. 2F) must be radically modified; it would appear that some of the laminated sequences described by Clayton and Brown (1958) belong to Gibbard's (1977) Watton Road and Moor Mill phases (Figs 3D and F), while others (not discussed by Gibbard) are of more ephemeral character, similar to those ice-marginal features described by Wooldridge (1953), Brown (1959) and Lawrence (1964) (Fig. 2G).

The first evidence of diversion is provided by the Anglian retreat outwash in the Vale of St Albans (Smug Oak Gravel) which correlates with the Black Park Terrace (Fig. 3H). Thereafter, the Thames, confluent with the diverted proto-Mole, pursued a course approximately along the line of the existing Lower Thames, via Dartford Heath and possibly Chatham (Fig. 3H). This route was succeeded by the Boyn Hill Terrace route (Fig. 2H).

It is clear that this new interpretation resolves the long-standing problem of correlating the terrace and glacial sequences, and effectively accounts for the origin and form of the Vale of St Albans and Finchley Depression. There are, however, four residual problems unanswered in the new model:

(1) *The Chiltern Drift.* Was there a glaciation prior to the Anglian stage which, though not actually diverting the Thames, nevertheless deposited a local till and significantly altered sediment-discharge characteristics in the early (Westland Green) Thames?

(2) *The Mid-Essex Depression.* At what stage was it cut and occupied by the Thames, and at what stage was it abandoned in preference for a more southerly route?

(3) *The origin of the Lower Lea.* The size of the Lower Lea valley south of Hoddesden, and its discordance to the formerly continuous synclinal Hampstead-Epping Ridge, remains unexplained. Neither the old model (Fig. 2) nor the new (Fig. 3) satisfactorily accounts for this important physiographic feature.

(4) *The 'Romford River' valley.* The significant depression running between Upminster and Maldon, into which the Chalky Boulder Clay descends at Hornchurch, also remains unexplained. Was this a temporary routeway adopted by the Thames during the Anglian stage, or was it perhaps a former course of the proto-Darent?

## The Chiltern Drift and its supposed equivalents

The Chiltern Drift (Wooldridge, 1938) and its probable equivalents (Fig. 2D) are neither lithologically nor stratigraphically related to the Chalky Boulder Clay, despite the numerous Chiltern Drift-Hanningfield Till correlations (Clayton, 1957, 1960, 1964; Wooldridge, 1957) which are now regarded as erroneous. These deposits occur at relatively high levels (largely between 105 m and 120 m) in three main areas—the Chiltern backslopes, the South Hertfordshire Plateau and Epping Forest Ridge (Fig. 1)—and are usually separated altitudinally from the lower lying

TABLE IV

*The nature of the Chiltern Drift and its probable equivalents*

| Area | Literature | Lithology | Stratigraphy | Height and morphology |
|---|---|---|---|---|
| (1) *Chiltern dipslope* (Welwyn) and Beaconsfield) | Barrow (1919), Sherlock and Noble (1912), Wooldridge and Linton (1955) | *Chiltern Drift.* A stony clay with mainly local erratics; on the south and south-east margins, Triassic debris is abundant. At Cowcroft (Chesham), contains thrust planes and overthrust structures considered by Barrow (1919) to be glaciotectonic in origin. | Overlies Upper Chalk; overlies Pebble Gravel with which it is mixed in certain localities. | On northern side of Vale of St Albans. Height range 115-155 m O.D. Southern limit marked by morphological break, below which lies the Leavesden Gravel Train. |
| (2) *Welwyn area* (Mardley Heath) | Wooldridge (1960), Wooldridge and Cornwall (1964) | 'Western' erratics, including microphytic olivine basalt, flow-banded rhyolite and quartzites. Clay matrix decalcified, eluviated and rubified. Rubification taken to indicate Braunlehm soil formation produced under humid temperate conditions. | Overlies Tertiary and Pebble Gravel formations | Height 108 m O.D. at Mardley Heath. Separate from Lowestoft Till tract (itself rising to 130 m); some altitudinal overlap therefore. |

TABLE IV (cont.)

| Area | Literature | Lithology | Stratigraphy | Height and morphology |
|---|---|---|---|---|
| (3) *South-east Hertfordshire* (Potter's Bar, etc) | Thomasson (1961) — | *Pebbly Clay Drift.* Silty or loamy clay matrix with thin sand or sandy clay bands in places. Clay matrix non-calcareous and rubified in places. Stone content includes sub-angular flint pebbles, red quartzites, sandstones, vein quartz and quartz pebbles. Pebbly Clay Drift differs from Pebble Gravel in morphology, texture and mineralogy. | Overlies London Clay and Pebble Gravel. Adjoins Lowestoft Till but "no clear case of Chalky Boulder Clay superimposed on Pebbly Clay Drift can be produced" (p. 295). | At a height of 108 m to 115 m O.D. on level or gently sloping ground. Max. thickness 3·4 m. Occupies plateau site similar to that of the Lowestoft Till, from which it is not easily distinguished morphologically. |
| (4) *Epping Forest Ridge* (High Beach, Jack's Hill, Coopersale Common) | Whitaker (1889), Warren (1910), Wells and Wooldridge (1923), Baker (1971) | *'Western Drift'.* Amorphous or massive clay matrix with variable lithology. Non-calcareous to its maximum depth. Erratic content includes Bunter quartzites, Cretaceous flints, Tertiary flints, quartz pebbles, and devitrified rhyolites. | Overlies solid Tertiary formations (Bagshot Beds and Claygate Beds) and Pebble Gravel. Possibly overlain by Lowestoft Till at one point (Baker, 1971). | Occurs on a plateau surface at 106-115 m O.D. Max. thickness 6·2 m. Abuts against Lowestoft Till tract along northern edge at 106 m. |

Chalky Boulder Clay. They are discontinuous and of limited extent, composed of heterogeneous, non-calcareous, weathered pebbly clays of assumed glacial origin, and have been described variously as Western Drift (Barrow, 1919; Wells and Wooldridge, 1923; Wooldridge, 1960; Wooldridge and Cornwall, 1964), Chiltern Drift (Wooldridge, 1938, 1957; Wooldridge and Linton, 1955), Pebbly Clay Drift (Thomasson, 1961), and 'stony clays' (Baker, 1971). (A summary of their properties and relationships is presented in Table IV.) Wells and Wooldridge (1923; p. 248) ascribed the Epping deposits, on the basis of their suite of western erratics, to an ice advance from the west which 'probably antedated by a considerable period the main glaciation of S.E. England' (p. 250). However, comparatively little data is available on these deposits, largely due to neglect, so that their age and origin remains unclear, and conclusions as to the possible existence of an early glaciation are necessarily tentative.

## Mode of origin

Wooldridge (1938) originally interpreted the Chiltern Drift as the highly weathered ground moraine of a local ice-cap on the Chiltern cuesta which subsequently moved south-eastwards into the London Basin and caused the initial diversion of the proto-Thames (Figs 2C and D). Manley (1951) refuted this hypothesis by arguing that the Chilterns are of insufficient elevation to have generated an ice-cap. The Chiltern Drift then came to be regarded as either the product of a continental ice-sheet (Wooldridge and Linton, 1955; Wooldridge, 1960), or a deposit of non-glacial origin.

A glacial derivation is certainly in accord with observations of structure which is typically amorphous, unstratified and massive. Such till-like characteristics are particularly well displayed in the Mardley Heath drift (Wooldridge and Cornwall, 1964). In addition, the Chiltern Drift at Cowcroft (near Chesham) contains large-scale thrust structures which suggest glacio-tectonics rather than landslipping or periglacial disturbance (Barrow, 1919).

The periglacial explanation was advanced by Thomasson (1961) who remarked that the occurrence of these deposits on the lower Chiltern backslopes suggested a solifluction origin. Hey (1965) noted that the erratic content of the Chiltern and Pebbly Clay drifts resembles that of the underlying Pebble Gravels, thereby indicating a secondary derivation by *in situ* weathering or solifluction. However, this interpretation does not account for its distinctive clay content (Wells and Wooldridge, 1923), nor for its separation from both Tertiary formations and Pebble Gravels by sand mineralogy and texture (Thomasson, 1961), nor for its stratigraphic distinction. Finally, arguments as to the unsuitability of relief conditions for a periglacial origin have been advanced for the Pebbly Clay Drift of south-east Hertfordshire (Thomasson, 1961), the Mardley Heath deposits (Wooldridge and Cornwall, 1964) and the Epping stony clays (Baker, 1971).

In spite of these arguments, recent research has tended to strengthen the periglacial hypothesis. The reinterpretation of part of the Hanningfield Till as solifluction debris (Bristow and Cox, 1973) has made a similar explanation for the Chiltern Drift more plausible. Of greater significance, however, is the establishment of a new evolutionary model for the proto-Thames (Gibbard, 1977; Green and McGregor, 1978) which has removed the requirement for a Chiltern Drift glaciation. Nevertheless, the controversy is far from being resolved, for suggestions of an early (possibly Baventian) glaciation supplying far-travelled erratics to the Westland Green proto-Thames (Green and McGregor, 1978) make it unwise to wholly discount the possibility of older glacial till on the Chiltern backslope, especially as Briggs *et al.* (1975) have indicated the possibility of a pre-Cromerian (Beestonian or Baventian) glaciation in western Britain, based on a Bunter-rich till-like deposit beneath Cromerian interglacial deposits at Sugworth, near Oxford. What has emerged is that if there was an early ice-advance, it appears to have had little impact on the drainage pattern. This would indicate either short duration or limited extent.

In the absence of conclusive sedimentological evidence, physiographic arguments must be employed to evaluate the likelihood of an early ice-advance. Three alternative working hypotheses may be considered: an ice-advance across the Chiltern Hills from the north-west; the intrusion of an ice-sheet into the London Basin from the west and finally, the arrival of ice from a north-easterly direction.

*Glacial advance from the north-west*

This is the logical development of the 'local ice-cap' hypothesis (Wooldridge, 1960) and suffers from similar shortcomings. It is significant that the Chiltern cuesta between the Goring Gap and Hitchin bears no sign of having been overridden by ice, even though the Hanningfield Till-Chiltern Drift correlation (Wooldridge, 1957; Clayton, 1957, 1960, 1964), and the correlation proposed in this paper, indicate extensive ice cover within the London Basin. As the present relief of the cuesta above the 125 m (Pebble Gravel) surface exceeds 120 m, it is logical to suggest that this represents a minimum figure for the relative relief of the escarpment at the time of glaciation. Glacial overriding would be expected to have imparted a distinctive set of erosional landforms, particularly near the crest. As none has been distinguished and the regolith of the cuesta appears free of far-travelled erratics (Avery, 1964), the hypothesis must be considered highly improbable.

In addition, the use of the arcuate shaped Borehamwood Channel (Fig. 2C) as evidence for a southward moving lobe is unjustified. The depression has a maximum floor elevation of 99 m (Wooldridge and Linton, 1955; p. 155), and although altimetric arguments indicate that its creation could slightly pre-date the eastward sloping Higher Gravel Train, such reasoning must be treated with caution in an area where local base-levels would have oscillated during glaciation. The suggestion that it represents the glacially deflected course of the proto-Thames is unconvincing in the face of alternative explanations that it represents (1) a meander of the Thames at, or slightly after, the Westland Green stage, or (2) an ice-marginal meltwater channel produced during the Anglian Glaciation.

There thus appears no morphological or sedimentological support for such an hypothesis.

*Glaciation from the west*

The presence of far-travelled erratics of 'western' provenance led Barrow (1919) to suggest an ice invasion from the west, an hypothesis which gained credibility when glacial deposits were reported from the Goring Gap (Hawkins, 1923). Although Wooldridge initially supported this view (Wells and Wooldridge, 1923), he later considered that the erratics could have been derived from pre-existing Thames gravels (Wooldridge, 1938, 1957), such as the Westland Green Gravels, which are now known to be rich in far-travelled material (Hey, 1965; Green and McGregor, 1978). Even so, it is of interest to note that Wooldridge indicated the occurrence of Chiltern Drift in the Middle Thames valley (Wooldridge and Linton, 1955; p. 116) (Fig. 1).

The major problem with this hypothesis is the lack of morphological or sedimentological evidence from the area between Henley and Maidenhead that can be used to support the idea of glacial activity. It is possible that ice may have penetrated a short distance into the Goring Gap, but there can be no support for the view that all of the Chiltern Drift was deposited by ice moving from this direction.

*Glaciation from the north-east*

The failure of the other two hypotheses make this suggestion worthy of consideration for the following reasons:

(1) The Chalk cuesta to the east of Hitchen has obviously been overridden by ice. Although covered with Chalky Boulder Clay, there is no reason why all the denudation should be ascribed to the Anglian glaciation. It can be argued that the great extent of the Anglian ice advance in the eastern London Basin reflects the pre-existing pattern of relief and ease of entry from the north-east. While warping due to the subsidence of the North Sea Basin is the most likely explanation, erosional lowering of the landsurface by an earlier glaciation is also possible, especially as the warping argument still applies.

(2) The variation in erratic content of the Chiltern Drift deposits and their apparent 'western' provenance is explicable in terms of the pre-existing pattern of erratic-bearing fluvial gravels (Wooldridge, 1938, 1957), and does not preclude an ice-advance from the north-east.

(3) The arrival of ice into the London Basin from the north-east need not militate against the idea of a minor western advance via the Goring Gap. It is possible to envisage a southward moving ice-sheet riding up against the main Chiltern escarpment between Goring and Hitchin, but only being able to penetrate into the London Basin in the east and west.

While an ice-sheet advance from the north-east could have provided a logical explanation for the initial diversion of the Thames in the double deflection model, recent work in the Vale of St Albans (Gibbard, 1977; Green and McGregor, 1978) has effectively excluded such an interpretation. However, there exists the possibility that this early ice-advance may have had a similar impact on the drainage pattern as the later Anglian glacial, temporarily initiating the changes subsequently accomplished by the Anglian phase. Such an hypothesis is not wholly implausible, as the course of the proto-Thames during the Pebble Gravel (Wooldridge, 1938; Wooldridge and Linton, 1955) and Westland Green Gravel (Hey, 1965) phases appears to lie well to the north of the Mid-Essex Depression (Figs. 2B and 3A). It is possible, therefore, that the proto-Thames was displaced southwards in Essex at this time.

## Relative age

The age distinction between the Chiltern Drift, Western Drift, etc., and the Chalky Boulder Clay is based on both morphological and lithological considerations in the absence of unequivocal stratigraphic relationships (Thomasson, 1961; Baker, 1971). Morphological relations are not as clear as is often suggested, for there exists a considerable altitudinal overlap. Chalky Boulder Clay lies at over 120 m O.D. on the South Hertfordshire Plateau, while patches of apparently undisturbed Western Drift are found as low as 105 m in west Essex. The overlap is most marked in south-east Hertfordshire, where both Pebbly Clay Drift and Chalky Boulder Clay occupy contiguous plateau sites between 105 m and 120 m, but where "no clear case of Chalky Boulder Clay superimposed on Pebbly Clay Drift can be produced" (Thomasson, 1961; p. 295). There is, however, physiographic evidence in the western Vale of St Albans to indicate that Chiltern Drift pre-dates the Chalky Boulder Clay. Here the fluviatile sequence is, in order of decreasing altitude and age: Higher Gravel Train, Lower Gravel Train/Harefield Terrace and Winter Hill Terrace. As the Winter Hill Terrace = Chalky Boulder Clay correlation is generally accepted and the Higher Gravel Train cuts into the Chiltern Drift to form what Wooldridge called a "terminal rise", it is clear that deposition of the Chiltern Drift long preceded the Chalky Boulder Clay glaciation.

Of particular significance is the prominent reddening and mottled coloration of the deposits at Welwyn, south-east Hertfordshire and Epping Forest, which was taken by Wooldridge and Cornwall (1964) to indicate Braunlehm soil development under humid, tropical conditions in the 'Great Interglacial' or Hoxnian, thereby implying a Lowestoft age for the deposits (Wooldridge and Cornwall, 1964). A Hoxnian date for this weathering activity can no longer be accepted, since it has recently been established that the Chalky Boulder Clay is also pre-Hoxnian and must, therefore, have been subjected to the same intensity of weathering and yet has no concomitant Braunlehm soil development.

Rather similar reddened soils have recently been discovered widely preserved within the middle Pleistocene sequences of north Essex and southern East Anglia (Rose *et al.*, 1976; Rose and Allen, 1977; Baker, 1977). These rubified *sols lessivés modaux* are developed on fluvial deposits (Kesgrave Sands and Gravels) laid down under periglacial conditions and are thought to be the product of pedogenesis under humid, temperate conditions during the Cromerian. This is of great relevance, as work on the Pleistocene floral record (Turner, 1975) indicates a major glacial advance in the Baventian, a suggestion supported by recent findings near Oxford (Briggs *et al.*, 1975) and the work of Hey (1976) which has shown that the first influx of exotic pebbles into East Anglia occurred at this time. Although glacial deposits of Baventian age have yet to be discovered in north Essex (Rose *et al.*, 1976; Baker, 1977), it is possible that the Chiltern Drift may be a relic of this early glacial phase which became reddened during the Cromerian or Anglian (Table III).

### Conclusion

Both the glacial and periglacial explanations of the Chiltern Drift have their merits but must await substantiation by further sedimentological investigations. While the balance of contemporary opinion appears to favour a periglacial origin, evidence has emerged which indicates that the London Basin may have experienced glaciation in the mid-Pleistocene. The most logical conclusion is that the Chiltern Drift, like the Hanningfield Till, is part solifluction deposit and part deeply weathered till. The till component is likely to be restricted to the north-east London Basin (Mardley Heath, etc.) to conform with an early ice-advance from this direction, which displaced the proto-Thames southwards to the Mid-Essex alignment.

## The proto-Thames in west and central Essex

Establishment of the proto-Thames course in Essex is dependent upon correlation between the Pleistocene stratigraphy of the Vale of St Albans (Gibbard, 1977) and that of north-east Essex (Rose *et al.*, 1976). Unfortunately this area was heavily glacierized and much evidence is concealed by Chalky Boulder Clay up to 35 m thick (Fig. 6). In addition, it must be recognized that considerable reworking of pre-existing gravels has occurred. The problem may be approached by considering the form of the sub-drift surface (Fig. 4) and the distribution of drift formations at the solid-drift interface (Fig. 5).

### Interpretation of the sub-drift surface

Previous sub-drift reconstructions for Hertfordshire and Essex have been made by Wooldridge and Henderson (1955), Clayton (1957, 1960), Brown (1959), Baker (1977) and Gibbard (1977). The present coverage (Fig. 4) is more comprehensive, and employs recent commercial site investigations and other borehole records collated by the Institute of Geological Sciences (see Acknowledgements). The form of the sub-drift surface is believed to broadly approximate to the pre-Anglian land surface, though the precise interpretation of the surface in terms of the palaeo-drainage system depends on a number of assumptions and conditions, which may or may not be satisfied at different points on the surface. These conditions may be enumerated as follows.

*Basal drift formations are not isochronous*

Figure 5 indicates that four distinct groups of drift deposits are in contact with the bedrock surface. In the north and on the Thames valley margin, Chalky Boulder

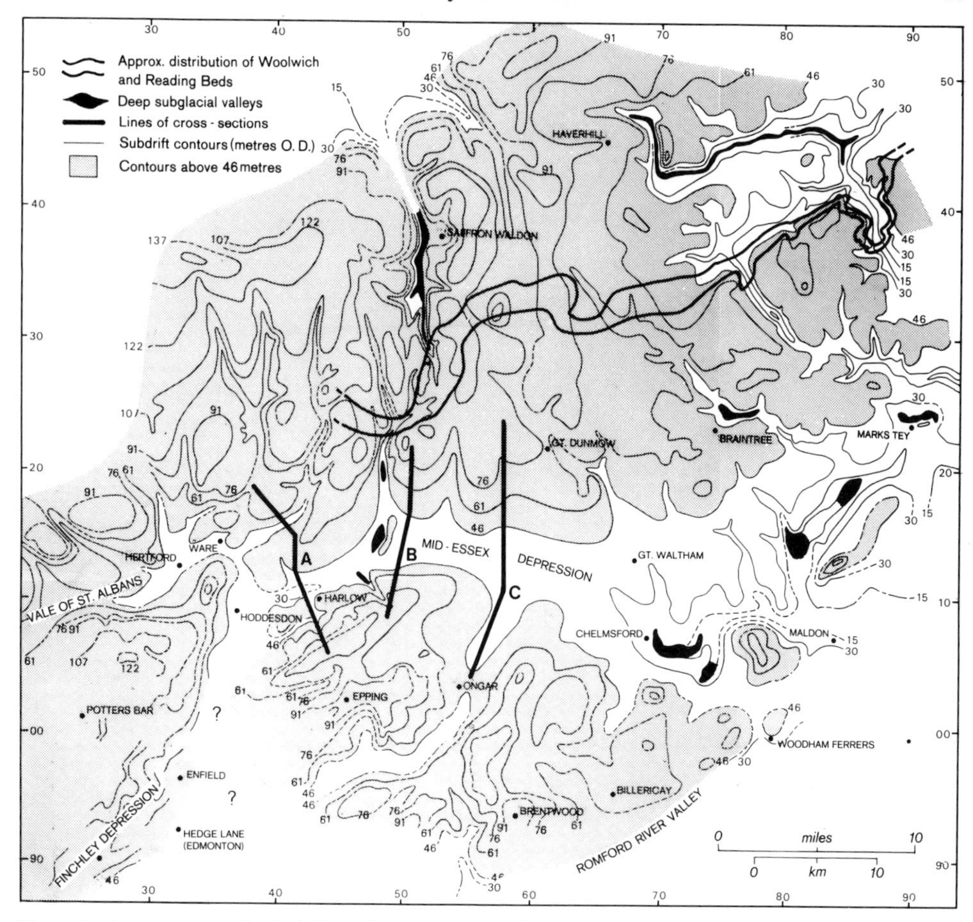


*Figure 4.* Contour map of sub-drift surface in east Hertfordshire and Essex.

Clay lies in direct contact with bedrock. The lower dipslope in the north-east is blanketed by stratified sands, while across the central tract, and along valley floors, undifferentiated sands and gravels (mainly gravels) predominate. In localized valley floors, buried channel deposits are encountered which mark the former courses of subglacial meltwater paths; these may be of three types—coarse gravels with variable interbedded sediments, uniform fine sands and silts, and till (Baker, 1977). Consequently, there are a variety of basal drift formations of differing age and origin; the sub-drift surface is therefore diachronous.

*Variability within the Chalky Boulder Clay*

Even beneath the apparently uniform sheet of till, the sub-drift surface may be diachronous. There are four principal sources of variability here (Baker, 1976, 1977). Much of the Chalky Boulder Clay consists of the lodgement till member of the Lowestoft Till formation. There are, however, occasional lower tills of identical character (Fig. 5; Table V) of which the Maldon Till (Clayton, 1957) and Ware Till (Gibbard, 1977) are but two examples. Some buried channels are entirely filled with

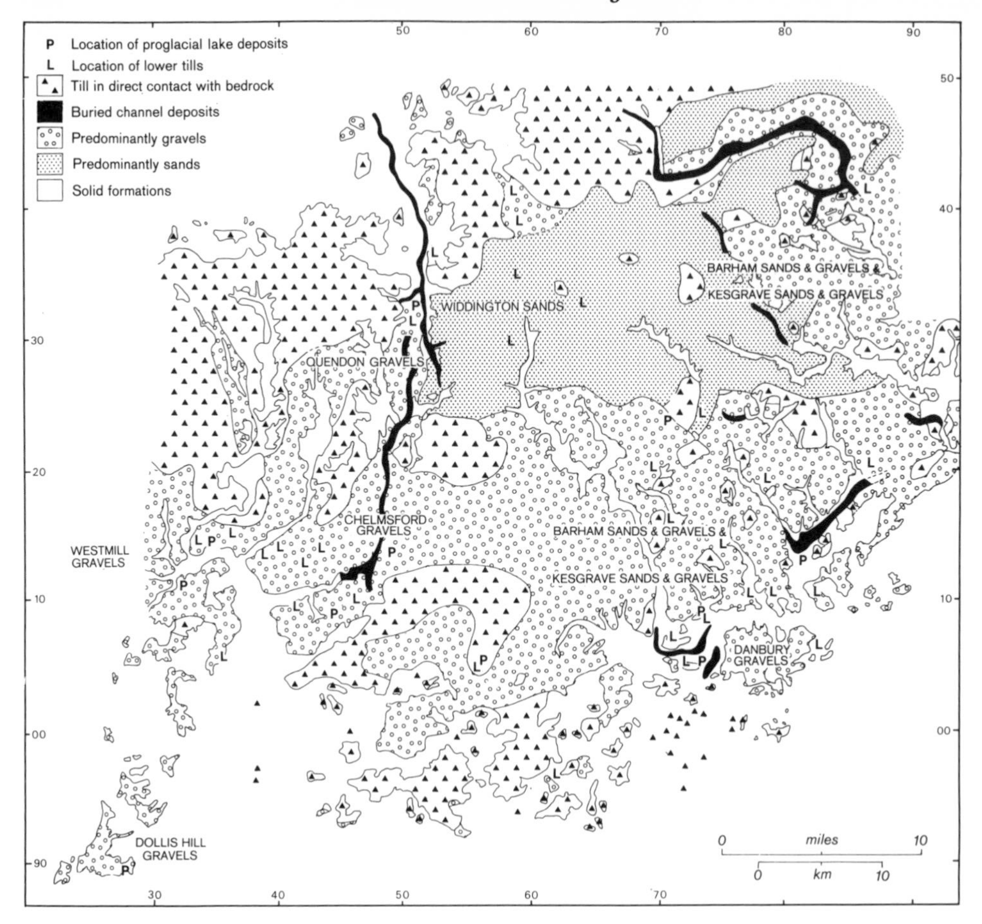


*Figure 5.* Map showing the nature of the basal drift unit at the solid-drift interface (area as for Fig. 4).

till; the Upper Stour valley and Upper Cam valley are examples (Fig. 6). Finally, there exist valley diamictons, either at the till sheet margin or as separate outliers, which represent the action of debris-sliding or earthflow under late Devensian periglacial conditions; these are particularly problematic as they impart an apparent 'pre-glacial' appearance to a valley cut post-depositionally in Chalky Boulder Clay (Baker, 1976).

### *Bedrock erosion during glaciation*

Previous authors (Wooldridge and Henderson, 1955; Clayton, 1957; Sparks, 1957; Brown, 1959; Rose *et al.*, 1976) are unanimous in assuming minimal glacial erosion, to the extent that the Chalky Boulder Clay is believed to have been emplaced without great disturbance to underlying sediments and without modifying the pre-glacial surface to any significant extent. This may be true in certain localities, but along buried channels (Woodland, 1970) (Fig. 5) evidence for sub-glacial scour is indisputable, while in areas of direct contact of ice with bedrock, the character of the overlying till points to at least some localized glacial erosion

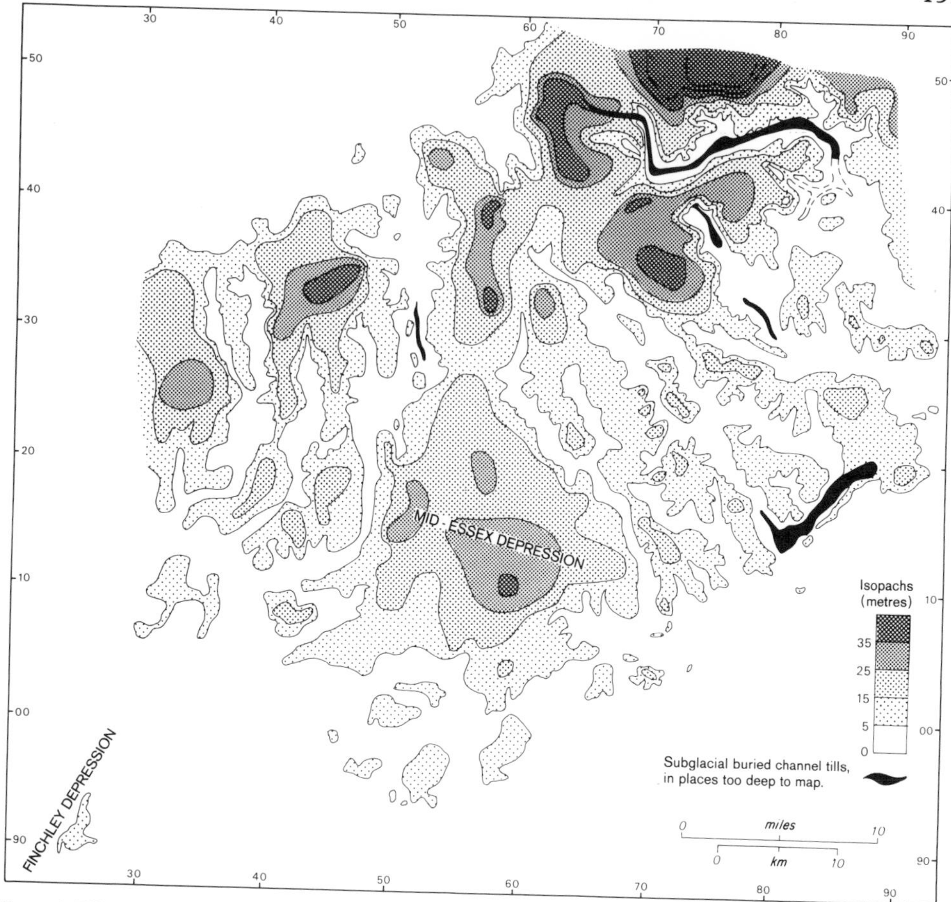


*Figure 6.* Till isopachyte map for eastern Hertfordshire and Essex (area as for Fig. 4).

(Baker, 1977). In these instances, the sub-drift surface can in no sense be regarded as the 'pre-glacial' surface.

*Tectonic stability*

Most authors assume that differential warping has not been a significant factor in southern East Anglia. Wooldridge and Henderson (1955; p. 30), however, argued for a NNW-SSE hingeline, which they called the 'Braintree Line', to the east of which subsidence of the North Sea Basin was believed to have modified the form of the sub-drift surface. This line was based on the distribution of the so-called '200 foot' (60 m) platform but has not been identified by subsequent geophysical investigations. Nevertheless, neither relative subsidence nor localized uplift should be discounted, even as recently as within the Ipswichian stage (West, 1972). Another source of variability may have been differential loading, both by ice and ground moraine, which can cause valley bulging of bedrock particularly within the London Clay tract, similar to that described by Kellaway (1972). Detailed interpretations of the sub-drift surface should take these possible factors into account.

*Chalk solution*

Individual, and anomalous, records of the sub-drift surface are occasionally encountered where drift lies directly on Chalk. One such example exists in north Essex (Baker, 1977) where a narrow column of collapsed fluvio-glacial sands infills a solution pipe, 14 m below the local Chalk surface. Similar structures are recorded by Thorez *et al.* (1971) from South Mimms, Hertfordshire. Misleading interpretations of the sub-drift surface can be formulated in such instances.

*Solid-drift differentiation*

Generally speaking, where drift sediments overlie Chalk or London Clay, the sub-drift surface is easily located in borehole records. However, where stratified drift overlies Woolwich and Reading Beds the solid-drift boundary is often indistinct (Whitaker *et al.*, 1878; Whitaker and Thresh, 1916) and the sub-drift surface accordingly difficult to establish. The sub-drift distribution of these beds is shown in Fig. 4.

In the light of these considerations, the form of the sub-drift surface should be interpreted with care, recognizing that it is only an approximate indicator of pre-glacial and palaeodrainage conditions. At best, it can reveal only broad lineations of pre-glacial topography in so far as glacial processes may have been preferentially directed or conditioned by that pre-existing surface.

## Elements of the sub-drift surface

The sub-drift surface is at its highest (152 m) on the upper backslope of the Chalk cuesta, south of Royston (Fig. 4). From here, sub-drift summit elevations decline southwards to 91 m on the margins of the Lower Thames valley, and eastwards to around 46 m. The surface is dissected along all principal river valleys, showing that much of the relief is, at least in the local sense, pre-glacial (i.e., pre-Anglian). Glacial overdeepening is evident in many of these valleys (Rib, Ash, Upper Cam, Stort, Chelmer, Upper Stour and Blackwater), indicating that sub-glacial meltwater paths were largely conditioned and directed by this palaeodrainage pattern.

Between the upper dipslope and the Bagshot-capped hills of the Thames valley margin, the sub-drift surface descends into the Mid-Essex Depression, a notable west-east trough which cuts diagonally across the structural grain of the Tertiary rocks (Fig. 4). Near Hertford bedrock surfaces lie at about 50 m (Gibbard, 1977) and decline progressively eastwards to 46 m at Harlow, 40 m at The Rodings and 30 m around Chelmsford. The Depression is heavily obscured by glacial drift, but north-south cross-sections reveal the morphological and stratigraphic relationships (Fig. 7). At the 61 m level, the Depression broadens progressively eastwards from 3·5 km at Hertford, to 5·5 km at Harlow, 7 km in the Pincey Brook area, 11 km in the Rodings, and 19 km through Chelmsford (Fig. 4).

Tributary valleys enter the Depression from both north and south. Of these, perhaps the most striking is that of the middle Roding north of Ongar. The sub-drift valley of the mid-Roding (Fig. 4) declines north-eastwards in sympathy with a main

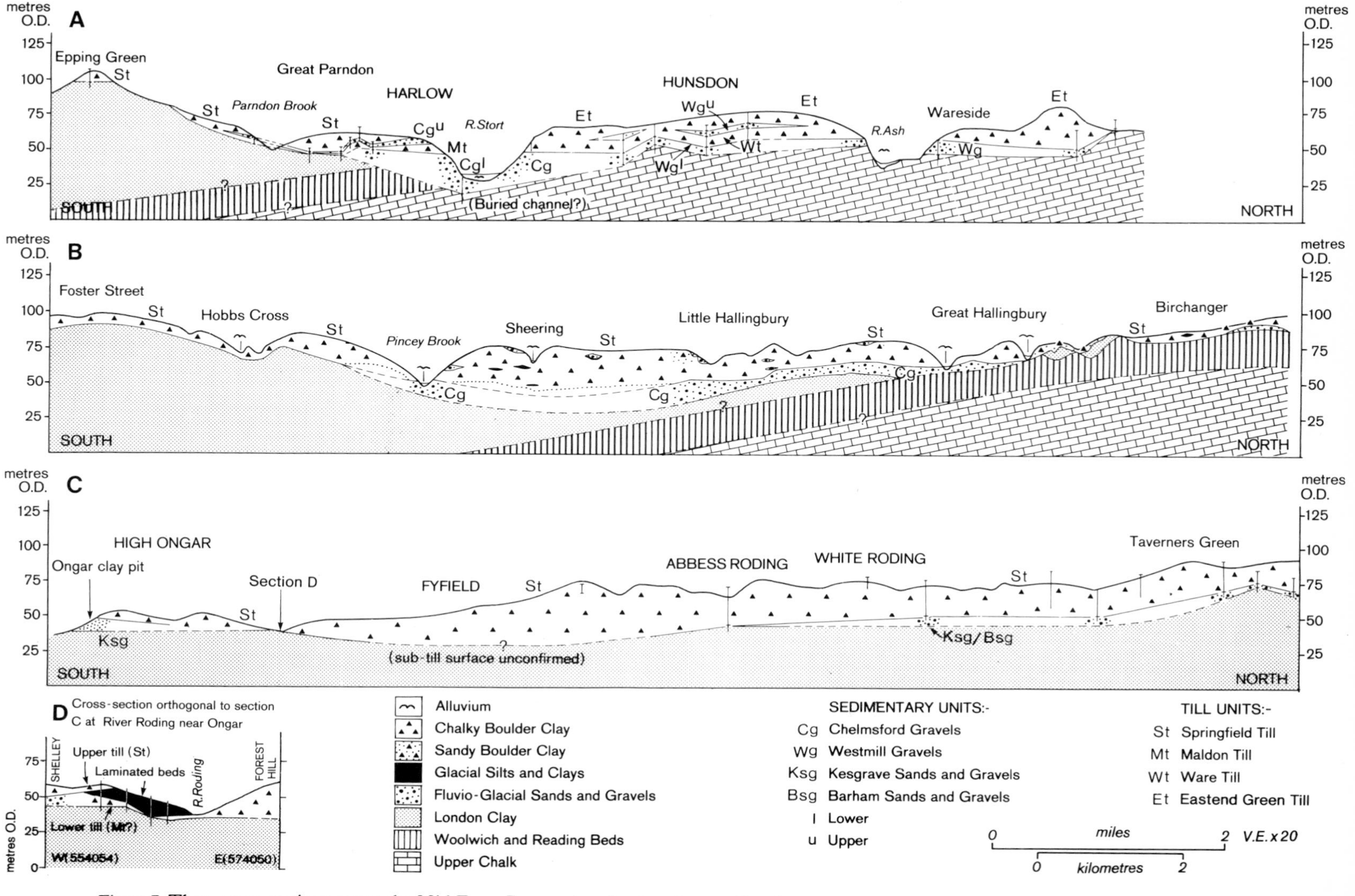


*Figure 7.* Three cross-sections across the Mid-Essex Depression. For location see Fig. 4.

eastward-flowing drainage line through the Mid-Essex Depression. The stratigraphy within this tributary valley is portrayed in Fig. 7D. These glacial sediments indicate that during glaciation, the mid-Roding was impounded in a proglacial lake against a former watershed near Ongar (to at least 62 m O.D.). The lake was subsequently overridden by ice and thick accumulations of till were emplaced in the Depression (Fig. 7C). Deglaciation resulted in the generation of a continuously south-flowing stream across the infilled Depression, effectively reversing the middle section of the Roding. Indeed, this part of the present Roding valley has a notably less mature appearance than the pre-glacial lower Roding, south-west of Ongar. The stratigraphy thus confirms earlier views on the origin of the Roding (Wooldridge and Henderson, 1955; Clayton, 1960).

Other principal features of the sub-drift surface (Fig. 4) include the Finchley Depression (between 61 m and 91 m around Hendon, Finchley and Enfield) and the so-called 'Romford River' valley south of Brentwood, where the sub-drift surface falls rapidly from 91 m at Brentwood to less than 24 m at Hornchurch (Wooldridge, 1957). At Woodham Ferrers (Fig. 4), the same depression is detected at about 30 m. The lower course of the early Anglian Roding valley may be presumed to have drained into the 'Romford River' (Fig. 8B).

## Stratified deposits in the Mid-Essex Depression and their implications

The higher margins of the Depression are free of stratified drift (Figs 5, 7B and C), Chalky Boulder Clay resting directly on London Clay. Within the central trough, however, stratified sub-till sands and gravels are evident (Figs 5, 7A-C), which reach 11 m in thickness, though 4-6 m is more usual. These are a continuation of the Westmill Gravels of Hertford and Ware (Gibbard, 1977) and the lower Chelmsford Gravels identified at Harlow by Clayton (1957), where they rest on a 2 km wide bench at 49 m. The gravels are overlain by a lower till (Maldon Till of Clayton) (Fig. 7A) similar to the Ware Till of Gibbard (1977). There is good reason, therefore, to believe that these lowest stratified deposits in the trough floor denote the continuous eastward passage of the proto-Thames just prior to the first ice advance of the Anglian glaciation, and that they correlate with the Westmill Lower Gravels and Dollis Hill Gravels (Fig. 3C). This is shown diagrammatically in Fig. 8B, the Mid-Essex Depression already having been excavated at the Lower Gravel Train stage (Fig. 8A).

Either contemporaneously with the Westmill Lower Gravel phase, or just prior to it, there is further evidence for eastward-flowing drainage within the Depression. In the Upper Cam-Stort watershed region, detailed drift mapping has revealed two distinct outwash formations (Baker, 1976, 1977). The older (Widdington Sands) consists of well-sorted and cross-bedded sand units, up to 13 m thick. Overall mean diameter ($+1{\cdot}34\ \phi$) is of medium sand grade with considerable vertical variability ($V_c = 77{\cdot}8\%$), though individual units are extremely well-sorted (mean $\sigma\phi = 0{\cdot}78$). The sand sequence culminates in a rubified palaeosol and leached coversand (preserved in solution-collapse structures) similar to those described by Rose and Allen (1977) from south-east Suffolk. There seems little doubt that these sands are to be correlated with the Kesgrave Sands and Gravels; indeed, such an interpretation of the Widdington stratigraphy is made by Rose *et al.* (1976). Consideration of the distribution of Widdington-type sands at the sub-drift surface (Fig. 5) suggests that they are continuous and widespread to the east and south-east of the Upper Cam, and that no equivalent sands exist to the west or south. It may be inferred, therefore, that the Widdington Sands were preferentially directed to the south-east (via the Chelmer valley) and did not pass to the south or south-west (via the Stort and Lea valleys). This interpretation is portrayed in Fig. 8B, and is consonant with Rose *et al.* (1976) in so far as their work suggests an eastward-flowing proto-Thames in the Mid-Essex Depression at the Kesgrave stage. We disagree, however, on the absolute chronology to be assigned to the Kesgrave Formation, preferring to regard it, at least in part, as early Anglian in age and fluvio-glacial in character.

Thus the pre-Anglian course of the proto-Thames via the Vale of St Albans and the Mid-Essex Depression appears to be established, the river subsequently being diverted by Anglian ice which twice advanced into the area.

## The first Anglian ice advance

At Ware and Harlow, the basal early Anglian gravels are overlain by a lower till—the Ware Till (Green, 1918; Barrow, 1919; Sherlock and Pocock, 1924; Clayton and Brown, 1958; Gibbard, 1977) and the Maldon Till (Irving and Irving, 1913; Clayton, 1957) respectively. The proximity of these two sites and their similarity of stratigraphy strongly suggests contemporaneity; indeed, a similar succession is found in the intervening Ash-Stort watershed (Figs 5, 7A). Some stratigraphers attach little significance to these lower tills, preferring to regard them as products of ephemeral fluctuations in the position of the Anglian ice-sheet. Indeed, these impersistent lower tills are lithologically identical to the main Chalky Boulder Clay sheet. It should be noted, however, that appreciable thicknesses of lower tills are recorded throughout Essex and east Hertfordshire (Table V), separated from the main till sheet by up to 13 m of stratified sediments. The concentration of lower till units in the Harlow-Hunsdon-Ware area is a notable feature (Figs 5, 7A). The most southerly occurrence of lower till is at Hoddesdon, 9 km south of Ware.

Although data are too scarce at present to confirm a separate widespread ice-advance prior to the main glaciation, it is nevertheless clear that a double thrust of ice occurred in the Harlow-Ware area of the Mid-Essex Depression. Consequently, we may extend the Ware Till ice limit (first advance) reconstructed by Gibbard (1977) (Fig. 3D) to include Hunsdon, Hoddesdon, Harlow, and possibly Ongar (Fig. 8C). Few fabric analyses are available for this poorly-exposed lower till, but measurements at Quendon reveal a north-west to south-east trend with high vector magnitudes, low variability and strong up-glacier dip. This is markedly different

TABLE V

*Lower till stratigraphy and thicknesses in east Hertfordshire and west and north Essex*

| Site | Lower till thickness (m) | Underlying deposits (m) | Overlying deposits (m) |
|---|---|---|---|
| Westmill (340160) | 2·4 | Gravel (6·7) | Gravel (11) |
| Ware (376143) | 3·6-9·5 | Gravel (7·6) | Gravel (2·1-4·6) and till (10) |
| Hunsdon (418143) | 7·3-7·6 | Gravel (4·9) | Gravel (3·0-3·4) and till (5·2-5·8) |
| Hunsdon (413131) | 9·7 | Gravel (7·6) | Gravel (4) |
| Gilston (439135) | 10·0 | Gravel (1·8) | Gravel (1·2) and till (3·6) |
| Harlow (4410) | 4·3-8·5 | Gravel | Gravel (9·1) and till (4) |
| Harlow (4209) | 1·2-2·44 | Gravel (2·14) | Gravel (1·0-4·6) and till (2·0-9·5) |
| Hoddesdon (352061) | 3·4-6·14 | Gravel (1·54) | Gravel (2·7-6·1) |
| Ongar (5605) | 1·8-7·0 | Bedrock (LC) | Laminated beds (5·2-12·8) and till (0-4·3) |
| Quendon (5231) | 1·8-7·6 | Sand and gravel (1-5) | Solifluxion clay (1-3), gravel (3-5·5) and till (2-3) |
| Wendens Ambo (527369) | 0·9-1·8 | Bedrock (UC) | Gravel (3-6) and till (1-3) |
| Thaxted (594307) | 4·3 | Bedrock (WRB) | Sand (9·1) and till (8·7) |
| Ashdon (595395) | 8·2 | Bedrock (UC) | Sand (9·1) and till (18·3) |
| Ashdon (598419) | 9·8 | Bedrock (UC) | Gravel (2·9) and till (2·6) |
| Radwinter (603363) | 9·8 | Sand and gravel (1·8) and basal till (1·5) | Sand and gravel (1·8) and till (12·8) |
| Little Sandford (651338) | 5·5 | Bedrock (WRB) | Sand and gravel (5·5) |

from fabrics in the later upper till body (Baker, 1976), and accords with the general direction of ice movement detected in the Ware Till around Hertford (West and Donner, 1956; Gibbard, 1977) (Fig. 8C).

## The first diversion of the Thames

It is apparent from the extent of this first ice advance, however brief it may have been, that the Mid-Essex Depression was occupied by ice and the proto-Thames course blocked, with the likelihood of diversion to a more southerly route. It is our contention that damming occurred forming a lake in the vicinity of Hoddesdon, Cheshunt and Enfield, with both Thames and Mole water impounded against the formerly continuous Hampstead-Epping ridge. The ridge was breached at its lowest col with overflow discharging along the line of the Lower Lea into the pre-existing valley of the 'Romford River' (Fig. 8C). On retreat of the ice this spillway route was

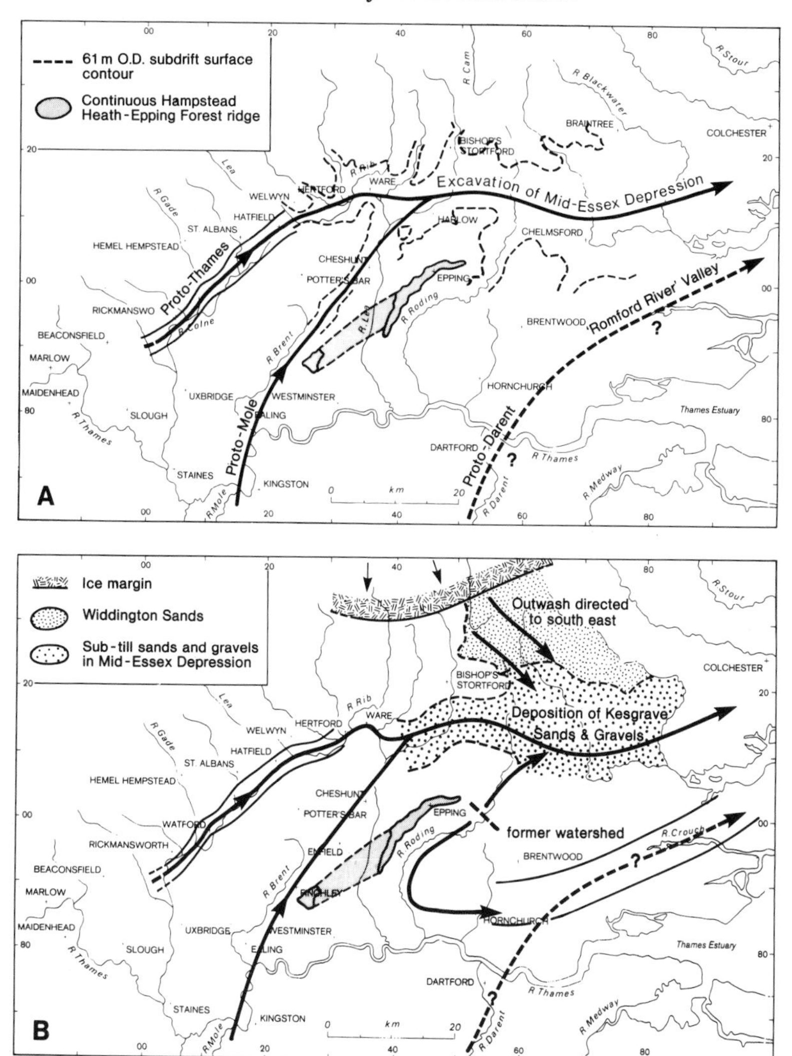


*Figure 8 A-F.* A new model for the evolution of the Thames and Lea drainage systems, in the Middle and Upper Pleistocene, based on the evidence presented in this paper. (A) Lower Gravel Train stage (Early Anglian). (B) First ice advance (early stage): Westmill Lower Gravel, Dollis Hill Gravel, Lower Chelmsford Gravels, Widdington Sands, Kesgrave Sands and Gravels (Anglian).

maintained, and the Lower Lea valley excavated (Fig. 8D). This reconstruction must be regarded as a working hypothesis at present, for no firm lithostratigraphic evidence has yet been adduced from the Lower Lea valley; indeed, very little glacial material has survived in this area. However, there are a number of lines of evidence that support this interpretation.

Such an hypothesis reiterates the earlier view (Hawkins, 1923) that the Thames flowed "up the Colne and down the Lea" (Fig. 2A). But whereas Hawkins considered this a "pre-glacial" course, we regard it as of mid-Anglian age and only temporary duration. One minimum estimate of the time interval separating

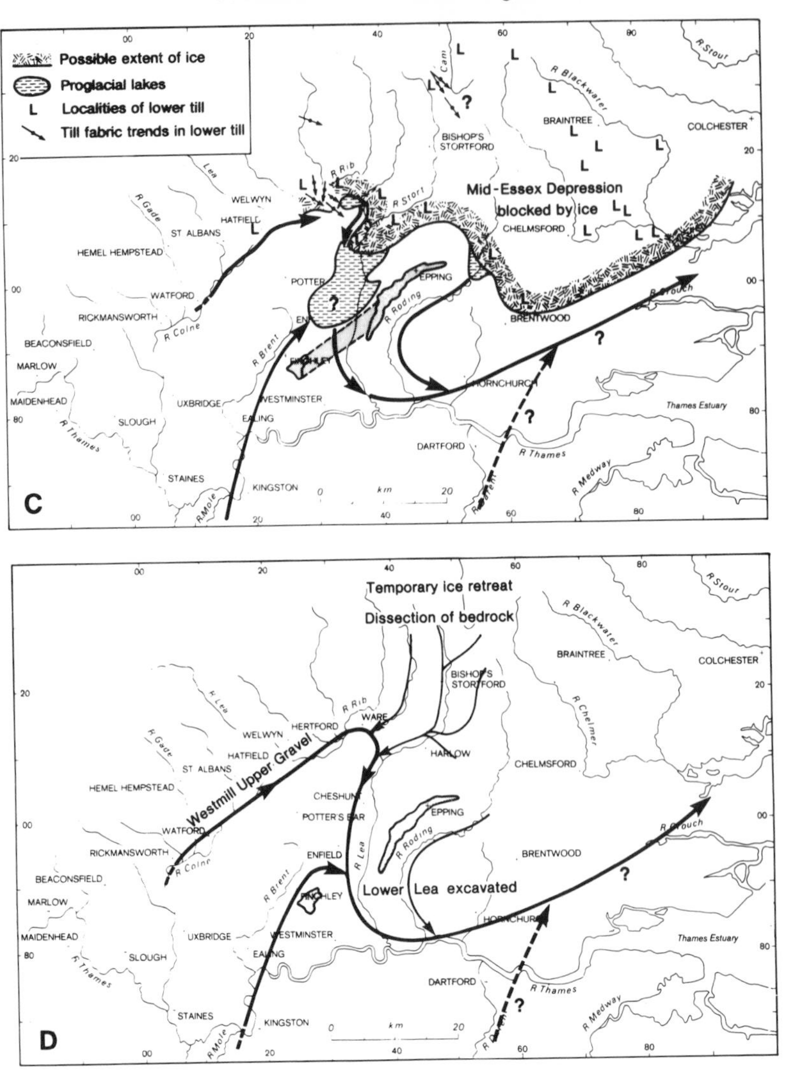


*Figure 8 (cont.).* (C) First ice advance (late stage: Quendon Till, Maldon Till, Ware Till, Watton Road Silts, Ongar Laminated Clays (Anglian). (D) Temporary ice retreat: Westmill Upper Gravel (Anglian).

deposition of a lower till (Quendon Till) and the main Lowestoft Till sheet, based on proglacial varves in the Newport area, is calculated as 5400 years (Embleton and King, 1975; Baker, 1977). A similar figure could apply in the Harlow area. Sherlock (unpublished) originally concurred with Hawkins in the notion of a Lower Lea course for the proto-Thames, but later abandoned the idea in favour of a more northerly route (Fig. 2A) after recognizing the Amwell 'gorge' (near Ware) as a glacial spillway cut by overflow from a proglacial lake held up at Ware (Sherlock, 1924). The hypothesis suggested here allows a Lower Lea route for the Thames to be retained while, at the same time, recognizing the Amwell 'gorge' to be proglacial in origin (Fig. 8C).

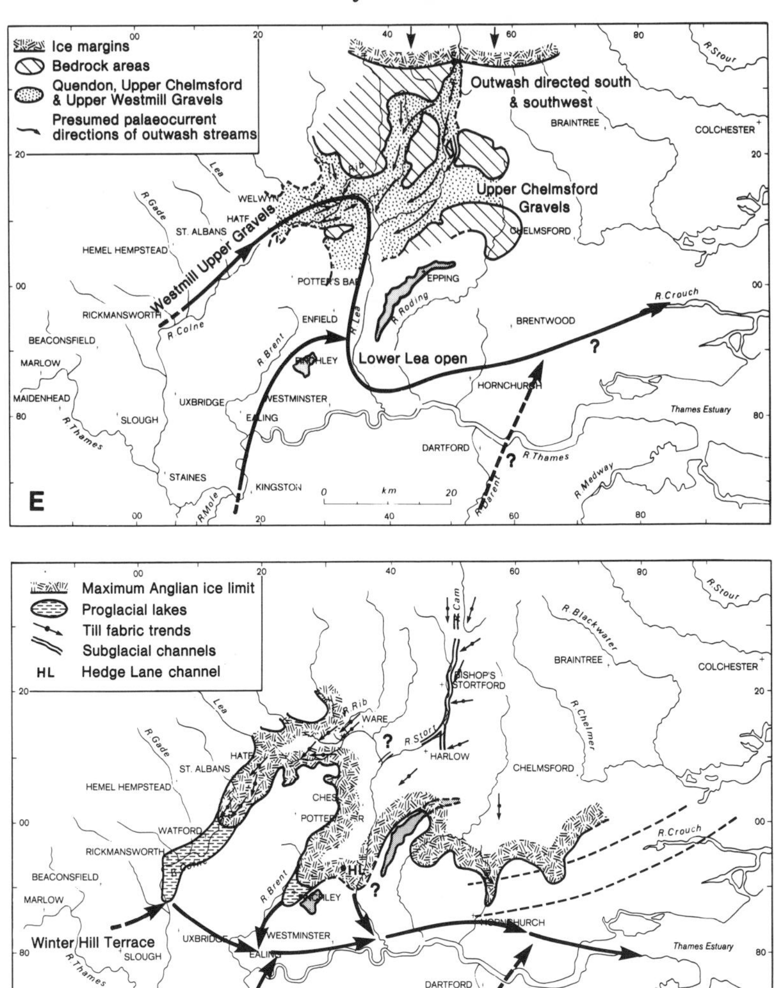


*Figure 8 (cont.).* (E) Second ice advance (early stage): Quendon Gravels, Upper Chelmsford Gravels, Westmill Upper Gravels (Anglian). (F) Second ice advance (late stage): Widdington Till, Eastend Green Till, Finchley Till, Moor Mill Laminated Clay, Coldfall Wood Laminated Clay, Winter Hill Terrace (Anglian).

An early diversion of the proto-Thames also agrees with the view that this river had migrated to the south of the Chelmsford-Colchester line prior to the arrival of the main Anglian ice-advance (Rose *et al.*, 1976; Rose and Allen, 1977). The explanation of such a displacement at the eastern end of the Mid-Essex Depression is difficult without recourse to river capture. However, if the Kesgrave Sands and Gravels are interpreted as fluvial and fluvio-glacial deposits of early Anglian age (see Table III), then it is possible to suggest that displacement was due to glacial interference, and that lack of sedimentary evidence for a mid-Anglian proto-Thames in east Essex was due to the prior deflection of the river down the Lower Lea valley.

## The second (Main) Anglian ice advance

Examination of the sub-drift surface (Fig. 4) and basal drift deposits (Fig. 5) in the Stort valley, reveal that the buried channel, in the vicinity of Sawbridgeworth, is cut deeply into the floor of the Mid-Essex Depression and is directed to the south-west (towards Harlow) and not towards the south-east, as would be expected if the main drainage within the Depression were still eastwards at the time of ice advance. This implies that sub-glacial meltwater could easily escape towards the south-west. While ponding in the Finchley Depression may explain this feature, it is more logical to argue that ease of escape was due to the prior existence of the Lower Lea valley.

Consideration of the advance outwash of the second ice-advance in the Stort is consistent with this view (Fig. 5). The younger of the two outwash formations mapped in the Upper Cam-Stort watershed (Baker, 1976, 1977) consists of coarse torrential gravels (Quendon Gravels) between 5 m and 24 m thick. Mean particle size ($-3{\cdot}90\ \phi$) is in the medium-to-coarse gravel category, with low overall variability ($V_c = 19{\cdot}6\%$), though individual samples display poor sorting (mean $\sigma\phi = 2{\cdot}20$). There is a general coarsening sequence upwards. The gravel-till contact is irregular, till lenses occurring in the gravel, and gravel itself incorporated within the till. This formation is interpreted as high-energy proximal outwash emanating from the advancing ice-front; little or no time lapse is envisaged before glacial overriding, the disturbance and assimilation of gravel material suggesting rapid sedimentation before permafrost could develop. A close parallel with the Barham Sands and Gravels (Rose *et al.*, 1976; Rose and Allen, 1977) is thus evident (see Table III), and we similarly regard these as the advance outwash of the second ice-advance and late Anglian in age.

The Quendon Gravels principally occur west of the Upper Cam valley (Fig. 5), and are preferentially directed south-westwards into the Ash valley and southwards (into the Stort valley) via three gaps cut through a dissected 91 m bedrock surface. From Bishop's Stortford, outwash gravels appear to be directed towards the south-west, in sympathy with the sub-drift contours and the course of the buried channel (Fig. 5). At Harlow (Fig. 7A) these gravels were identified by Clayton (1957) as upper Chelmsford Gravels, sandwiched between the Maldon and Springfield Tills; they would appear to correlate, therefore, with the Westmill Upper Gravels at Hertford (Gibbard, 1977). Thus the distribution of advance outwash prior to the second ice-advance, and the course of the sub-glacial buried channels, both indicate that the Lower Lea valley was open at this stage, the Mid-Essex routeway already having been abandoned. This is summarized in Fig. 8E. Palaeocurrent analyses of the lower and upper Chelmsford Gravels at Harlow, together with fabric analysis of the Maldon Till, would provide valuable data to confirm or modify this sequence of events.

## The second diversion

Progressive ice advance following deposition of the Quendon Gravels was responsible for lodgement of the main body of Chalky Boulder Clay—the

Springfield Till of Clayton (1957). Available till fabric measurements demonstrate that ice crossed the Chalk backslope under the influence of a unidirectional energy source (Rose, 1974), veering systematically from a NNE-SSW trend in the Upper Cam to an ENE-WSW trend in west Essex (Fig. 8F). Vector magnitudes are strong with low variability, and the fabric dips with a strong polarity in the up-glacier direction (Baker, 1976); strength of fabric declines south-westwards (Rose, 1974; Baker, 1977). From west Essex, distributary ice lobes penetrated the Vale of St Albans, Finchley Depression, the Lower Lea, Lower Roding and Ingrebourne valleys (Fig. 8F). Maximum till thicknesses at interfluve sites also decline south-westwards, from 21 m (west of Bishop's Stortford) to 19·2 m (Widford), 13·7 m (Ware) and 12·8 m (south of Hertford), while within the Mid-Essex Depression, thicknesses of over 26 m (Pincey Brook) and 35 m (The Rodings) are recorded (Fig. 6). The Mid-Essex Depression was infilled at this stage, although it had been abandoned previously by the proto-Thames. In the Finchley Depression and Vale of St Albans, proglacial impounding occurred (Gibbard, 1977, 1979), causing major diversions into the Lower Thames route via Westminster (Fig. 8F).

The precise extent of ice in the Lower Lea is unknown. The most southerly till is found at an elevation of 60 m at Chingford (Thompson, 1915). On this evidence alone, Wells and Wooldridge (1923) concluded that Chalky Boulder Clay was laid down on a floor some 46 m above the present valley bottom, but added that "it is possible that the boulder clay formerly filled the lower parts of the valley floor from which it has since been removed." That this may indeed be the case can be gauged from two considerations:

(1) Equivalent till is found at elevations much lower than 60 m in the Roding and Ingrebourne valleys, i.e. 43 m and 24 m respectively.

(2) An isolated and anomalous till-filled channel exists at Hedge Lane, Edmonton, with its base below 2·7 m O.D. (Hayward, 1957) (Fig. 4). It may be that this deposit is a till erratic swept post-depositionally into a channel of Taplow Terrace age, but the possibility of an *in situ* sub-glacial buried channel should not be overlooked.

That the outlier of glacial drifts at Finchley is preserved, while equivalent material in the Lower Lea valley has been destroyed, may result merely from its relatively protected interfluve site. Thus, the lack of sedimentary evidence in the Lower Lea does not preclude the former presence of a morainic ridge there also.

At Hornchurch, an ice lobe descended the Ingrebourne valley to an elevation of 24 m (Holmes, 1892; Wooldridge, 1957) partially infilling the former 'Romford River' valley (Fig. 8F). If the proto-Thames formerly occupied this valley (Wooldridge and Linton, 1955), it is reasonable to suppose that the river was expelled at this phase of the Anglian glaciation. Holmes (1894), however, regarded the Lower Thames and Romford rivers as contemporaneous, with the former eventually capturing the latter. In either event, the immediate post-glacial course of the newly-diverted Thames is clearly registered in the Black Park and Boyn Hill Terraces (Fig. 2H), with the Lower Lea well-established as a north-bank tributary, 4-5 km west of its present course (Fig. 2H). A substantial change in catchment area and consequent reduction in discharge was responsible for the Lower Lea's misfit appearance within an overly-wide valley. That this valley is, in addition, discordant to the Hampstead-Epping structural line adds weight to the hypothesis of glacial overflow and temporary occupation by the confluent proto-Thames and proto-Wey-Mole, as advocated here.

## Conclusion

The drainage development model for the Lower Thames advanced by Wooldridge (1938) was, to a large extent, based on the spatial relationship of deposits and surfaces. This model, with its double deflection of the proto-Thames, proved widely acceptable for over thirty years, even though an increasing number of defects emerged with the passage of time. The recent advent of more rigorous and detailed lithostratigraphic investigations have provided significant new evidence concerning both the fluvial and glacial sequences (Hey, 1965; Baker, 1971; Bristow and Cox, 1973; Rose *et al.*, 1976; Rose and Allen, 1977; Gibbard, 1977, 1979; Green and McGregor, 1978). This work has conclusively shown that while the proto-Thames routeway via the Vale of St Albans and Mid-Essex Depression can be substantiated, the subsequent evolutionary history is different from that envisaged by Wooldridge. Thus, the traditional model of two separate glacial diversions has been replaced by a single, but rather complex, diversion caused by the Anglian ice-sheet. Of particular significance is the rejection of the Finchley Depression as a routeway of the ancestral Thames and its reinterpretation as the valley of the proto-Wey-Mole (Gibbard, 1979).

While much has been resolved within the central London Basin, further investigation is required in the glacial and fluvio-glacial terrains of Essex, particularly with regard to the age and origin of the Kesgrave Sands and Gravels (Rose *et al.*, 1976), and the Chiltern Drift and its equivalents. The possibility that an early ice-advance entered the London Basin from the north-east and displaced the proto-Thames to a position along the line of the Mid-Essex Depression cannot be excluded and must await the re-examination of these long-neglected deposits. This chapter has sought to provide new evidence as to the middle Pleistocene evolution of this area. The nature and distribution of drift deposits and the configuration of the sub-drift surface suggest a double advance of the Anglian ice-sheet and the displacement of the proto-Thames prior to the arrival of the main (second) ice-advance. It is argued that glacial impounding led to overflow across the Hampstead-Epping Ridge, thereby creating the ancestral Lower Lea valley which continued to be utilized by the proto-Thames drainage system until the main Anglian ice-sheet blocked both the Thames and Wey-Mole valleys and caused the diversions described by Gibbard (1979). The envisaged sequence of events can be summarized as follows:

(1) The early Anglian course of the proto-Thames lay along the lines of the Vale of St Albans and Mid-Essex Depression (Fig. 8A).

(2) The first advance of the Anglian ice-sheet (Ware Till and Maldon Till) resulted in blocking of the Mid-Essex Depression. Impounding of both the Thames

and Wey-Mole resulted in overflow across Hampstead-Epping Ridge to the Romford River (Figs. 8B and C).

(3) During a subsequent retreat phase (Westmill Upper Gravel, Winter Hill Terrace) the proto-Thames system continued to utilize the Lower Lea-Romford River route, with development of south-westward oriented drainage in the western part of the Mid-Essex Depression (Fig. 8D).

(4) Re-advance of the Anglian ice-sheet progressively interfered with the drainage system. At an early stage pro-glacial outwash was widespread, much passing south-westwards to the proto-Thames and thence down the Lower Lea (Fig. 8E). Further advance led to sub-glacial flows in a similar direction along the previously established tributary networks and resulted in over-deepening of the sub-drift surface (e.g. Stort buried channel) (Fig. 8F).

(5) Further re-advance resulted in lobes penetrating the Vale of St Albans, Finchley Depression and Lower Lea valley. The proto-Thames routeway was blocked and impounding within the Vale of St Albans and Finchley Depression resulted in spillway development as described by Gibbard (1979) (Fig. 3G).

(6) Deglaciation saw the Thames established along its present course, and the initiation of the Lower Brent and Colne systems (Fig. 3H). The drainage of the abandoned and drift-covered proto-Thames course became integrated to form the Lea system, with a watershed that probably included the Finchley and Aldenham moraines and a drift divide within the Mid-Essex Depression. The present position of the Colne-Lea divide in the Vale of St Albans indicates recent drainage abstraction in favour of the Colne system. Elsewhere in northern Essex drainage became established on the drift cover but closely paralleling sub-drift pre-Anglian valleys.

It is recognized that much of this model is based on circumstantial evidence and will have to be verified by detailed sedimentological investigation. Nevertheless it does represent a return to previous suggestions (Hawkins, 1923; Dury, 1958) and explains the size and discordance of the Lower Lea valley, the unusual shape of the Lea drainage system and the nature of the sub-drift surface morphology in the western part of the Mid-Essex Depression. With the ever-increasing emphasis placed upon lithostratigraphy by Quaternary research workers, fundamental morphological criteria, such as the size of the Lower Lea valley, should not be overlooked. We believe that the model of drainage diversion proposed in this chapter is both a necessary and sufficient explanation for the existence and form of the present Lower Lea valley.

## Acknowledgements

The authors wish to thank the Institute of Geological Sciences (East Anglia and South-east England Unit) for kindly allowing access to borehole data. Discussion and correspondence with Dr P. Gibbard, Mr R. W. Hey and Mr J. Rose on the interpretation of the drift stratigraphy is also gratefully acknowledged. Finally, the authors wish to thank the secretarial and technical staff of the Geography Department, London School of Economics, for their assistance, and in particular Jane Shepherd for drawing all the maps and diagrams.

## References

Allender, R. and Hollyer, S. E. (1973). The sand and gravel resources of the country around Shotley and Felixstowe. Description of 1:25000 resource sheet TM23, *Rep. Inst. Geol. Sci.* **73/13**, 71 pp.

Avery, B. W. (1964). Soils and land use of the district round Aylesbury and Hemel Hempstead, *Mem. Soil Surv. U.K.*

Baden-Powell, D. F. W. (1948). The Chalky Boulder Clays of Norfolk and Suffolk, *Geol. Mag.* **85**, 279-96.

Baden-Powell, D. F. W. (1951). The age of the interglacial deposits at Swanscombe, *Geol. Mag.* **88**, 344-54.

Baker, C. A. (1971). A contribution to the glacial stratigraphy of West Essex, *Essex Naturalist* **32**, 318-30.

Baker, C. A. (1976). Periglacial phenomena in the Upper Cam Valley, North Essex, *Proc. Geol. Ass.* **87**, 285-306.

Baker, C. A. (1977). Quaternary stratigraphy and environments in the Upper Cam Valley, Ph.D. Thesis, University of London (unpublished).

Barrow, G. (1919). Some future work for the Geologists' Association, *Proc. Geol. Ass.* **30**, 36-48.

Briggs, D. J., Gilbertson, D. D., Goudie, A. S., Osbourne, P. J., Osmaston, H. A., Pettit, M. E., Shotton, F. W. and Stuart, A. J. (1975). New interglacial site at Sugworth, *Nature* **257**, 477-9.

Bristow, C. R. and Cox, F. C. (1973). The Gipping Till: a reappraisal of East Anglian glacial stratigraphy, *J. Geol. Soc. Lond.* **129**, 1-37.

Brown, J. C. (1959). The sub-glacial surface in East Hertfordshire, *Trans. Inst. Br. Geogr.* **26**, 37-50.

Bull, A. J. (1942). Pleistocene chronology, *Proc. Geol. Ass.* **53**, 1-20.

Clayton, K. M. (1957). Some aspects of the glacial deposits of Essex, *Proc. Geol. Ass.* **68**, 1-21.

Clayton, K. M. (1960). The landforms of part of Southern Essex, *Trans. Inst. Br. Geogr.* **28**, 55-74.

Clayton, K. M. (1964). The glacial geomorphology of Southern Essex, *in* K. M. Clayton (ed.), "Guide to London Excursions". 20th I.G.U. Congress, London.

Clayton, K. M. (1969). Post-war research on the geomorphology of south-east England, *Area* **1** (2), 9-12.

Clayton, K. M. and Brown, J. C. (1958). The glacial deposits around Hertford, *Proc. Geol. Ass.* **69**, 103-19.

Dury, G. H. (1958). Contribution to discussion of Clayton and Brown (1958), *Proc. Geol. Ass.* **69**, 268.

Embleton, C. E. and King, C. A. M. (1975). "Glacial geomorphology". Arnold, London.

Gibbard, P. L. (1977). The Pleistocene history of the Vale of St Albans, *Phil. Trans. Roy. Soc. Lond.* **B280**, 445-83.

Gibbard, P. L. (1979). Middle Pleistocene drainage in the Thames Valley, *Geol. Mag.* **116**, 35-44.

Gibbard, P. L. and Aalto, M. M. (1977). A Hoxnian interglacial site at Fishers Green, Stevenage, Hertfordshire, *New Phytol.* **72**, 505-23.

Green, J. F. N. (1918). Excursion to Ware, *Proc. Geol. Ass.* **29**, 42-45.

Green, C. P. and McGregor, D. F. M. (1978). Pleistocene Gravel Trains and the River Thames, *Proc. Geol. Ass.* **89**, 143-56.

Gregory, J. W. (1922). "The Evolution of the Essex Rivers". Benham, Colchester.

Hare, F. K. (1947). The Geomorphology of parts of the Middle Thames, *Proc. Geol. Ass.* **58**, 294-339.

Hawkins, H. L. (1923). The relation of the River Thames to the London Basin, *Rep. Br. Ass.* **1922**, 365-6.

Hayward, J. F. (1957). Borehole records from the Lea Valley in the neighbourhood of Edmonton, Middlesex, *Proc. Geol. Ass.* **68**, 39-44.

Hey, R. W. (1965). Highly quartzose pebble gravels in the London Basin, *Proc. Geol. Ass.* **76**, 403-20.

Hey, R. W. (1976). Provenance of far-travelled pebbles in the pre-Anglian Pleistocene of East Anglia, *Proc. Geol. Ass.* **87**, 69-81.

Holmes, T. V. (1892). The new railway from Gray's Thurrock to Romford, *Q. J. Geol. Soc. Lond.* **48**, 365-72.

Holmes, T. V. (1894). Further notes on some sections on the new railway from Romford to Upminster, *Q. J. Geol. Soc. Lond.* **50**, 443-52.

Irving, A. and Irving P. A. (1913). The Harlow Boulder Clay and its place in the glacial sequence of Eastern England, *Rep. Br. Ass.* 480-1.

Kellaway, G. A. (1972). Development of non-diastrophic Pleistocene structures in relation to climate and physical relief in Britain, 24th Session of the International Geology Conference, Canada (1972), Vol. 12, pp. 134-46.

Lawrence, G. R. P. (1964). Some proglacial features near Finchley and Potters Bar, *Proc. Geol. Ass.* **75**, 15-29.

Linton, D. L. (1969). The formative years in geomorphological research in South-east England, *Area* **1** (2), 1-8.

Manley, G. (1951). The range of variation of the British climate, *Geog. J.* **117**, 43-68.

Mitchell, G. D., Penny, L. F., Shotton, F. W. and West, R. G. (1973). A correlation of Quaternary deposits in the British Isles, *Sp. Rep. Geol. Soc. Lond.* **4**.

Perrin, R. M. S., Davies, H. and Fysh, M. D. (1973). Lithology of the Chalky Boulder Clay, *Nature Phys. Sci.* **245**, 101-4.

Rose, J. (1974). Small-scale variability of some sedimentary properties of lodgement till and slumped till, *Proc. Geol. Ass.* **85**, 239-58.

Rose, J. and Allen, P. (1977). Middle Pleistocene stratigraphy in South-east Suffolk, *J. Geol. Soc. Lond.* **133**, 83-102.

Rose, J., Allen, P. and Hey, R. W. (1976). Middle Pleistocene stratigraphy in Southern East Anglia, *Nature* **263**, 492-4.

Rose, J., Allen, P. and Wymer, J. J. (1978a). Weekend Field Meeting in South East Suffolk, *Proc. Geol. Ass.* **89**, 81-90.

Rose, J., Sturdy, R. G., Allen, P. and Whiteman, C. A. (1978b). Middle Pleistocene sediments and palaeosols near Chelmsford, Essex, *Proc. Geol. Ass.* **89**, 91-6.

Sealy, K. R. and Sealy, C. E. (1956). The terraces of the middle Thames, *Proc. Geol. Ass.* **67**, 369-92.

Sherlock, R. L. (1924). On the superficial deposits of South Herts and South Bucks etc., *Proc. Geol. Ass.* **35**, 19-28.

Sherlock, R. L. and Noble, A. N. (1912). On the glacial origin of the Clay-with-Flints and on a former course of the Thames, *Q. J. Geol. Soc. Lond.* **68**, 199-212.

Sherlock, R. L. and Pocock, R. W. (1924). The geology of the country around Hertford, *Mem. Geol. Survey U.K.*

Sparks, B. W. (1957). The evolution of the relief of the Cam Valley, *Geog. J.* **123**, 188-207.

Sparks, B. W., West, R. G., Williams, R. B. G. and Ranson, M. (1969). Hoxnian interglacial deposits near Hatfield, Herts, *Proc. Geol. Ass.* **80**, 243-67.

Thomasson, A. J. (1961). Some aspects of the drift deposits and geomorphology of South-east Hertfordshire, *Proc. Geol. Ass.* **72**, 287-302.

Thompson, P. G. (1915). Note on the occurrence of Chalky Boulder Clay at Chingford, *Essex Naturalist* **18**, 2-4.

Thorez, J., Bullock, P., Catt, J. A. and Weir, A. H. (1971). The petrography and origin of deposits filling solution pipes in chalk near South Mimms, Hertfordshire, *Geol. Mag.* **108**, 413-23.

Turner, C. (1970). The Middle Pleistocene deposits at Marks Tey, Essex, *Phil. Trans. Roy. Soc.* **B257**, 373-440.

Turner, C. (1973). Eastern England, *in* Mitchell, G. F., Penny, L. F., Shotton, F. W. and West, R. G. (eds), "A Correlation of Quaternary deposits in the British Isles". Sp. Rep. Geol. Soc. Lond. **4**.

Turner, C. (1975). The correlation and duration of Middle Pleistocene interglacial periods in north-west Europe, *in* K. W. Butzer and G. L. Isaac (eds), "After the Australopithecines: stratigraphy, ecology and culture change in the Middle Pleistocene". Mouton, The Hague.

Warren, S. H. (1910). Excursion to the Loughton district of Epping Forest, *Proc. Geol. Ass.* **21**, 451.

Warren, S. H. (1912). On a late glacial stage in the valley of the River Lea, subsequent to the epoch of river-drift Man, *Q. J. Geol. Soc. Lond.* **68**, 213-51.

Wells, A. K. and Wooldridge, S. W. (1923). Notes on the geology of Epping Forest, *Proc. Geol. Ass.* **34**, 244-52.

West, R. G. (1968). "Pleistocene Geology and Biology". Longmans, London.

West, R. G. (1972). Relative land-sea-level changes in south-eastern England during the Pleistocene, *Phil. Trans. Roy. Soc. Lond.* **A272**, 87-98.

West, R. G. (1973). A state of confusion in Norfolk Pleistocene stratigraphy, *Bull. Geol. Soc. Norfolk* **23**, 3-9.

West, R. G. and Donner, J. J. (1956). The glaciations of East Anglia and the east Midlands: a differentiation based on stone orientation measurements of the tills, *Q. J. Geol. Soc. Lond.* **112**, 69-91.

Whitaker, W. (1889). The geology of London and part of the Thames Basin, *Mem. Geol. Surv. U.K.*

Whitaker, W. and Thresh, J. C. (1916). The water supply of Essex from underground sources, *Mem. Geol. Surv. U.K.*

Whitaker, W., Penning, W. H., Dalton, W. H. and Bennet, F. J. (1878). The geology of part of NW Essex, with parts of Cambridgeshire and Suffolk and the NE part of Hertfordshire, *Mem. Geol. Surv. U.K.*

Woodland, A. W. (1970). The buried tunnel-valleys of East Anglia, *Proc. Yorks. Geol. Soc.* **37**, 521-77.

Wooldridge, S. W. (1923). The minor structures of the London Basin, *Proc. Geol. Ass.* **34**, 175-190.

Wooldridge, S. W. (1927). The Pliocene period in the London Basin, *Proc. Geol. Ass.* **38**, 49-132.

Wooldridge, S. W. (1938). The glaciation of the London Basin and the evolution of the Lower Thames drainage system, *Q. J. Geol. Soc. Lond.* **94**, 627-67.

Wooldridge, S. W. (1953). Some marginal features of the Chalky Boulder Clay ice-sheet in Hertfordshire, *Proc. Geol. Ass.* **64**, 208-31.

Wooldridge, S. W. (1957). Some aspects of the physiography of the Thames in relation to the Ice Age and Early Man, *Proc. Prehist. Soc.* **23**, 1-19.

Wooldridge, S. W. (1960). The Pleistocene succession in the London Basin, *Proc. Geol. Ass.* **71**, 113-29.

Wooldridge, S. W. and Cornwall, I. W. (1964). A contribution to a new datum for the prehistory of the Thames Valley, *Bull. Inst. Arch. Univ. Lond.* **4**, 223-32.

Wooldridge, S. W. and Henderson, H. C. K. (1955). Some aspects of the physiography of the eastern part of the London Basin, *Trans. Inst. Br. Geogr.* **21**, 19-31.

Wooldridge, S. W. and Linton, D. L. (1939). Structure, surface and drainage in South-east England, *Inst. Br. Geogr. Sp. Pub.* No. 10.

Wooldridge, S. W. and Linton, D. L. (1955). "Structure, surface and drainage in south-east England". George Philip, London.

Zeuner, F. E. (1959). "The Pleistocene Period". Hutchinson, London.

SEVEN

# Quaternary evolution of the River Thames

C. P. GREEN AND D. F. M. MCGREGOR
*Department of Geography, Bedford College, University of London*

## Introduction

The evolution of the River Thames during the Quaternary has been a subject of scientific enquiry for more than 150 years. The subject has been examined from two main points of view. On the one hand, attempts have been made to understand the origin and stratigraphic succession of the Quaternary sediments in the Thames Basin. On the other hand, geographical studies of landforms and deposits have sought to trace the successive positions of the Thames and its tributaries. The main sources of evidence have been the fossil record in the Quaternary sediments, the lithology and structure of the sediments and the surface and bedrock relief of the river valleys. During 150 years of investigation, a large amount of factual information has been recorded. Attention has tended to focus successively on different problems and corresponding stages of research can be identified. Increasingly precise methods have led to changes in the quality of information but the early work is the essential basis of present-day studies. Research has often dealt with local problems related to the stratigraphic significance of individual localities, such as Swanscombe or Ilford; or with regional problems of landform development or stratigraphy, such as the origin of the Goring Gap, or the diversion of the Thames from its early course through the Vale of St Albans.

The aims of this chapter are two-fold: (1) To trace and draw together various strands of research which, in the past, have tended to remain either thematically or geographically separate. (2) To suggest, as a necessary prerequisite for stratigraphic interpretation, a scheme of terrace development based on environmental interpretation of the terraces and terrace sediments.

## Early concepts and research

The earliest fields of sustained scientific research seem to have been investigations of the mammal and shell faunas in the brickearths, sands and gravels to the east of London (Morris, 1838; Brown, 1838, 1840; Cotton, 1847). The development of ideas during the nineteenth century is described and documented by Whitaker in his

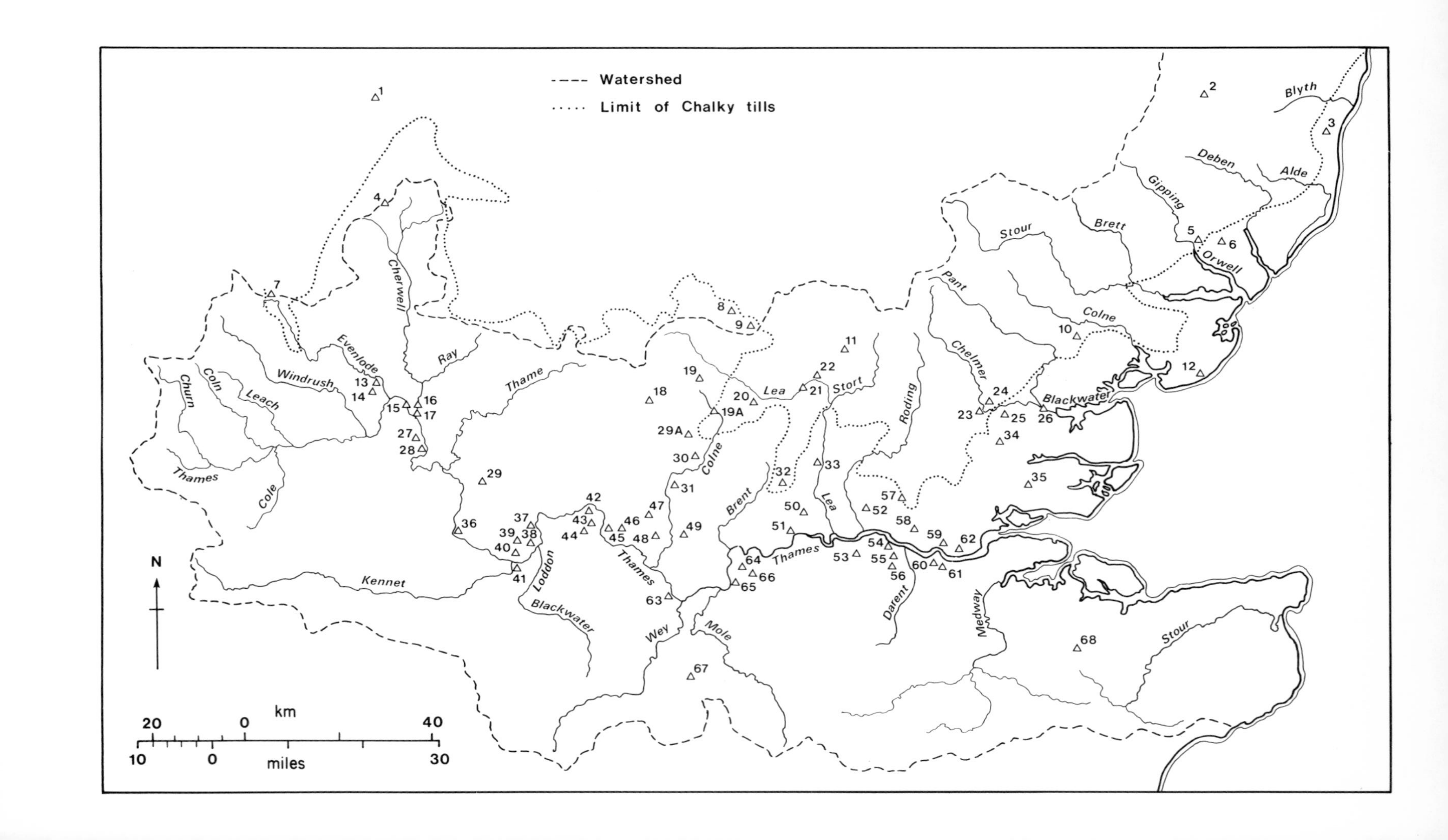
Watershed
Limit of Chalky tills
Blyth
Alde
Deben
Gipping
Brett
Orwell
Stour
Pant
Colne
Chelmer
Blackwater
Roding
Stort
Lea
Brent
Colne
Thames
Darent
Medway
Stour
Mole
Wey
Thames
Loddon
Blackwater
Kennet
Thame
Ray
Cherwell
Evenlode
Windrush
Leach
Coln
Churn
Thames
Cole
N
km
miles
20
0
40
10
0
30

*Figure 1.* The Thames basin and southern East Anglia, showing key localities, the main rivers and the southern limit of chalky tills.

| | | | | | |
|---|---|---|---|---|---|
| Aveley | 58 | Hitchin | 8 | Richmond | 64 |
| Binfield | 39 | Hornchurch | 57 | Rothamsted | 19 |
| Black Park | 47 | Hoxne | 2 | St Albans | 19A |
| Boyn Hill | 44 | Ilford | 52 | Shiplake | 38 |
| Caversham | 40 | Ipswich | 5 | Shooter's Hill | 53 |
| Chelmsford | 23 | Iver | 48 | Springfield | 24 |
| Chertsey | 63 | Kingston | 65 | Stevenage | 9 |
| Clacton on Sea | 12 | Kesgrave | 6 | Stoke Newington | 50 |
| Crayford | 55 | Leavesden | 29A | Sugworth | 27 |
| Dartford | 56 | Lenham | 68 | Summertown | 16 |
| Danbury | 25 | Little Heath | 18 | Swanscombe/Barnfield | 60 |
| Erith | 54 | Long Hanborough | 13 | Taplow | 45 |
| Fenny Compton | 4 | Lynch Hill | 46 | Tilbury | 62 |
| Finchley | 32 | Maldon | 26 | Trafalgar Square | 51 |
| Freeland | 14 | Marks Tey | 10 | Wallingford | 29 |
| Furze Platt | 43 | Moreton in Marsh | 7 | Ware | 22 |
| Goring | 36 | Netley Heath | 67 | Watford | 30 |
| Grays | 59 | Northfleet/Ebbsfleet | 61 | Westland Green | 11 |
| Hanningfield | 34 | Oxford | 17 | Westleton | 3 |
| Harefield | 31 | Ponders End | 33 | Wimbledon | 66 |
| Hatfield | 20 | Radley | 28 | Winter Hill | 42 |
| Henley | 37 | Rayleigh | 35 | Wolston | 1 |
| Hertford | 21 | Reading | 41 | Wolvercote | 15 |

important memoir on the geology of London (1889). Research at this time developed in two main directions. On the one hand, continuing interest in the fossil record led to the discovery and description prior to 1914 of many of the key stratigraphic sites (Fig. 1). Study of the fossil record received a great stimulus from the realization in the 1860s that evidence of early man, in the form of flint implements, occurs in the 'River Drift' in association with the remains of extinct animals (Evans, 1872). Prior to 1914, the work of French archaeologists in the valley of the Somme (Commont, 1910) had indicated the potential stratigraphic value of implement typology, and this concept was to influence the interpretation of terrace sequences in this country until comparatively recently. On the other hand, in deposits relatively remote from the present river, where Palaeolithic artifacts were not found, attention during the period up to 1914 tended to focus on the composition, provenance and distribution of the river gravels (White, 1895, 1897; Salter, 1898) and it was at this time that glacial deposits were first recognized in the basin of the Thames (Walker, 1871). Much useful information was recorded between 1890 and 1930 in Reports on the excursions of the Geologists' Association, and between 1905 and 1930 the Geological Survey published maps and Memoirs for large parts of the middle Thames and London areas. During the same period, the earliest stages in the development of the Thames were a particular focus of attention, leading to a succession of papers on the supposed Pliocene deposits and the Pebble Gravels (Leach, 1912; Gilbert, 1919; Bury, 1922).

## S. W. Wooldridge: the context and influence of his work

From 1927, when Wooldridge published his now classic paper on the Pliocene of the London basin (Wooldridge, 1927), until 1960 when he offered what proved to be his last published summary of Pleistocene events in the London Basin, the geomorphological concepts that he expounded and the succession of Pleistocene events that he worked out, assumed an increasingly important place in Quaternary studies of the Thames.

### Geomorphological method

Wooldridge argued that earlier workers, who had given due weight to fossil and lithological evidence and to the distribution of terrace deposits, had paid insufficient attention to the down-valley continuity and relative elevations of the terrace deposits. He recognized that a comprehensive morphostratigraphic scheme would contribute significantly to an understanding of the Thames' terraces and their deposits. He repeatedly expressed his confidence in the methods of geomorphology, being particularly concerned to trace the morphological expression of the terraces along the valley (Wooldridge, 1928) and to recognize a vertical sequence of terraces that could be related to fossil evidence and to the glacial sequence. The value of such concepts had been recognized by some earlier workers, such as H. J. O. White, but Wooldridge sought to accommodate a much more substantial volume of evidence in his scheme of terrace development, and he provided a regional scheme in which to evaluate local problems that had previously been considered in relative isolation.

In the work of Wooldridge on the Quaternary evolution of the Thames, and in the work of those who during his lifetime were inspired by him, it is possible to distinguish a special concern for four separate but closely related problems, namely, (1) the early course of the Thames, (2) the diversion of the Thames from this early course, (3) the terrace succession, and (4) the glacial succession. It is appropriate here to consider briefly the work of Wooldridge on these problems and to indicate his debt, frequently acknowledged by him, both to earlier workers and to his contemporaries, and his influence on the course of research.

## The early course of the Thames

Wooldridge in his 1927 paper used a skilful combination of stratigraphic and geomorphological methods to resolve the 'Pebble Gravel' problem, which had exercised geologists for the previous twenty years. Wooldridge proposed that the bulk of the Pebble Gravel comprised a single unit, separate from the Pliocene deposits. He suggested a plausible stratigraphic position for it, and he discussed its relation to the early course of the Thames and its Wealden tributaries (Fig. 2). The idea of an early course of the Thames passing north of the present course was not however originated by Wooldridge, for both Salter (1905) and Sherlock (1924) had suggested the possibility. The effectiveness of Wooldridge's proposals is clearly indicated however by the fact that almost no new work on the Pebble Gravels was undertaken until Hey began to re-examine them in the 1960s (Warren, 1957; Hey, 1965).

At a relatively early date, Wooldridge recognized the possibility that the Thames had at one time extended from the neighbourhood of Ware "through the broad depression which leads eastwards from the valley of the lower Stort and so into the Chelmer lowlands" (Saner and Wooldridge, 1929). This idea, elements of which appear in the work of Gregory (1922) on the rivers of Essex, is developed more fully (Fig. 2) by Wooldridge in a later paper (Wooldridge and Henderson, 1955).

## Diversion of the Thames

In some respects the work of Wooldridge (1938) on the diversion of the Thames from its early course is more representative of his fundamental contribution to Quaternary studies than his work on the early course itself, in so far as it depends much more completely on the geomorphological evidence. Wooldridge based his hypothesis on the topographic evidence of terraces and on terrace gradients at levels above the Geological Survey's 100 foot or Boyn Hill Terrace. He extended the terrace sequence upward, towards the Pebble Gravels, by identifying a Winter Hill Terrace and Lower and Higher Gravel Trains, all of which he regarded as composite in nature. This work included an appraisal of the stratigraphic position of the aban-

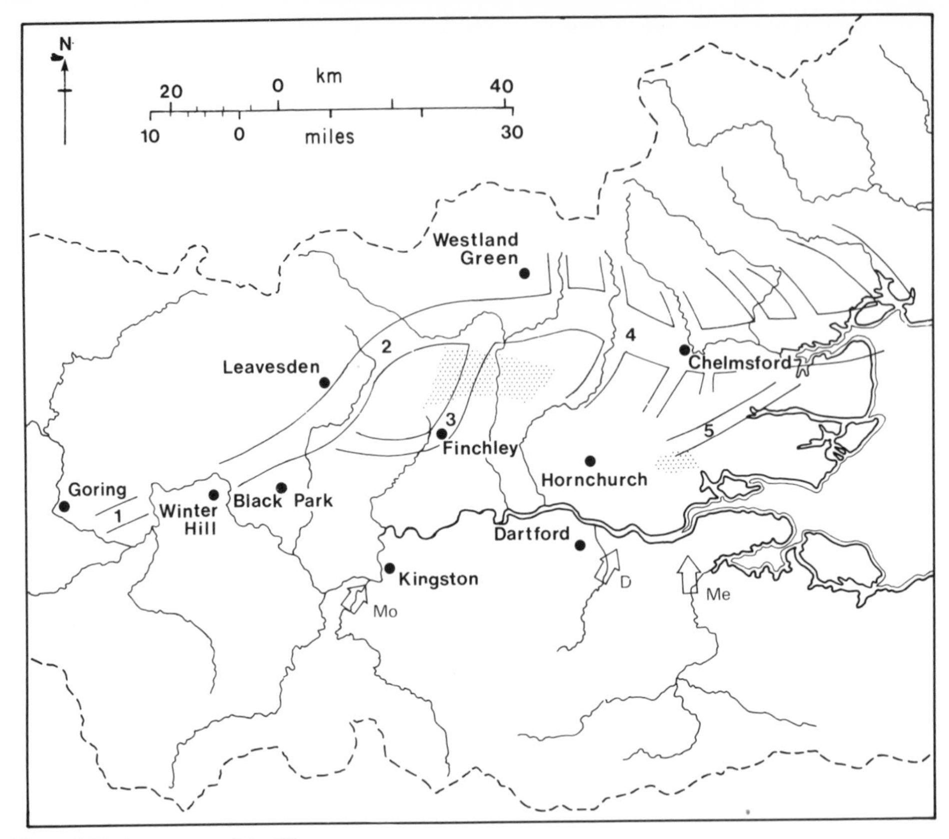


*Figure 2.* Early courses of the Thames.

(1) Caversham Channel: identified by Treacher (1926); latest sediments appear to form part of the Winter Hill Terrace.

(2) Vale of St Albans: recognized by Sherlock (1924) as course of the proto-Thames.

(3) Finchley Depression: recognized by Wooldridge as possible right bank tributary of proto-Thames and thought by him to have been later occupied, during the Gravel Train and Winter Hill stages, by the Thames itself.

(4) Mid-Essex Depression: identified by Wooldridge and Henderson (1955) as the probable course of the Thames at the Lower Gravel Train stage.

(5) The Romford River: depression first noted by Holmes (1894) and speculatively identified by Wooldridge and Linton (1955) as the course of the Thames at the Black Park stage.

Stipple: areas in which Lower Greensand chert is common; considered by Wooldridge (1927) to mark courses of right bank tributaries of the proto-Thames at the Pebble Gravel stage. Arrows: possible sources of Lower Greensand chert. Mo: Mole, D: Darent, Me: Medway.

doned valley of the Thames between Caversham and Henley (Fig. 2). This feature was first recognized by Treacher (1926) and Wooldridge suggested that it was last occupied by the Thames during the Winter Hill stage. Again it should be noted that the recognition of these earlier and more elevated terraces is foreshadowed in the work of White (1908).

Wooldridge attributes the diversion of the Thames to the presence on two occasions of glacial ice in the valley of the river. On the first occasion the river was diverted from the Vale of St Albans between Watford and Ware into a more southerly course through the Finchley Depression. On the second occasion the river was diverted from the Finchley Depression to the present course between Staines and the sea. Wooldridge believed that in the period immediately following this second diversion, the river occupied the low ground in Essex between the Danbury and Rayleigh ridges and flowed thence towards the area now occupied by the lower estuary of the Blackwater (Fig. 2).

## The terrace sequence

The arrangement of the Quaternary deposits of the River Thames in a series of terraces was generally recognized by the 1870s (Whitaker, 1889). Before this date a wide range of alternative explanations of the 'River Drift' had been proposed. Thereafter the terrace concept was usually accepted and it was commonly assumed that greater terrace elevation indicated greater antiquity. The nomenclature of a three-fold terrace sequence was formalized by Bromehead (1912). This sequence was adopted by the Geological Survey and is regarded as the basis of all subsequent schemes in the Thames Valley (Fig. 3). It should be noted however that the four-fold scheme proposed by Pocock (1908) for the upper Thames area is the basis of subsequent work there.

In a previous paragraph the work of Wooldridge which extended the terrace sequence above the Boyn Hill level has been described. At lower levels the most important developments in the period up to 1939 arose not from geomorphological investigations, but from stratigraphic work on the fossil record. In the basin of the Upper Thames, the work of Sandford (1924, 1926) around Oxford developed and refined Pocock's earlier scheme. In the middle and lower valley of the Thames, the new results came from the work of Burchell on the Ebbsfleet Channel Series (1933, 1935), from work on implement typology (Lacaille and Oakley, 1936), from the work of Warren (1916) in the Lea valley, and from work at Swanscombe (Marston, 1937).

A stratigraphic synthesis of some of this new material was published in 1936 by King and Oakley. Their paper, which followed a pattern established by Breuil (1932) in his work on the terraces of the Somme, attached a new importance to glacio-eustatic fluctuation of sea-level during the Pleistocene. The influence of this paper is clear in the treatment of Pleistocene events in the first edition of *Structure, Surface and Drainage* published in 1939. As far as the terraces of the Middle and Lower Thames are concerned, that treatment was not radically changed in any of Wooldridge's later work on this subject.

In Wooldridge's examination of this material, two emphases are apparent. Firstly, he argues that the terrace sequence records reliably the episodes of the glacial sequence. Secondly, he stresses the morphological continuity of the terrace features. This latter idea was tested in a number of instrumental surveys inspired by him, of which the most influential were undertaken in the Middle Thames area (Hare, 1947; Sealy and Sealy, 1956). In both these studies general morphological characteristics of the terraces were noted, including topographic boundaries,

| 1912 | 1921 | 1927 | 1929 | 1932 | 1936 | 1938 | 1945 | 1947 | 1956 | 1965 | Present terms | Stage |
|---|---|---|---|---|---|---|---|---|---|---|---|---|
| Pliocene | | | | | | | | | | | Lenham Beds<br>Rothamsted Sandstone | Pliocene<br>Waltonian |
| Pebble Gravels | | 500 foot Pebble Gravels<br>400 foot Pebble Gravels | | | | | | | | Westland Green Gravel | 500 foot Pebble Gravels<br>400 foot Pebble Gravels<br>Westland Green Gravel | |
| Plateau and Glacial Gravels | | | Winter Hill | Binfield | | Higher Gravel Train<br>Lower Gravel Train | Finchley Leaf<br>Kingston Leaf | Harefield<br>Black Park | Rassler<br>Upper Winter Hill<br>Lower Winter Hill | | Higher Gravel Train<br>Harefield / Rassler (Lower Gravel Train)<br>Upper Winter Hill<br>Lower Winter Hill<br>Kingston Leaf<br>Black Park | ? Late Cromerian<br>Late Anglian |
| Boyn Hill | L.Boyn Hill or U.Taplow | or Furze Platt or Iver | | | v L.Barnfield and viii M.Barnfield<br>viiia M.Barnfield or ixa Iver | | | Lynch Hill | | | Boyn Hill<br>Lynch Hill | Hoxnian<br>Hoxnian or Wolstonian |
| Taplow | | | | | xi Taplow | | | | Upper Taplow<br>Lower Taplow | | Upper Taplow<br>Lower Taplow | Ilfordian or Ipswichian |
| Floodplain | Upper Floodplain<br>Lower Floodplain | | | | xiv Ponders End<br>xvi Halling<br>xva, xviia | | 1st, 2nd, 3rd, P | | | | Upper Floodplain<br>Lower Floodplain | Ipswichian<br>Devensian |
| Alluvium | | | | | xviib Tilbury | | | | | | Alluvium | Flandrian |

*Figure 3.* The terrace sequence : the development of the nomenclature since 1912. Dates indicate the following sources: 1921 Dewey and Bromehead; 1927 Wooldridge; 1929 Saner and Wooldridge; 1932 Ross; 1936 King and Oakley; 1938 Wooldridge; 1945 Zeuner; 1947 Hare; 1956 Sealy and Sealy; 1965 Hey. Dashed lines indicate subdivisions which either are of local significance only, or are disputed, or superseded.

P: relationship of Ponders End sediments in Zeuner (1945); 1st: first sunk channel of Zeuner (1945); 2nd: second sunk channel; 3rd: third sunk channel.

Roman numerals signify the stages of King and Oakley (1936). An attempt is made to show the position in the terrace sequence, as supposed by King and Oakley, of the sediments from which the stages are named. Pre-Boyn Hill stages (i-iv), brickearths (vii, xii), periglacial deposits (x, xvb) and certain erosional stages (vi, ix, xiii) are omitted.

The correlations indicated in the final column reflect the authors present opinion.

longitudinal gradients and irregularities of cross-profile. In addition, qualitative observations were made on the varying character of the terrace sediments. It is clear that neither Hare nor the Sealys regarded the morphological approach as a complete answer to the deciphering of the terrace sequence. Frequent reference is made by them to the disturbed nature of the terrace deposits and to the evidence of post-depositional effects. Nevertheless, their work tended to confirm the terrace succession already outlined by Wooldridge, while elaborating it to the extent described by him in his later papers (Wooldridge and Linton, 1955; Wooldridge, 1958, 1960).

## The glacial sequence

By 1938 the previously influential concept of a monoglacial Pleistocene in the British Isles had long been seriously discredited. In developing a multiglacial Pleistocene succession for the Thames area, Wooldridge (1938) acknowledged the importance of the multiglacial succession proposed by Solomon (1932) for East Anglia, in which four glacial episodes were identified.

Wooldridge recognized the deposits of two glaciations in the basin of the Middle and Lower Thames (Fig. 4), and equated these with the first and second glaciations of Solomon's succession (Fig. 5). The earlier episode, called by Wooldridge the Chiltern Glaciation, he recognized on the basis of stony clays occupying the dip slope of the Chilterns. Both Sherlock and Noble (1912) and Barrow (1919) had argued a glacial origin for these clays, and Wooldridge proposed that the deposits represented a glaciation, later than the Pebble Gravel stage, and responsible for the diversion of the Thames from its original course through the Vale of St Albans. Although the relationship of the Gravel Trains to this glacial episode is not entirely unambiguous in Wooldridge's account, a close relationship is implied.

The chalky till, which is widespread in Essex and which extends into the Vale of St Albans and the Finchley Depression, had been recognized as a glacial deposit at an early date (Walker, 1871; Hicks, 1891) and had been described in detail by the Geological Survey. Wooldridge (1938) referred the chalky till to his Eastern Drift Glaciation which he placed in the interval between the Winter Hill and Boyn Hill stages, and which he held responsible for the diversion of the Thames from the Finchley Depression. In deciding the stratigraphic position of the chalky till Wooldridge attached great importance to observations made by Holmes (1892, 1894) in south Essex. Holmes had found chalky till overlain by gravel which appeared to him to be part of the 100 foot terrace of the Thames, thus determining the position of the chalky till in the terrace succession. Wooldridge and Linton (1955) describe this finding as "a vital fixed point in the Pleistocene stratigraphy of the Thames valley and one upon which most of the correlations since proposed have essentially rested".

The sedimentology of the tills and the details of their stratigraphy were never described by Wooldridge himself, but he clearly influenced the work of Clayton and Brown (Clayton, 1957; Clayton and Brown, 1958; Brown, 1959; Clayton, 1960)

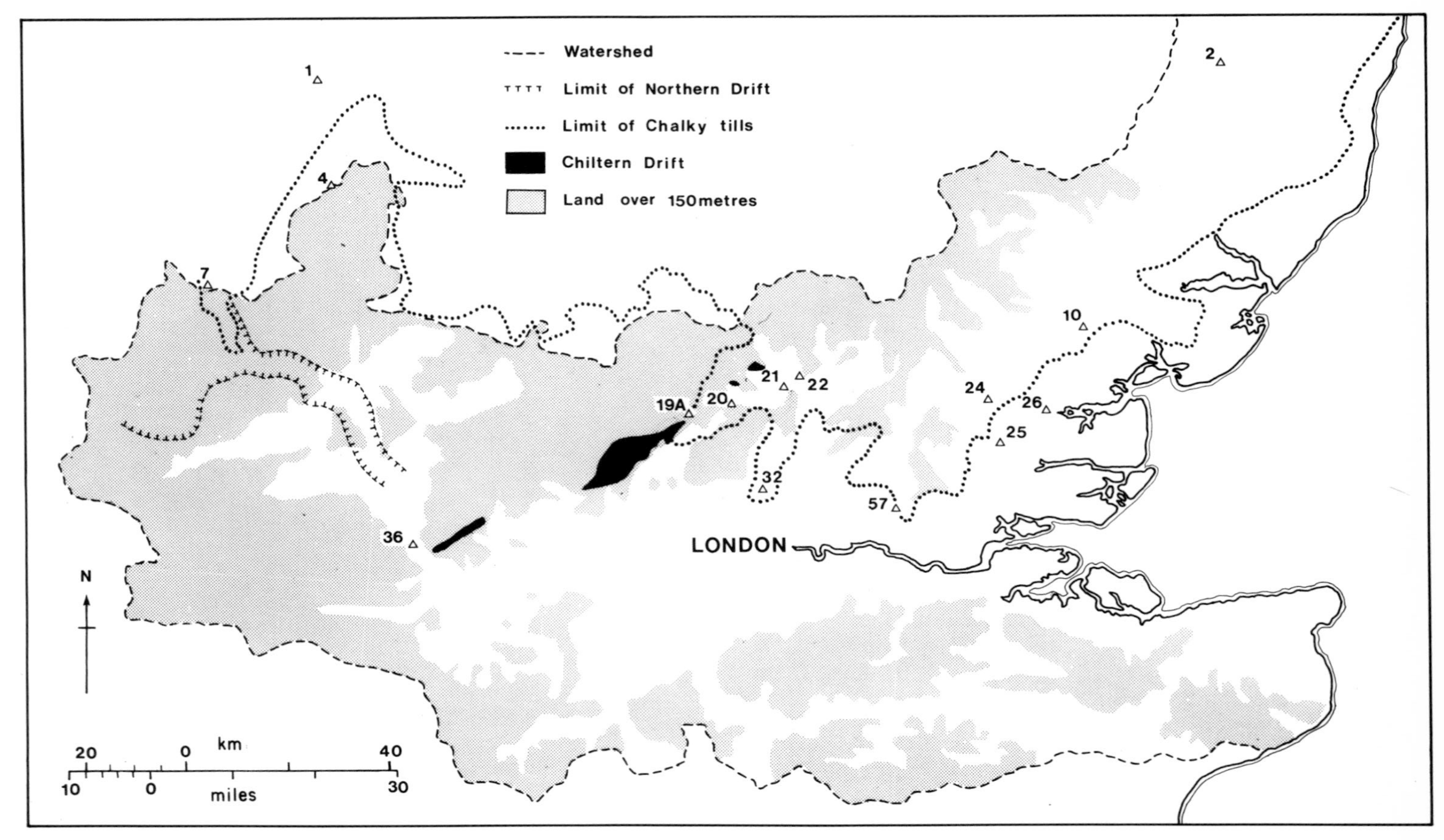


*Figure 4.* Glacial deposits in the Thames Basin and southern East Anglia. Contour at 150 m. Sites as in Fig. 1.

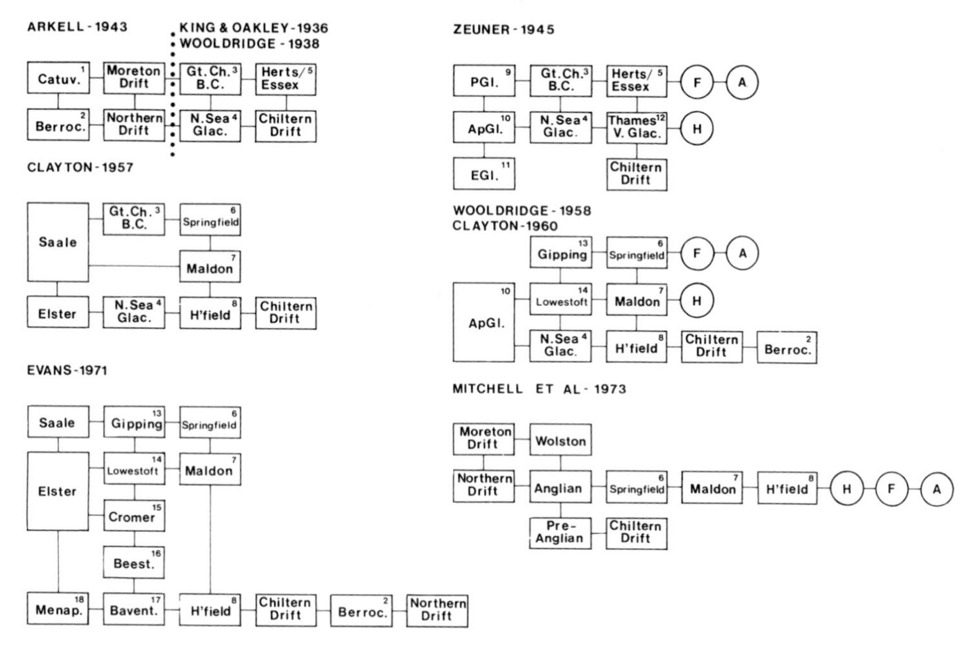


*Figure 5.* The glacial succession in the Thames Basin and its correlation.
F : Finchley Till; H : Hornchurch Till; A : till in Vale of St Albans.
Nomenclature and sources: (1) Catuvellaunian glaciation, (2) Berrocian glaciation (Arkell, 1943); (3) Great Chalky Boulder Clay, (4) North Sea glaciation (Solomon, 1932); (5) undifferentiated chalky tills of Hertfordshire and Essex, (6) Springfield Till, (7) Maldon Till, (8) Hanningfield Till (Clayton, 1957); (9) Penultimate glaciation, (10) Antepenultimate glaciation, (11) Early glaciation, (12) Thames valley glaciation (Zeuner, 1945); (13) Gipping glaciation, (14) Lowestoft glaciation (Baden-Powell, (1948); (15) Cromer Till (West and Donner, 1956); (16) Beestonian stage, (17) Baventian stage, (18) Menapian (West, 1963).

in which evidence of three tills is presented and in which correlation with the glacial succession of the Middle Thames area is attempted. Clayton in 1957, without reference to the Hornchurch evidence, suggested that the later tills in Essex (his Springfield and Maldon Tills) are of Saale age, which in terms of the correlations that were generally accepted at that time, placed them after, rather than before, the formation of the Boyn Hill Terrace. Wooldridge (1958) did not accept this suggestion. Clayton equated the earlier till in Essex (his Hanningfield Till) with the Chiltern Drift and regarded these deposits as the product of the Elster glaciation. Evidence of more than one glacial episode at sites in the Thames Basin was also presented by West and Donner (1956) on the basis of till fabrics. The episodes were equated by them with the Lowestoft and Gipping stages of East Anglia. A post-Boyn Hill date was therefore implied for the later episode. In 1960 Clayton correlated the Springfield Till, including drifts in the Vale of St Albans and at Finchley, with the Gipping stage, and the Maldon Till, including the Hornchurch Drift, with the Lowestoft stage. The Hanningfield Till he correlated with the Cromer Till and Norwich brickearth.

This uncertainty regarding the broader relationships of the glacial sequence in the Middle and Lower Thames valley has close parallels in the Upper Thames Basin. In most discussions of the glacial sequence in the latter area, two fundamental points of agreement can be recognized. Firstly, that ice penetrated the Upper Thames Basin via the Moreton Gap and upper Evenlode valley on two occasions; and secondly, that on the second occasion, outwash gravel extended down the Evenlode valley to form the Wolvercote Terrace. Prior to the work of Shotton (1953) it was generally believed (Arkell, 1943, 1947; Wills, 1948) that the second glacial episode, represented by chalky till of eastern derivation, pre-dated the Great Interglacial (the Hoxnian interglacial of the modern terminology). Following the work of Shotton, the second glacial episode has generally been placed after the Hoxnian interglacial, mainly because tills, correlated with the chalky tills near Moreton, are found near Birmingham resting on organic deposits of Hoxnian aspect (Shotton, Banham and Bishop, 1977).

*Pleistocene history of the Middle and Lower Thames*

It is evident from this account that by 1939, the year in which *Structure, Surface and Drainage* first appeared, Wooldridge had already assembled all the essential components for a Pleistocene history of the Middle and Lower Thames, although his interest did not extend into the basin of the Upper Thames. Wooldridge had himself provided an interpretation of the early course and diversion of the Thames. To this he now added an account of the glacial and terrace sequences based on the concept of a multiglacial Pleistocene and owing its inspiration in part to the stratigraphic work of Solomon (1932) and King and Oakley (1936).

Wooldridge's main achievement in *Structure, Surface and Drainage* and in all subsequent revisions of the ideas presented there was the synthesis of the components into a convincing morphostratigraphic scheme for the Pleistocene, primarily of the Thames Basin, but potentially of the whole of extraglacial Britain. The foundation of this scheme was the belief that the surface of each river terrace could be treated as a stratigraphic horizon and that a stratigraphy based on this principle would prove superior to one based on the fossil record. In its historical context this approach has usually been seen as a major achievement in the development of geomorphology as an independent field of study (Linton, 1969). It can also be seen, however, as an original attempt to cope with the complex stratigraphical problems arising from the recognition of a multiglacial Pleistocene characterized by repeated glacio-eustatic fluctuations of sea-level.

## Recent developments

Since 1960 effort has been increasingly devoted to examination or re-examination of evidence in greater detail. Such studies have been mainly concerned with the Quaternary sediments of the Thames Basin and rarely with the detailed morphology of landforms.

## The terrace gravels

The papers of Hey (1965) and Walder (1967) are indicative of the change of emphasis from a morphological approach to more detailed studies of the terrace sediments. They also indicate a development of quantitative analysis in the study of gravel deposits. Hey was able to subdivide the 400 foot Pebble Gravel into an earlier 400 foot stage and a later Westland Green stage. He showed that substantial proportions of far-travelled material, mainly Triassic pebbles from the Midlands, appeared for the first time in the gravels of the Thames at the Westland Green stage, and that at that stage the Thames still occupied a course through the Vale of St Albans.

Walder (1967), differentiating between the Upper and Lower Winter Hill Terraces near Reading on the basis of pebble composition, indicated the possibility of relatively precise subdivision of the terrace stages using these methods. The present writers (Green and McGregor, 1978; McGregor and Green, 1978) have sought to refine further the study of gravel composition and shape by using intercomponent ratios rather than simple proportions of components, and by using a more elaborate shape criterion. Studies of gravels in the Middle Thames area and in the Vale of St Albans show that the Thames was not, as Wooldridge believed, diverted from its course through the Vale of St Albans into the Finchley Depression in the period immediately preceding the Higher Gravel Train stage, but continued to occupy the Vale throughout Gravel Train times.

## The glacial deposits

Since 1960, work on the chalky tills of southern East Anglia (Baker, 1971; Bristow and Cox, 1973; Perrin *et al.*, 1974; Rose and Allen, 1977), including tills in the basin of the Thames, has failed to establish any satisfactory basis for their subdivision, and a single glacial phase of Anglian age is now usually recognized in this region. Using sedimentological techniques, Rose (1974) and Gibbard (1977) have described in some detail the conditions near the edge of the Anglian ice, and Gibbard has suggested that the course of the Thames through the Vale of St Albans was blocked by Anglian ice and that as a result the river was diverted into its present course.

The reality of pre-Anglian glaciation in this region remains unproven. Kellaway *et al.* (1973) suggest a pre-Anglian age for the Chiltern Drift, but little has been added in recent years to earlier qualitative descriptions of this deposit; although Thomasson (1961) suggests that stony clays overlying the Pebble Gravel are most probably a "very ancient till".

The reality of pre-Anglian glaciation in the Thames Basin as a whole has received indirect attention in various studies of the terrace succession. The Westland Green gravels which represent a substantial influx into the Thames Basin of far-travelled material, almost certainly of glacial origin, are placed by Kellaway *et al.* (1973) in the pre-Anglian period. These gravels have been tentatively correlated by Hey (1965) with the highest parts of the Northern Drift of the Upper Thames area. A tentative correlation has also been proposed by Evans (1971) between the Lower Gravel Train of the Middle Thames and parts of the Northern Drift complex, which he regards as pre-Anglian (Baventian) in age. More recently Green and McGregor (1978) have suggested that as many as three pre-Anglian glacial episodes may be represented as influxes of far-travelled material into the gravels of the Middle Thames area. In the context of these opinions, the Northern Drift might be regarded as evidence of pre-Anglian glaciation. Such an interpretation is a restatement in modern terms of the views of Wills (1948) and Arkell (1943). It is not supported however by current interpretations of the evidence in the Midlands, and West (1963) and Shotton (1973) place the Northern Drift of the Upper Thames in the Anglian.

## The fossil record

Investigations into the fossil record, including Palaeolithic archaeology, have also developed in various important respects in the last twenty-five years. The study of pollen has made possible a reappraisal of many of the key localities, including to the east of London, Clacton (Pike and Godwin, 1953) and Ilford (West *et al.*, 1964); and in London itself, Trafalgar Square (Franks, 1960). Pollen analysis has also contributed to the significance of a number of newly discovered sites, notably at Hatfield, where the recognition of Hoxnian organic deposits overlying chalky till shows conclusively that the chalky till of the Vale of St Albans is of Anglian age or earlier; and at Sugworth Farm near Abingdon, where organic sediments of probable Cromerian age have been identified for the first time in the Thames valley (Briggs, *et al.*, 1975). These organic sediments rest on gravel containing pebbles derived from the Northern Drift, thus suggesting that the latter is of pre-Cromerian age. In addition controlled archaeological investigations at sites yielding Palaeolithic material (Waechter and Conway, 1971; Singer *et al.*, 1973; Kerney and Sieveking, 1977) have provided a wealth of new information on local depositional environments of the terrace sediments. Implement typology however, once a cornerstone of stratigraphic studies of the terrace sequence, is seen now to have serious limitations for stratigraphic purposes (Wymer, 1974).

## Towards a new model

Research findings during the last twenty years reflect the introduction of more rigorous research methods, a better understanding of geomorphological and sedimentological processes and of the processes of climatic change, and to a lesser extent the examination of new localities. Much new information has been obtained but problems of interpretation remain. Interpretation of the evidence has always been conditioned by prevailing views on the nature of the Quaternary record in the Thames valley. In this connection, influential developments were the recognition, about one hundred years ago, that the 'River Drift' could be subdivided into terraces, and the subsequent recognition that these terraces could be related to the stages of a multiglacial Pleistocene. As a result of the latter observation, study of the terrace sequence during the last fifty years has often had as an underlying aim the stratigraphic interpretation of the evidence.

Recent review papers (Evans, 1971; Brown, 1975; Hey, 1976; Clayton, 1977) show however that the terrace sequence recognized by Wooldridge has not been greatly modified in the last forty years, and that stratigraphic interpretation of it has also made surprisingly little progress. Only Evans (1971) has offered a radical reinterpretation of the terrace stratigraphy by relating it to a stratigraphy based on oceanic and astronomical evidence.

## Problems of stratigraphical interpretation

The problems of stratigraphical interpretation arise from several causes. Firstly, localities yielding good biological evidence of their stratigraphic position are few in number. Secondly, there are critical areas wherein evidence either is lacking or has not been studied; for example the built-up area of London and around the Goring Gap. Interpretative schemes have tended therefore to relate either to the Upper Thames, the Middle Thames, or the Lower Thames. Correlative links between these areas have been at best tenuous. Thirdly, the distribution of terrace remnants is fragmentary and the arrangement of the terrace sediments is obviously complex.

The complexity of the terrace sequence arises from two causes: (1) Terrace sediments are variable due to down-valley effects such as selective entrainment and comminution, and cross-valley effects such as meandering and the access of colluvial material from the valley side (McGregor and Green, 1978). (2) Terraces and terrace sediments of several types, reflecting a variety of environmental conditions, may be represented in the sequence.

## Factors influencing terrace development

The main environmental factors that might influence terrace formation and the significance that has been attached to them are described below.

### *Glacial isostasy*

Although glacial ice has unquestionably penetrated into the Thames Basin, no attention appears to have been given to the possiblity that river behaviour might have been modified by associated isostatic effects.

### *Tectonics*

Continuing subsidence of the North Sea area during the Pleistocene is now generally accepted and this process has been invoked to explain apparent anomalies of the terrace sequence, for example in Kent (Coleman, 1952) and in Essex (Kerney, 1971) In addition, local uplift has been suggested by West (1972) to explain the apparently anomalous elevation of supposed Ipswichian deposits at Ilford, and by Holmes (1965) to explain the diversion of the Thames from the Vale of St Albans.

*Sea-level*

Following the work of King and Oakley (1936), the interpretation of the Thames terraces has relied to an important extent on eustatic explanations. Aggradational terraces have been related to rising sea level and therefore to climatic amelioration. Erosional benches, especially those with steep seaward gradients, have been related to low sea-levels and often by inference to glacial stages.

*Climate*

In addition to the glacio-eustatic effects associated with climatic change, the possible significance of hydrological effects has also been recognized. Zeuner (1945) in particular argued that "in districts removed from the influence of the changing sea-level", climatic deterioration led to aggradation of the valley bottom, whereas climatic amelioration led to erosion. Such erosion was thought by Zeuner to produce a broad valley floor which would become the rock bench beneath a subsequent aggradational terrace.

## Previous interpretation

Despite the complexity of the terrace sequence, interpretation of the relationship between environmental conditions and terrace formation has been largely guided by the relatively uncomplicated ideas noted above. Wooldridge and Linton (1955) for example make only the most perfunctory reference to the environmental conditions in which the formation of terraces and terrace sediments occurs, confining themselves almost entirely to brief comments on the behaviour of sea-level.

Little attention has been paid to the detailed geomorphological implications of the biological and lithological evidence. Consequently little has been done to establish the nature of river activity at various stages in the Pleistocene, or to relate the pattern of river activity to the stratigraphic evidence. No comprehensive scheme of terrace development based on a geomorphological interpretation of the environmental evidence has gained general approval. Wymer (1968) has suggested such a scheme but the present state of uncertainty can be illustrated in a comparison between his views and those of Clayton (1977).

Wymer visualizes that rivers acquired greatly increased competence and capacity under periglacial climatic conditions and he proposes that broad valley floors were cut in these conditions at a level determined by the low sea-level of the glacial phase. Wymer regards the gravels overlying such valley floors as broadly contemporary with them. He emphasizes the substantial movement of colluvial material from the valley side into the valley bottom at this time. Finer deposits (brickearths) overlying the gravels are related by Wymer to bare ground conditions during the period of climatic amelioration marking the onset of the subsequent interglacial stage. During the main part of the interglacial stage Wymer argues that rivers occupied much reduced channels in the deposits of the previous periglacial stage, that erosional activity was negligible, but that well-stratified channel deposits and flood loams accumulated. With the onset of renewed periglacial conditions, the competence and capacity of the rivers were again increased, interglacial deposits were largely destroyed, or reworked and mixed with pre-existing periglacial sediments. As the influence of the low sea-level was felt progressively further upstream, valleys were cut below the largely reworked sediments at the interglacial level, leaving them as a terrace, and new valley floors were formed in relation to the new low sea-level.

Clayton (1977) appears to have a very different view of terrace formation, suggesting "that many terraces are not formed by aggradation of braided rivers, but by cut and fill by meandering channels" and "that most British river terraces are the dissected cut and fill floodplains of streams not unlike those forming floodplains today".

## Problems of geomorphological interpretation

Some of the main problems with which a scheme of terrace development must deal are next outlined.

(1) In some periods during the Pleistocene, the erosional activity of rivers led to valley deepening, but at other periods it led to valley widening and the formation of a broad valley floor.

(2) If the present geomorphological activity of the Thames and its tributaries is regarded as representative of their normal interglacial behaviour, and in terms of the environmental evidence this seems reasonable, then little valley development, either widening or deepening, has taken place in interglacial conditions.

(3) Substantial erosion, whether valley widening or valley deepening, appears usually to have been associated with the deposition of sands and gravels. The arrangement of these sediments indicates deposition by rivers having a braided channel habit. These sediments are unlike those now being transported and deposited by the Thames and its tributaries, which are mainly fine-grained and represent the channel and floodplain deposits of meandering rivers. Evidence of braided channel patterns and increased competence are not necessarily associated with evidence of steeper channel gradients and have therefore usually been explained in terms of increased discharge related to periglacial climatic conditions.

(4) Aggradations appear to have taken place under both interglacial and periglacial conditions.

(5) Sediments suggestive of both periglacial and interglacial conditions are found in various relationships within the deposits of a single morphological terrace, and sometimes within a single section.

(6) Alluvial sediments apparently of a single age may be found within the height range of more than one morphological terrace.

(7) Alluvial sediments of one age may be eroded at a later date to form the bench beneath later sediments.

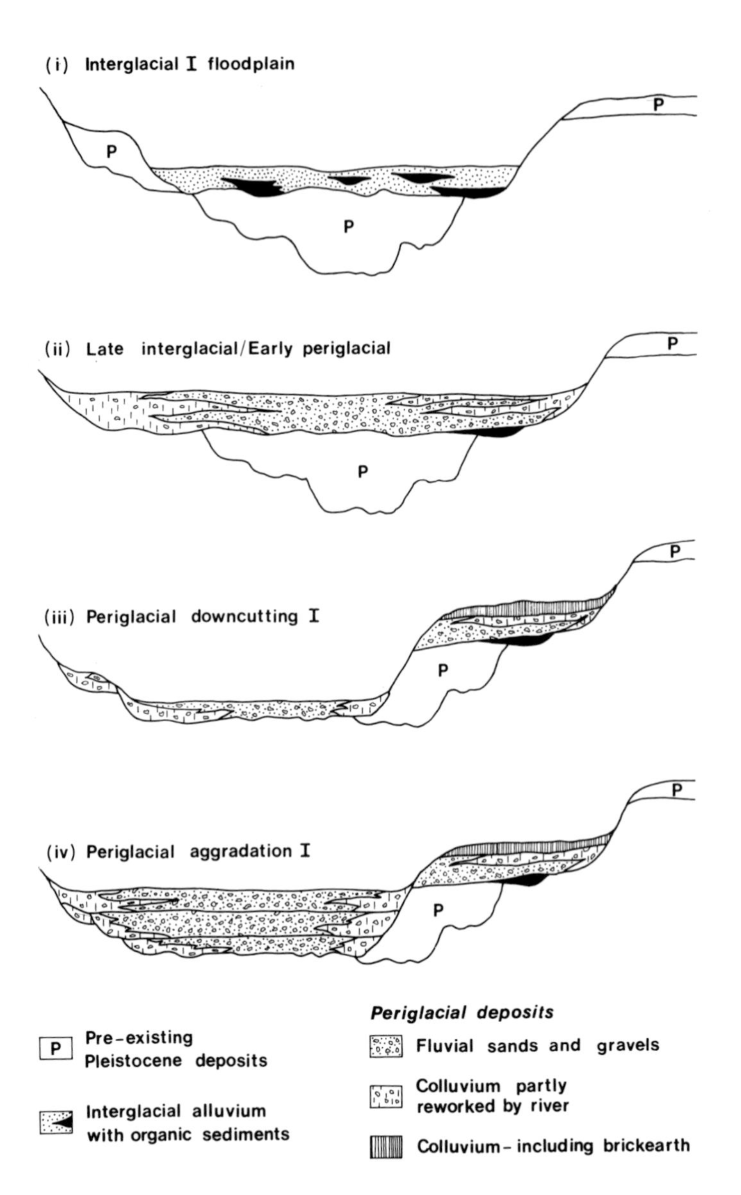


*Figure 6.* Scheme of terrace development.

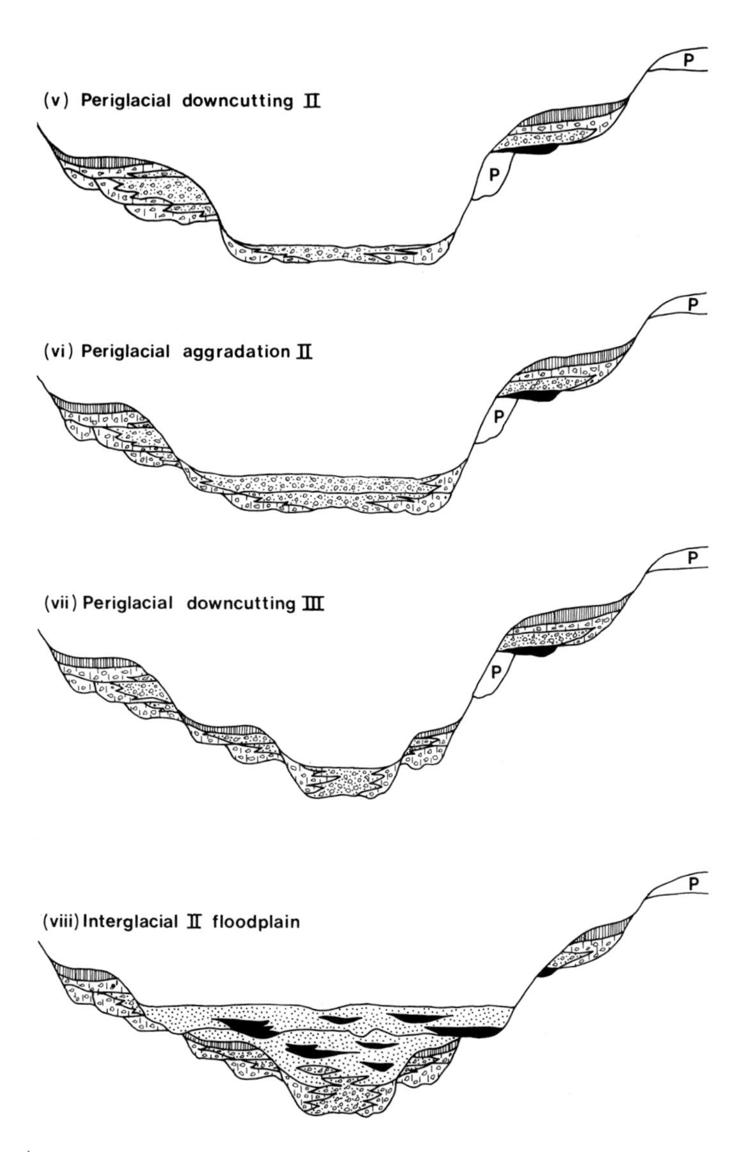
(v) Periglacial downcutting II
P
P
(vi) Periglacial aggradation II
P
P
(vii) Periglacial downcutting III
P
P
(viii) Interglacial II floodplain
P

*Figure 6.* (cont.)

## Development of the Thames Terrace model

In the final part of this paper a scheme of terrace development is proposed. This scheme which is shown in Fig. 6 has affinities with that of Wymer rather than Clayton but seeks to take into account more fully the great complexity of the terrace sequence.

In Fig. 6(i) an interglacial floodplain is illustrated. The alluvium is the product of meander activity and ranges from bedload material, mainly sand, to silt and clay deposits of overbank origin. Organic deposits are commonly present. The maximum depth of the alluvium laid down under full interglacial conditions, in a situation in which the river is neither aggrading nor degrading its channel, is probably equivalent to the difference between maximum channel scour and a level above the bankfull stage. The actual depth is therefore a function of channel size and in the Middle and Lower Thames might be several metres. This depth may be increased where minor environmental fluctuations cause minor phases of aggradation or downcutting at this time. In Fig. 6(i) the interglacial alluvium is shown resting on sediments which are the product of aggradation in the early part of the interglacial, and less extensively on a bedrock bench formed by the meander activity of the river during the interglacial period.

As a result of climatic deterioration marking the transition from interglacial to periglacial conditions, the discharge and load characteristics of the river change and cause a transformation to the braiding habit. Two results are apparent (Fig. 6(ii)). Firstly the pre-existing interglacial alluvium suffers substantial erosion. The thinner interglacial sequences are removed altogether and the thicker ones are truncated. Poorly-structured sands and gravels are deposited over the whole valley floor. Secondly, the valley floor is widened by lateral erosion, so that in many places the sands and gravels of this stage rest directly on a newly created rock bench. At this time large amounts of colluvial material may be supplied from the valley side, so that the valley floor sediments often grade from poorly sorted, poorly structured deposits near the valley side, to well structured sands and gravels near the valley axis (Briggs, 1976).

As a result either of further change in the hydrological conditions, or of falling base level associated with extensive glaciation in higher latitudes, the river subsequently cuts down (Fig. 6(iii)) and portions of the previous valley floor are preserved as a terrace. If downcutting is uninterrupted, a relatively narrow valley form may result, but if the process is checked, a broad valley floor is likely to develop. Deposition of sand and gravel persists in the new valley bottom. Periglacial slope processes may remain active, and may both contribute to the new valley floor deposits and continue to modify the character of the recently created terrace. Brickearth may form as a colluvial deposit on the terrace surface in situations where suitable source outcrops of fine-grained sediment are present in nearby terrace bluffs or on the valley side. Such material may be mixed with wind-blown dust, and to some extent with sediments of the terrace itself. If downcutting is discontinuous, a number of valley side benches may be created.

Under continuing periglacial conditions, further change of either hydrological conditions or base level may lead to aggradation, and considerable thicknesses of sand and gravel may be built up in the valley bottom (Fig. 6(iv)). Subsequent downcutting may leave portions of the surface of such an aggradational accumulation as a terrace (Fig. 6(v)). During a prolonged periglacial phase, both the aggradational and downcutting activity may be episodic in character, allowing scope for lateral erosion and the development of valley side benches, cut either in the solid rock or in pre-existing Pleistocene deposits (Figs. 6(iii)-(vii)).

Climatic amelioration marking the return from periglacial to interglacial conditions is accompanied by aggradation in response to either rising base level or changed hydrological conditions or both (Fig. 6(viii)). The latter part of this aggradational phase may be accomplished under full interglacial conditions, and the presence of fine-grained alluvium may indicate the re-establishment of a meandering channel habit. At the culminating level of this aggradation, assuming a continuation of interglacial conditions, the situation illustrated in Fig. 6(i) recurs.

Figure 6 shows that rock benches may be formed in a variety of environmental circumstances, and that the sediments overlying a rock bench may be either contemporary with its formation, or later, or both. Thus terrace form may disguise a number of stratigraphic sequences. Figure 6 is based on the premise, not necessarily correct, that interglacial floodplain deposits of later date are not found at levels above those of earlier date, nevertheless it shows that greater elevation is not invariably indicative of greater age.

## Conclusion

This paper has sought to establish the consensus of opinion regarding the Quaternary evolution of the Thames, not in terms of the detailed evidence, but by identifying the questions that have been asked, the extent to which these questions have been answered and, perhaps most important, the questions that have not been asked. We conclude that the most important questions that have been neglected in the last forty years, rather surprisingly considering the influence of Wooldridge during this period, relate to the geomorphological activity of rivers during the Pleistocene and to the environmental factors that influenced geomorphological processes. The lack of an adequate geomorphological interpretation of terrace formation undoubtedly hinders stratigraphic studies, and it can be argued that further refinement of the stratigraphic scheme must await a better understanding of the processes responsible for terrace formation. The early work of Wooldridge, emphasizing the orderly nature of terrace morphology, illustrates this point, but subsequent progress has been slow. As Clayton (1977) notes, the climatic approach to river terrace development, expounded by Zeuner (1945) and more fully discussed by Wymer (1968), has been largely ignored. In the scheme of terrace formation suggested in this paper, the emphasis is on the diversity of the processes involved, and on the corresponding diversity of interpretations that must be considered in studies of terraces and terrace sediments.

## References

Arkell, W. J. (1943). The Pleistocene rocks at Trebetherick Point, North Cornwall: their interpretation and correlation. *Proc. Geol. Ass.* **54**, 141-170.

Arkell, W. J. (1947). "The Geology of Oxford". Clarendon, Oxford.

Baden-Powell, D. F. W. (1948). The chalky boulder clays of Norfolk and Suffolk, *Geol. Mag.* **85**, 279-296.

Baker, C. A. (1971). A contribution to the glacial stratigraphy of western Essex, *Essex Naturalist* **32**, 317-330.

Barrow, G. (1919). Some future work for the Geologists' Association, *Proc. Geol. Ass.* **30**, 1-48.

Breuil, H. (1932). D l'importance de la solifluxion dans l'ĕtude des terrains quaternaires de la France et des pays voisins, *Revue Gĕogr. phys. Gĕol. dyn.* **7**, 269-331.

Briggs, D. J. (1976). Some Quaternary problems in the Oxford area, *in* Roe, D. (ed.), "Quaternary Research Association Field Guide to the Oxford Region".

Briggs, D. J., Gilbertson, D. D., Goudie, A. S., Osborne, P. J., Osmaston, H. A., Pettit, M. E., Shotton, F. W. and Stuart, A. J. (1975). New interglacial site at Sugworth, *Nature, Lond.* **257**, 477-479.

Bristow, C. R. and Cox, F. C. (1973). The Gipping Till: a reappraisal of East Anglian glacial stratigraphy, *J. Geol. Soc. Lond.* **129**, 1-37.

Bromehead, C. E. N. (1912). On diversions of the Bourne, Chertsey, *Summ. Prog. Geol. Surv. U.K.* for 1911, Appendix II, 74-77. H.M.S.O., London.

Brown, E. H. (1975). The Quaternary terraces of the River Thames, *in* Macar, P., "L'ĕvolution Quaternaire des bassins fluviaux de la Mer du Nord Mĕridionale", pp. 241-251. Societĕ Gĕologique de Belgique, Liege.

Brown, J. (1838). Discovery of a large pair of fossil horns in Essex, *Ann. Mag. Nat. Hist.* (New Series) **2**, 163-164.

Brown, J. (1840). Notice of a fluvio-marine deposit containing mammalian remains occurring in Clacton on the Essex coast, *Ann. Mag. Nat. Hist.* (New Series) **4**, 197-201.

Brown, J. C. (1959). The sub-glacial surface in eastern Hertfordshire and its relation to the valley pattern, *Trans. Inst. Br. Geogr.* **26**, 37-50.

Burchell, J. P. T. (1933). The Northfleet 50-foot submergence later than the coombe-rock of post-Early Mousterian times, *Archaeologia* **83**, 67-92.

Burchell, J. P. T. (1935). Evidence of a further glacial episode within the valley of the lower Thames, *Geol. Mag.* **72**, 90-91.

Bury, H. (1922). Some high level gravels of north-east Hampshire, *Proc. Geol. Ass.* **33**, 81-103.

Clayton, K. M. (1957). Some aspects of the glacial deposits of Essex, *Proc. Geol. Ass.* **68**, 1-21.

Clayton, K. M. (1960). The landforms of parts of southern Essex, *Trans. Inst. Br. Geogr.* **28**, 55-74.

Clayton, K. M. (1977). River terraces, *in* Shotton, F. W., "British Quaternary studies, Recent Advances", pp. 153-168. Oxford University Press.

Clayton, K. M. and Brown, J. C. (1958). The glacial deposits around Hertford, *Proc. Geol. Ass.* **63**, 103-119.

Coleman, A. (1952). Some aspects of the development of the lower Stour, Kent, *Proc. Geol. Ass.* **63**, 63-86.

Commont, V. (1910). Note préliminaire sur les terraces fluviatiles de la vallée de la Somme, *Annls Soc. géol. N.* **39**, 185-210.

Cotton, R. P. (1847). On the Pliocene deposits of the valley of the Thames at Ilford, *Ann. Mag. Nat. Hist.* (New Series) **20**, 164-169.

Dewey, H. and Bromehead, C. E. N. (1921). The geology of South London, *Mem. Geol. Surv. U.K.*

Evans, J. (1872). "The ancient stone implements, weapons and ornaments of Great Britain". Longmans, London.

Evans, P. (1971). Towards a Pleistocene time-scale, *in* "The Phanerozoic time-scale—a supplement". Part 2. Geol. Soc. Lond. Sp. Pub. No. 5, pp. 123-356.

Franks, J. W. (1960). Interglacial deposits at Trafalgar Square, London, *New Phytol.* **59**, 145-52.

Gibbard, P. L. (1977). Pleistocene history of the Vale of St Albans, *Phil. Trans. Roy. Soc.* **B280**, 445-483.

Gilbert, C. J. (1919). On the occurrence of extensive deposits of high-level sands and gravels resting upon Chalk at Little Heath, near Berkhamsted, *Q. J. Geol. Soc. Lond.* **75**, 32-50.

Green, C. P. and McGregor, D. F. M. (1978). Pleistocene Gravel Trains of the River Thames, *Proc. Geol. Ass.* **89**, 143-156.

Gregory, J. W. (1922). "The evolution of the Essex rivers and of the lower Thames". Benham, Colchester.

Hare, F. K. (1947). The geomorphology of a part of the middle Thames, *Proc. Geol. Ass.* **58**, 294-339.

Hey, R. W. (1965). Highly quartzose Pebble Gravels in the London Basin, *Proc. Geol. Ass.* **76**, 403-420.

Hey, R. W. (1976). The terraces of the middle and lower Thames, *Studia Soc. Sci. torun.* **8** (Sectio C), 115-122.

Hicks, H. (1891). On some recently exposed sections in the glacial deposits at Hendon, *Q. J. Geol. Soc. Lond.* **47**, 575-584.

Holmes, A. (1965). "Principles of Physical Geology", 2nd edn. Nelson, London.

Holmes, T. V. (1892). The new railway from Greys Thurrock to Romford. Sections between Upminster and Romford, *Q. J. Geol. Soc. Lond.* **48**, 365-372.

Holmes, T. V. (1894). Further notes on some sections on the new railway from Romford to Upminster and on the relations of the Thames valley beds to the boulder clay, *Q. J. Geol. Soc. Lond.* **50**, 443-452.

Kellaway, G. A., Worssam, B. C., Holmes, S. C. A., Kerney, M. P. and Shephard-Thorn, E. R. (1973). South-east England, *in* Mitchell, G. F., Penny, L. F., Shotton, F. W. and West, R. G., "A correlation of Quaternary deposits in the British Isles". Geol. Soc. Lond. Sp. Rep. No. 4.

Kerney, M. P. (1971). Interglacial deposits in Barnfield pit, Swanscombe and their molluscan fauna, *J. Geol. Soc. Lond.* **127**, 69-86.

Kerney, M. P. and Sieveking, G. deG. (1977). Northfleet, *in* Shephard-Thorn, E. R. and Wymer, J. J., "South-east England and the Thames Valley". Tenth INQUA Congress; guidebook for excursion A5, pp. 44-46.

King, W. B. R. and Oakley, K. P. (1936). The Pleistocene succession in the lower part of the Thames valley, *Proc. prehist. Soc.* **2**, 52-76.

Lacaille, A. D. and Oakley, K. P. (1936). The Palaeolithic sequence at Iver, Bucks, *Antiquaries J.* **16**, 420-443.

Leach, A. L. (1912). On the geology of Shooter's Hill, Kent, *Proc. Geol. Ass.* **23**, 112-124.

Linton, D. L. (1969). The formative years in geomorphological research in south-east England, *Area* **1**, 1-8.

Marston, A. T. (1937). The Swanscombe skull, *J. Roy. anthrop. Inst.* **67**, 339-406.

McGregor, D. F. M. and Green, C. P. (1978). Gravels of the River Thames as a guide to Pleistocene catchment changes, *Boreas* **7**, 197-203.

Mitchell, G. F., Penny, L. F., Shotton, F. W. and West, R. G. (1973). "A correlation of Quaternary deposits in the British Isles". Geol. Soc. Lond. Sp. Rep. No. 4.

Morris, J. (1838). On the deposits containing Carnivora and other mammalia in the valley of the Thames, *Ann. Mag. Nat. Hist.* (New Series) **2**, 540-546.

Perrin, R. M. S., Davies, H. and Fysh, M. D. (1973). The lithology of the Chalky Boulder Clay, *Nature phys. Sci.* **245**, 101-104.

Pike, K. and Godwin, H. (1953). The interglacial at Clacton-on-sea, Essex, *Q. J. Geol. Soc. Lond.* **108**, 261-272.

Pocock, T. I. (1908). The geology of the country around Oxford, *Mem. Geol. Surv. U.K.*

Rose, J. (1974). Small scale spatial variability of some sedimentary properties of lodgement and slumped till, *Proc. Geol. Ass.* **85**, 223-237.

Rose, J. and Allen, P. (1977). Middle Pleistocene stratigraphy in south-east Suffolk, *J. Geol. Soc. Lond.* **133**, 83-102.

Ross, B. R. M. (1932). The physiographic evolution of the Kennett-Thames, *Rep. Br. Ass. Advmt Sci.* for 1931 (Section C) p. 386.

Salter, A. E. (1898). Pebbly and other gravels in southern England, *Proc. Geol. Ass.* **15**, 264-286

Salter, A. E. (1905). On the superficial deposits of central and parts of southern England, *Proc. Geol. Ass.* **19**, 1-56.

Sandford, K. S. (1924). The river gravels of the Oxford district, *Q. J. Geol. Soc. Lond.* **80**, 113-179.

Sandford, K. S. (1926). Pleistocene deposits, *in* Pringle, I., "The geology of the country around Oxford". *Mem. Geol. Surv. U.K.*

Saner, B. R. and Wooldridge, S. W. (1929). River development in Essex, *Essex Naturalist* **22**, 244-250.

Sealy, K. R. and Sealy, C. E. (1956). The terraces of the middle Thames, *Proc. Geol. Ass.* **67**, 369-392.

Sherlock, R. L. (1924). The superficial deposits of south Buckinghamshire and south Hertfordshire, and the old course of the Thames, *Proc. Geol. Ass.* **35**, 1-28.

Sherlock, R. L. and Noble, A. H. (1912). On the origin of the Clay-with-Flints of Buckinghamshire and on a former course of the Thames, *Q. J. Geol. Soc. Lond.* **68**, 199-212.

Shotton, F. W. (1953). Pleistocene deposits of the area between Coventry, Rugby and Leamington and their bearing on the topographic development of the Midlands, *Phil. Trans. Roy. Soc.* **237B**, 209-260.

Shotton, F. W. (1973). English Midlands, *in* Mitchell, G. F., Penny, L. F., Shotton, F. W. and West, R. G., "A correlation of Quaternary deposits in the British Isles". Geol. Soc. Lond. Sp. Rep. No. 4.

Shotton, F. W., Banham, P. H. and Bishop, W. W. (1977). Glacial-interglacial stratigraphy of the Quaternary in Midland and eastern England, *in* Shotton, F. W., "British Quaternary studies, Recent Advances", pp. 267-282. Oxford University Press.

Singer, R., Wymer, J. J., Gladfelter, B. and Wolff, R. (1973). Excavation of the Clactonian industry of the golf course, Clacton-on-sea, Essex, *Proc. prehist. Soc.* **39**, 6-74.

Solomon, J. D. (1932). The glacial succession on the north Norfolk coast, *Proc. Geol. Ass.* **43**, 241-271.

Sparks, B. W., West, R. G., Williams, R. B. G. and Ransom, M. (1969). Hoxnian interglacial deposits near Hatfield, Herts, *Proc. Geol. Ass.* **80**, 243-267.

Thomasson, A. J. (1961). Some aspects of the drift deposits and geomorphology of south-east Hertfordshire, *Proc. Geol. Ass.* **72**, 287-302.

Treacher, L. (1926). Excursion to Shiplake, *Proc. Geol. Ass.* **37**, 440-442.

Waechter, J. d'A. and Conway, B. W. (1971). Swanscombe 1971, *Proc. Roy. anthrop. Inst. 1971*, 73-85.

Walder, P. S. (1967). The composition of the Thames gravels near Reading, Berkshire, *Proc. Geol. Ass.* **78**, 107-119.

Walker, H. (1871). On the glacial drifts of north London, *Proc. Geol. Ass.* **2**, 289-298.

Warren, S. H. (1916). The late glacial stage of the Lea valley, *Q. J. Geol. Soc. Lond.* **71**, 164-182.

Warren, S. H. (1957). On the early Pebble Gravels of the Thames Basin from the Hertfordshire-Essex border to Clacton-on-sea, *Geol. Mag.* **94**, 40-46.

West, R. G. (1963). Problems of the British Quaternary, *Proc. Geol. Ass.* **74**, 147-186.

West, R. G. (1972). Relative land-sea-level changes in south-eastern England during the Pleistocene, *Phil. Trans. Roy. Soc.* **272A**, 87-98.

West, R. G. and Donner, J. J. (1956). The glaciations of East Anglia and the Midlands: a differentiation based on stone orientation measurements of the tills, *Q. J. Geol. Soc. Lond.* **112**, 69-91.

West, R. G., Lambert, C. A. and Sparks, B. W. (1964). Interglacial deposits at Ilford, Essex, *Phil. Trans. Roy. Soc.* **247B**, 185-212.

Whitaker, W. (1889). The geology of London and of part of the Thames Valley, *Mem. Geol. Surv. U.K.*

White, H. J. O. (1895). On the distribution and relations of the Westleton and glacial gravels in parts of Oxfordshire and Berkshire, *Proc. Geol. Ass.* **14**, 11-30.

White, H. J. O. (1897). On the origins of the high-level gravel with Triassic debris adjoining the valley of the Upper Thames, *Proc. Geol. Ass.* **15**, 157-174.

White, H. J. O. (1908). The geology of the country around Henley, *Mem. Geol. Surv. U.K.*

Wills, L. J. (1948). "The palaeogeography of the Midlands". Liverpool University Press.

Wooldridge, S. W. (1927). The Pliocene history of the London Basin, *Proc. Geol. Ass.* **38**, 49-132.

Wooldridge, S. W. (1928). The 200-ft platform in the London Basin, *Proc. Geol. Ass.* **39**, 1-26.

Wooldridge, S. W. (1938). The glaciation of the London Basin and the evolution of the Lower Thames drainage system, *Q. J. Geol. Soc. Lond.* **94**, 627-667.

Wooldridge, S. W. (1958). Some aspects of the physiography of the Thames valley in relation to the Ice Age and early man, *Proc. prehist. Soc.* **23**, 1-19.

Wooldridge, S. W. (1960). The Pleistocene succession in the London Basin, *Proc. Geol. Ass.* **71**, 113-129.

Wooldridge, S. W. and Henderson, H. C. K. (1955). Some aspects of the physiography of the eastern part of the London Basin, *Trans. Inst. Br. Geogr.* **21**, 19-31.

Wooldridge, S. W. and Linton, D. L. (1939). Structure, surface and drainage in south-east England, *Trans. Inst. Br. Geogr.* No. 10, 124 pp.

Wooldridge, S. W. and Linton, D. L. (1955). "Structure, surface and drainage in south-east England". George Philip, London.

Wymer, J. J. (1968). "Lower Palaeolithic archaeology in Britain: as represented by the Thames valley". John Baker, London.

Wymer, J. J. (1974). Clactonian and Acheulian industries in Britain. Their chronology and significance, *Proc. Geol. Ass.* **85**, 391-421.

Zeuner, F. E. (1945). The Pleistocene period: its climate, chronology and faunal successions, *Ray. Soc. Pub.* No. 130.

EIGHT

# Late Quaternary palaeohydrology: the Kennet Valley case-study

G. HENRY CHEETHAM
*Department of Geography, University of Reading*

## Introduction

Although the work of Wooldridge and Linton (1939) was pioneering in its attempt to explain the widespread occurrence of river terraces in southern Britain by reference to drainage development on a regional scale, their work can be fairly criticised for the general failure to explain the environmental implications. Since the publication of their research a considerable amount of modern process data has amassed which now enables us to make a more realistic evaluation of terrace forming events.

Palaeoenvironmental analysis is not a straightforward task for which there is a single all-encompassing technique. Reconstruction is piecemeal, dependent not so much on an adequate information base of modern process data, but restricted more by the availability of field evidence. Consequently, it is appropriate to adopt a case-study approach to analyse in depth the available palaeoenvironmental evidence from a specified area and attempt to integrate such information into a meaningful reconstruction.

The Kennet Valley in Berkshire is suitable for this purpose as various complementary types of field evidence are available. In addition to being a key area in the study of Wooldridge and Linton, the Kennet Valley was also in part the subject of the pioneering palaeohydrological studies of Dury who described the present-day River Kennet as underfit (Dury, 1964; p. A46).

Dury (1965) concluded that the underfit state of rivers in Lowland Britain indicates at least a twenty times decrease in bankfull discharge since late Quaternary times. Tinkler (1971), however, considered this difference to be considerably overstated according to a re-examination of active meandering valleys. He concluded that a three to five times difference would be a more realistic interpretation.

## Study area

### Location and geology

The River Kennet rises on the dip slope of the Chalk escarpment west of Marlborough in Wiltshire and flows in an easterly direction as far as Reading where it joins the Thames (Fig. 1). The catchment area consists of Upper Cretaceous and Eocene strata. Lower and Middle Chalk outcrop in the upper part of the catchment towards the drainage divide, where there are also restricted outcrops of Upper Greensand. In the middle and lower parts of the catchment, these are overlain by Upper Chalk, Reading Beds, London Clay and Lower Bagshot Beds, although frequently obscured by more recent surficial deposits. The Lower Bagshot Beds are essentially sandy deposits, whereas the Reading Beds are lithologically much more varied, but generally composed of sand in the lower part of the sequence and clay in the upper.

The Quaternary history of the Kennet Valley is evidenced by a sequence of gravel spreads and river terraces, originally recognized in part by Osborne White (1907) and subsequently mapped in more detail by Thomas (1957, 1961) who attempted to correlate the Kennet Terraces with those of the Middle Thames. This chapter is primarily concerned with the lower part of the alluvial sequence, namely the Lower Taplow and Floodplain Terraces of Thomas (1961), approximately 10 m and 2 m above the floodplain respectively.

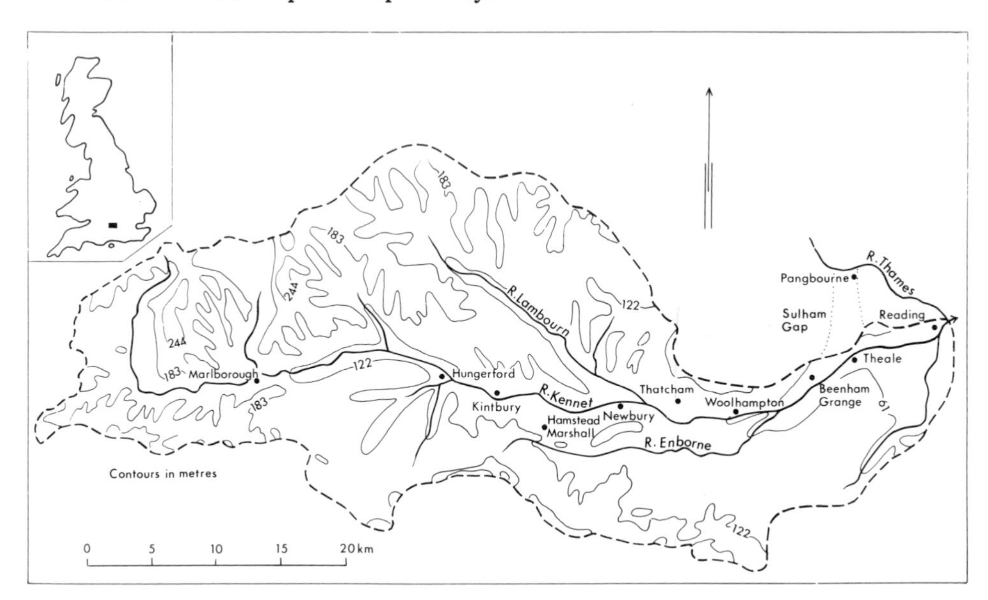


*Figure 1.* The Kennet catchment: location and sites.

### Terrace morphology

Valley bottom terraces of the Kennet are largely confined to the area downstream of Hungerford (Fig. 1) where the valley broadens out. Two distinct terraces were identified from aerial photographs and field surveys.

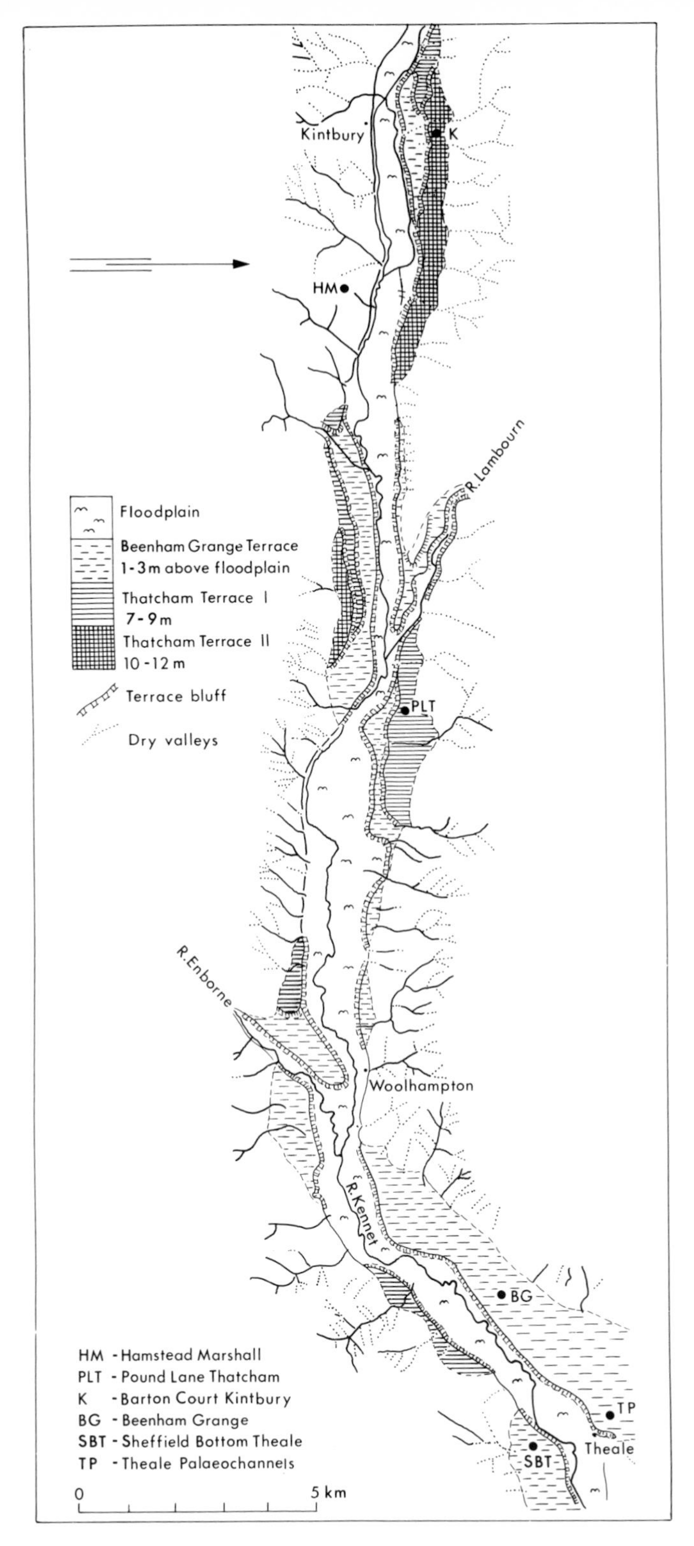


*Figure 2.* Valley bottom terraces of the Kennet between Kintbury and Theale.

The higher terrace displays minor variations in height sufficient to warrant sub-division into an upper and lower part, 10-12 m and 7-9 m above the floodplain respectively. The upper is morphologically distinct from the lower on account of steeper in-valley slopes.

The lowest terrace is situated approximately 2 m above the floodplain. In the eastern end of the valley, separation of this low terrace from the floodplain proved to be difficult on a purely altimetric basis. Although this terrace surface can be traced as a continuous spread of gravel in this area (Fig. 2), it becomes morphologically indistinguishable from the floodplain at the easternmost end of the valley. This is because the terrace has a much lower gradient than the floodplain. Since the terrace is largely composed of gravels and the floodplain of peat and tufa deposits, separation of the two was based on contrasting pedological development reflecting the different underlying material.

Thomas (1961) attempted to correlate the Kennet Terraces to the Thames Valley and accordingly applied Thames Valley terminology. However, in view of the speculative nature of many of these correlations, in the present instance it was considered more appropriate to use local terminology as listed below. Type locality descriptions are detailed in Cheetham (1975) and Chartres (1975).

| *Terrace* | *Height above floodplain (m)* |
|---|---|
| Floodplain | - |
| Beenham Grange | 1-3 |
| Thatcham I | 7-9 |
| Thatcham II | 10-12 |

## Hydrology of the modern Kennet

To provide a present-day comparative basis for palaeodischarge estimates, hydrological records of the River Kennet at the Theale gauging station (SU 646704) were examined. Records date from October 1961. For the subsequent thirteen-year period of record, the mean annual flood is 40 $m^3 s^{-1}$ with a recurrence interval of approximately 2 years and bankfull discharge is 35 $m^3 s^{-1}$ assuming a 1·56 year recurrence interval.

## Terrace sedimentology

Terrace sedimentology in the present context is a useful palaeohydrological tool for two reasons: firstly, it enables one to specify the general nature of the fluvial environment; and secondly, by detailed analysis, it is possible to identify the type of channel which deposited the sediments. To this end, empirical data such as the discriminant relationship between meanders and braids based on channel slope and bankfull discharge (Leopold and Wolman, 1957) can be used to form the basis of a palaeohydrological reconstruction. This may then be refined by additional techniques, such as the analysis of palaeochannels and drainage densities.

## Thatcham I Terrace

The terrace deposits of Thatcham I were well exposed by deep excavations at Pound Lane Thatcham (SU 499673). The sediments consist largely of sandy gravels capped by 50-100 cm silt loam, and range in thickness from 2·5 m to 4·0 m reflecting a gently undulating basal unconformity with the Tertiary. However, subtle matrix variations occur in the gravels. Characteristic sedimentological features are illustrated on the profile section (Fig. 3). At the base of the section between 202 cm and 293 cm there are three couplets of alternating openwork gravel overlain by sandy gravel. The final cycle is overlain by a 12 cm thick bed of medium sand, then 110 cm of massive sandy gravel which exhibits no indications of structure or bedding. Finally, the sequence is capped by a sandy unit, within which a textural break at 50 cm separates a siltier topsoil. Fabric studies down the profile indicated that cryoturbation had not affected the sediments to any great extent, and that the massive sandy-gravel unit is essentially a primary alluvial facies (Cheetham, 1975), although micromorphological analysis indicated some disturbance in the top part of the profile (Chartres, 1975).

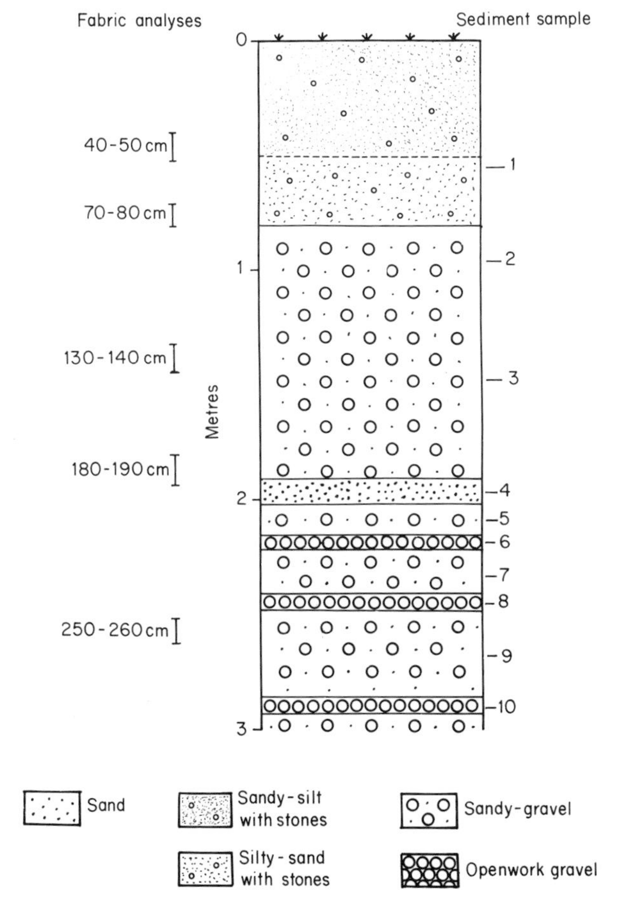


*Figure 3.* Characteristic profile section through Thatcham Terrace I at Pound Lane Thatcham.

This type of alluvial sequence bears strong resemblances to the internal structure of braided stream longitudinal bars (McDonald and Banerjee, 1971; Smith, 1974). Specifically, the openwork gravel cycles in the lower part of the sequence are indicative of a stage-sensitive emergent island which subsequently grew to a mature bar by lateral sedimentation producing a massive sandy-gravel facies. Subsequently, this was capped by sand which was probably fed to the upper surface of the bar at higher discharge.

## Thatcham II Terrace

Thatcham II Terrace, as exposed at the Barton Court gravel pit, Kintbury (SU 384683) consists of the following sequence of deposits (in stratigraphic order from the surface):

Unit 1 Sandy silt, up to 1·5 m thick;
Unit 2 Massive silty clay gravel, 2-4 m thick;
Unit 3 Bedded sandy gravel, 2-4 m thick;
Unit 4 Chalky rubble, >2 m thick.

The poorly sorted silty-clay of Unit 2 is interpreted as a solifluction deposit (Cheetham, 1975). It occurs as an extensive downslope sheet overriding sediments of a more obvious fluvial origin (Unit 3) with which it frequently has a sharp contact (Fig. 4). In the upper part of the solifluction deposit, Chartres (1975)

*Figure 4.* Solifluction gravel overlying fluvial gravel separated by a sharp, gently sloping contact at Barton Court, Kintbury.

identified palaeoargillic features indicative of a palaeosol, which was subsequently buried by a sandy silt deposit believed to be of aeolian, or possibly slopewash, origin.

The unit of greatest interest to the present study is Unit 3, the sandy gravel. This unit exhibits bedding features indicative of fluvial deposition. As at Pound Lane, Thatcham, the sediments are characterized by interbedded gravel facies which have distinctive matrix properties. At Kintbury, however, the organization of these facies was found to be much more complex. Three main types of gravel facies could be identified (Fig. 5), namely coarse sandy gravel, clayey gravel and openwork gravel, interbedded with occasional sand and silt lenses.

The dominant facies is the sandy gravel. Within it are thin, discontinuous beds and lenses of openwork gravel and silty-clay matrix gravel. In many cases the openwork gravels occur as small channel-like lenses. Overall, the section shows very little systematic sedimentological organization, although the sandy gravel tends to be slightly coarser at the base of the sequence.

Such a facies arrangement clearly suggests a complex alluvial system. The juxtaposition of sandy gravels, openwork gravels and clayey gravels are indicative of a system in which stage changes had the effect of isolating certain areas from the main flow and produced stage-sensitive regions where sediment winnowing was operative. Such sedimentological features are entirely compatible with braided stream processes.

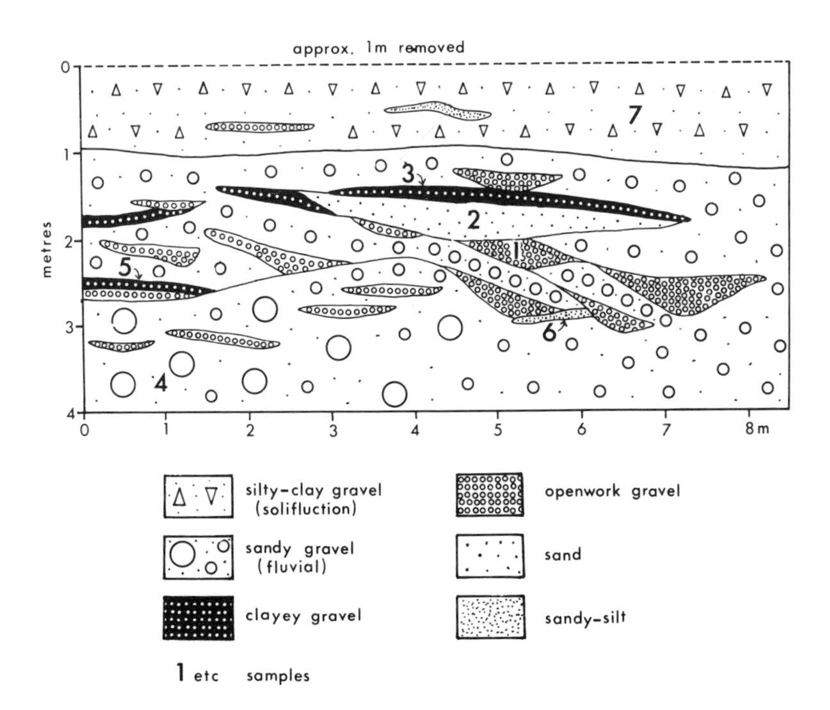


*Figure 5.* Section through the sediments of Thatcham Terrace II at Barton Court, Kintbury. Numbers 1 to 7 refer to the samples.

## Beenham Grange Terrace

The sediments of the Beenham Grange Terrace were studied in detail at Sheffield Bottom gravel pit, Theale (SU 654699) situated at the eastern end of the Kennet Valley (Fig. 2) where 2-5 m of essentially sandy gravels rest on Reading Beds. As regards the age of this terrace, the analysis of mollusca supported by radio-carbon dates from the contemporary floodplain (Cheetham, 1975) demonstrated that the Beenham Grange Terrace had been abandoned as an active alluvial surface by early Flandrian times.

In common with other terrace exposures in the Kennet Valley, the gravels were characterized by *matrix facies.* Lateral tracing of these facies along the sections was hindered by talus, hence it was decided to examine three closely-spaced profile sections arranged as shown on Fig. 6.

These three sections show a complex interrelationship of a wide variety of gravel facies. Unlike previous sections described, the facies are a little more subtle, although five types are clearly identifiable: (1) sandy gravel; (2) coarse gravel with sand; (3) low matrix content sandy gravel; (4) low matrix content sandy clay gravel; and (5) openwork gravel. The only consistent arrangement displayed

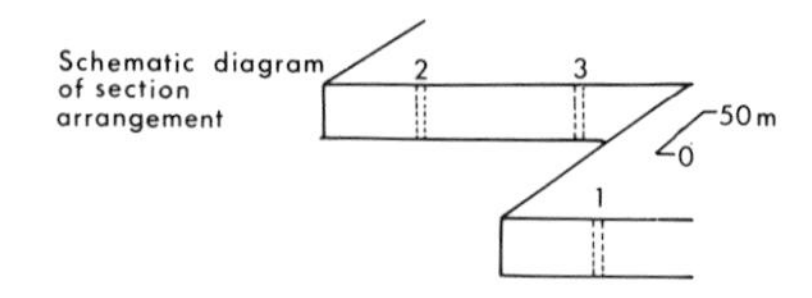


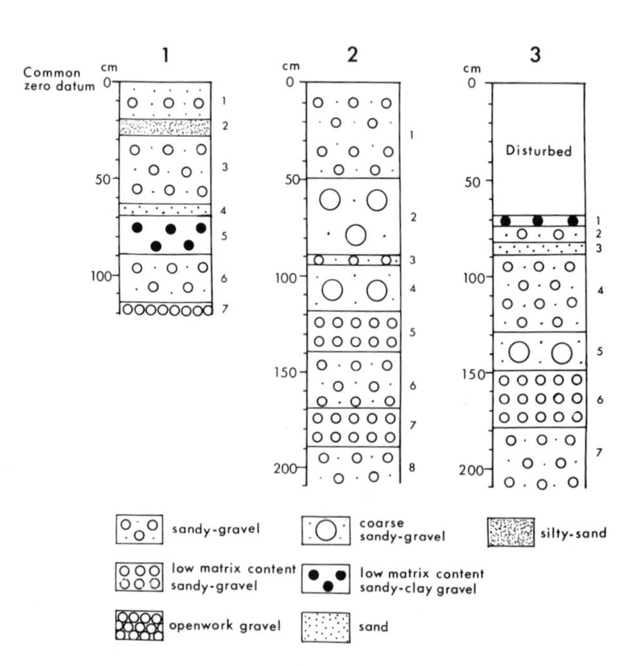


*Figure 6.* Profile sections through the Beenham Grange Terrace at Sheffield Bottom, Theale. Numbers 1 to 7 refer to the samples.

by these sequences is the sandy gravel facies overlying low matrix content sandy or sandy clay gravel.

Particle size analysis (Cheetham, 1975) showed that the basal sandy gravel and overlying openwork gravel of Section 1 are practically identical except for the virtual absence of sand and a higher clay content of the latter. This would suggest a period of winnowing followed by the deposition of fine suspended material, such as produced by discharge fluctuations in a partially abandoned area. This sequence is then overlain by a succession of sand, sandy gravel, silty sand and finally poorly sorted sandy gravel which probably reflects spatial changes in channel pattern.

Section 2, however, presents a slightly different situation. The sequence is composed predominantly of a coarser gravel with less indication from the matrix contents of abandonment and winnowing processes. This section more readily suggests sedimentation of a major aggrading channel.

Section 3 is similar to Section 2 as regards textural irregularities but there are notable differences. The gravel fractions are less coarse and the 'fines' percentages are greater. Consequently, a peripheral position to a major channel is envisaged.

The general picture indicated by these sections is a complex alluvial environment of shifting channels producing abandoned areas. Sections 2 and 3 suggest an active channel position, whereas Section 1 is more consistent with a partially abandoned channel showing possible effects of re-occupation. The upper part of the sequence changes character which perhaps reflects channel reorganization. Such features characterize a braided stream environment. On the surface of this terrace 2·5 km to the north-west, braided palaeochannels are in fact preserved. These will be examined in a later section to refine the general palaeohydrological conclusions based on the incidence of braiding.

## Palaeohydrological implications

Leopold and Wolman (1957) demonstrated that meandering and braided channels can be separated by the discriminant function $S=0{\cdot}012\ Q^{-0{\cdot}44}$ where $S$ is the channel slope in m $m^{-1}$ and $Q$ is bankfull discharge in $m^3\ s^{-1}$. For a given slope, braided channels have a higher discharge than meandering channels, thus this relationship provides a methodology for estimating the minimum bankfull discharge of a braided stream. Cant and Walker (1976) adopted this method to reconstruct the palaeohydrology of the Devonian Battery Point Sandstone in Quebec. The preceding analysis of the sedimentology of the Kennet Valley Terraces also provided distinct indications of braided stream sedimentation, hence the Leopold and Wolman relationship can usefully be applied to attempt a palaeohydrological reconstruction. Channel slope can be approximated by terrace slope allowing for a low channel sinuosity, arbitrarily taken as 1·2.

Since Leopold and Wolman's relationship is not a perfect discriminant line, their data were re-examined to enable probability limits of palaeodischarge estimates to be specified. The relationship between bankfull discharge, $Q_{bf}$, and channel slope, $S$, obtained by least-squares regression, for their braided channel data only is:

$$Q_{bf} = 0{\cdot}000585\, S^{-2{\cdot}01},$$

where $r = 0{\cdot}948$, SE $= 0{\cdot}3964$ m$^3$ s$^{-1}$ in $\log_{10}$ units.

The application of this relationship in terms of mean and minimum bankfull discharge within a 95 per cent probability range is summarized on Table I, together with estimates from Leopold and Wolman's discriminant function. The latter indicates that bankfull discharge of the River Kennet was at least three to four times greater than the present-day Kennet, when each terrace was deposited. However, according to the relationship based solely on braided channels, there is a 95 per cent probability that the bankfull discharges associated with the terraces were approximately two times greater than the modern river, although mean estimates are considerably greater. These general probability limits on the palaeo-discharge of the Kennet can be refined by further analysis first of preserved palaeochannels and then drainage densities.

TABLE I

*Palaeodischarge calculations for the Kennet Valley Terraces based on the data of Leopold and Wolman (1957)*

| Terrace | Slope | Palaeo-bankfull discharge (× greater than modern Kennet) | | |
|---|---|---|---|---|
| | | (i) $S = 0{\cdot}012\, Q_{bf}^{-0{\cdot}44}$ | (ii) $Q_{bf} = 0{\cdot}000585\, S^{-2{\cdot}01}$ | |
| | | | Mean | Mean − 2 SE |
| Thatcham | 0·0015 | 3·2 | 11·2 | 1·8 |
| Beenham Grange | 0·0014 | 3·8 | 12·9 | 2·1 |

## Theale palaeochannels

At the eastern end of the Kennet Valley, a series of palaeochannels are preserved on the surface of the Beenham Grange Terrace (Fig. 2). Because they are infilled with finer sediment than the surrounding alluvial surface, they appear as distinct crop patterns on aerial photographs and were first noted during archaeological investigations by Wymer (1968). Further aerial survey revealed various morphological types (Figs 7, 8 and 9) and extensive ground survey by augering, aided by a pipeline cross-section, confirmed their braided form (Cheetham, 1976).

*Figure 7.* Bronze age ring ditches adjacent to a large palaeochannel on the Beenham Grange Terrace (at SU 624703) near Theale.

## Palaeochannel morphology

Essentially two main types of channel were identified. The first and most obvious type is the relatively narrow (up to 10 m wide) low sinuosity channel formed around large islands (Fig. 8). The second type consists of a complex series of channels and bars, arranged within one wide channel belt (Fig. 9). A similar morphological distinction was also observed in a present-day braided stream environment in north Norway (Cheetham, 1979), where the complex channel and bar networks were seen to be part of the active channel belt, and the narrower and better defined channels formed part of a high stage or flood-routing network. On the basis of this present-day analogy, it is suggested that the well-defined channels of Fig. 8 are flood channels associated with the active channel belt composed of a channel and bar complex of the type shown in Fig. 9.

Attention was focused primarily on the latter in one particular area where a pipeline cross-section cut across the terrace surface exposing the entire depth of the silt-filled palaeochannels immediately west of Theale Green School (SU 633709).

*Figure 8.* Braided palaeochannels on the Beenham Grange Terrace (at SU 618692) near Theale.

*Figure 9.* Complex series of braided channels and low relief bars preserved on the surface of the Beenham Grange Terrace at Theale Green School (SU 633709).

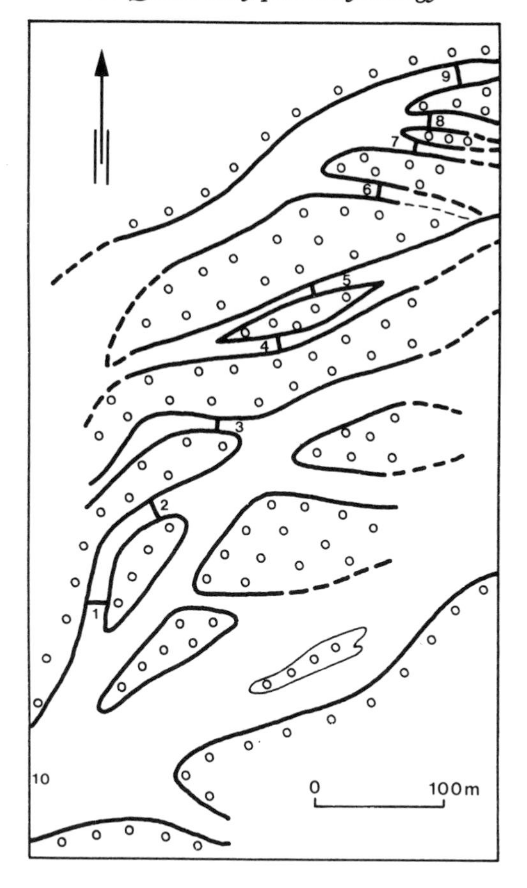


*Figure 10.* Plan geometry of palaeochannels immediately west of Theale Green School (SU 633709). Numbers refer to cross-sections subsequently corrected to true channel width on Table II.

Augering on a 10 m × 10 m grid around this cross-section in the adjacent area was undertaken to establish the three-dimensional form of these channels. From this a complex series of low-relief channels and bars emerged, arranged as longitudinal bars within a wide channel belt as shown on Fig. 10. This figure illustrates the major channels and higher relief bars as they would appear close to bankfull stage.

## Cross-sectional analysis

The maximum observed dimensions can be taken as a reasonable approximation of bankfull conditions immediately prior to abandonment. From the fundamental parameters of width, depth and energy gradient approximated by terrace gradient, one can estimate mean flow velocity, discharge, and bed shear stress. In addition, the basal gravel was sampled at the maximum depth of each channel along the cross-section to obtain a measure of the traction load. Palaeohydraulic reconstruction is normally related to conditions at the threshold of sediment motion, but in the present case it is preferable to adopt an alternative approach. It is more meaningful to calculate flow conditions in relation to bankfull discharge since one can then estimate whether bankfull conditions were sufficient to transport bedload.

A summary of the palaeochannel measurements and palaeohydraulic calculations is listed on Table II. The width-depth ratios are generally low, but since all the cross-sections are located on distributaries, this is consistent with observations of active braided channels (Cheetham, 1979). Their low width-depth ratios are indicative of unstable channel conditions similar to active distributaries which were suddenly abandoned, a characteristic feature of a braided stream environment.

TABLE II

*Summary of Theale palaeochannel data*

| Channel no. | $D_{65}$ (mm) | $d_{max}$ (m) | $W$ (m) | $R$ (m) | $w/d$ | $A$ ($m^2$) | $\bar{V}$ ($m\,s^{-1}$) | $Q_{bf}$ ($m^3\,s^{-1}$) | $\tau_o$ ($Nm^{-2}$) | $\theta$ | $\theta^1$ |
|---|---|---|---|---|---|---|---|---|---|---|---|
| 1 | 27 | 0·91 | 18·3 | 0·87 | 21·0 | 16·1 | 1·26 | 20·3 | 12·5 | 0·029 | 0·026 |
| 2 | 20 | 0·80 | 25·6 | 0·73 | 35·1 | 19·7 | 1·16 | 22·9 | 11·0 | 0·035 | 0·032 |
| 3 | 15 | 0·60 | 9·1 | 0·66 | 13·8 | 6·3 | 1·10 | 6·9 | 8·2 | 0·035 | 0·032 |
| 4 | 17 | 0·76 | 6·1 | 0·61 | 10·0 | 4·1 | 1·06 | 4·4 | 10·4 | 0·038 | 0·035 |
| 5 | 16 | 0·62 | 7·6 | 0·54 | 14·1 | 4·3 | 0·99 | 4·3 | 8·5 | 0·034 | 0·031 |
| 6 | 19 | 0·84 | 13·7 | 0·73 | 18·8 | 10·7 | 1·16 | 12·4 | 11·6 | 0·038 | 0·035 |
| 7 | 17 | 0·76 | 6·1 | 0·62 | 9·8 | 4·2 | 1·09 | 4·5 | 10·4 | 0·038 | 0·035 |
| 8 | 15 | 0·61 | 6·7 | 0·51 | 13·1 | 3·7 | 0·97 | 3·6 | 8·4 | 0·035 | 0·032 |
| 9 | 17 | 0·75 | 9·1 | 0·65 | 14·0 | 6·3 | 1·09 | 6·9 | 10·3 | 0·039 | 0·036 |
| 10 | - | 0·90 | 70·0 | 0·70 | 100·0 | 49·0 | 1·13 | 55·4 | - | - | - |

$D_{65}$, grain size (65th percentile of gravel fraction); $d_{max}$, maximum channel depth; $W$, width; $R$, hydraulic radius; $w/d$, width-depth ratio; $A$, cross-sectional area; $\bar{V}$, mean velocity; $Q_{bf}$, bankfull discharge; $\tau_o$, bed shear stress; $\theta$, Shields' entrainment function; $\theta^1$, corrected value of Shields' entrainment function.

Mean sectional velocity was calculated for each palaeochannel from the Darcy-Weisbach relationship:

$$\bar{V} = \sqrt{\left(\frac{8dgS}{f}\right)}$$

where $\bar{V}$ is the mean flow velocity, $d$ is the depth of flow, $g$ is acceleration due to gravity, $S$ is a measure of the slope, and $f$ is the Darcy-Weisbach friction factor, taken as 0·06 as an appropriate value for flow over a gravel bed.

The calculation of palaeochannel unit bed shear stress, $\tau_0$, is based on the Du Boys equation:

$$\tau_0 = \gamma\, dS$$

where $\gamma$ is the specific weight of water, $d$ is the mean channel depth or hydraulic radius, and $S$ is the energy gradient.

Shield's entrainment function was calculated from the 65th percentile, $D_{65}$, of the gravel fraction, and subsequently corrected in the light of Neill's (1968) experimental data. The entrainment threshold was taken as 0·047 in accordance with Meyer-Peter and Müller (1948) for grain size mixtures and particle sizes in excess of 2 mm.

Thus, the calculated values of Shields' criterion, $\theta$, for the palaeochannels (Table II) indicated no bedload motion at bankfull stage. Stresses approximately double would be required to create bed mobilization.

## Discussion

The apparent conformity of the palaeochannels with the underlying sediments suggest they formed part of the closing stages in the deposition of the Beenham Grange Terrace. They were not related to an isolated catastrophic event because of the general absence of high boundary shear features such as gravel ripples or dunes indicative of high sediment transport rates (Thiel, 1932; Meyer-Peter and Müller, 1948). They are more compatible with moderate magnitude discharges, but to maintain a braided pattern, such discharges would need to be of a relatively frequent recurrence. Kirkby (1972) demonstrated from an analysis of stable and unstable channels that a river which is unable to transport its bed material will tend to develop a meandering form, rather than remain braided. This process is apparently dependent on sediment mobility. Since bankfull discharge of the palaeochannels appears to be insufficient to create bedload motion, moderate flood discharge of a frequent recurrence would thus be necessary to maintain a braided pattern. This question is examined further in the following analysis of drainage density changes.

## Drainage density and dry valleys

Dry valley networks occur extensively throughout the Kennet catchment area on all the major lithologies, although they are most prominent on the Chalk. The extent to which they are relict features related to former hydrological conditions is clearly a point which needs to be examined in detail. Gregory and Walling (1968) expressed the view that "dry valleys . . . . . may be completely independent of the present drainage net." (p.66). If such is the case, then valley axis densities are a potentially very useful palaeohydrological tool, particularly because drainage densities are a sensitive indicator of hydrological processes operating in a catchment area. Significantly, many of the dry valleys of the Kennet catchment grade into the Beenham Grange Terrace, rather than to the present river (Fig. 2).

Accordingly, it is intended to examine channel densities and valley axis densities of the Kennet catchment in the light of present-day hydrological records. In addition, it is instructive to examine the spatial distribution of these densities in relation to lithology, particularly the extent to which valley axis densities are controlled by the geology. If there is very little relationship between the two, such densities probably developed under different climatic conditions, such as in a permafrost environment where the influence of lithology is overriden.

## Data sources

Drainage density is usually based either on 'blue-line' networks or highest contour crenulations shown on published maps, supported by 'spot-checking' in the field. However, there are various problems of interpretation. In many studies, drainage density is considered to be a static quantity, but in reality, it is a dynamic feature which varies with time in response to precipitation events (Gregory and Walling, 1968). It is thus important to be consistent in the selection of data source so that spatial variations can be examined.

In the present study, attention is focused primarily on valley axis density. The 'blue-line' network is also considered, chiefly as an index of contemporary conditions and of the controls on drainage density. Both valley axis density and channel density are based on Ordnance Survey maps at a scale of 1 to 25000. Drainage density was then measured on a grid basis so as to produce a map of spatial variations within the catchment area.

## Results

Channel density and valley axis density maps (Figs 11a and 11b) were constructed downstream to Theale, where present-day hydrological records are available. Two important points emerge from this analysis: (1) there is a considerable difference between present-day drainage densities and valley axis densities, and (2) there is a distinct contrast between the spatial organization of present-day channel densities and valley axis densities.

The present-day Kennet catchment area downstream to Theale has a mean drainage density of 0·45 km km$^{-2}$. This low density is hardly surprising since the catchment area is dominated by highly permeable lithologies. The influence of lithology on the present-day network is best illustrated by comparing the channel density map (Fig. 11a) with the accompanying geological map (Fig. 11c). The upper and middle parts of the catchment area are characterized by very low densities, where Chalk is the dominant lithology. In the lower part of the catchment where Tertiary sediments assume greater importance, drainage densities generally rise to between 1·0 and 1·5 km km$^{-2}$, particularly in the valley of the southern tributary, the Enborne, where London Clay outcrops widely.

The mean valley axis density downstream to Theale is 2·34 km km$^{-2}$ which is more than five times that of the 'blue-line' density. 'Blue-line' densities generally indicate moderately low flow conditions; Ordnance Survey practice is to map stream channels at their 'normal winter level' (Gardiner, 1974), a definition which is unfortunately open to a wide range of interpretation. The important point is whether valley axis densities are the product of present-day valley forming processes. This is a very difficult question to answer conclusively, except to say that there

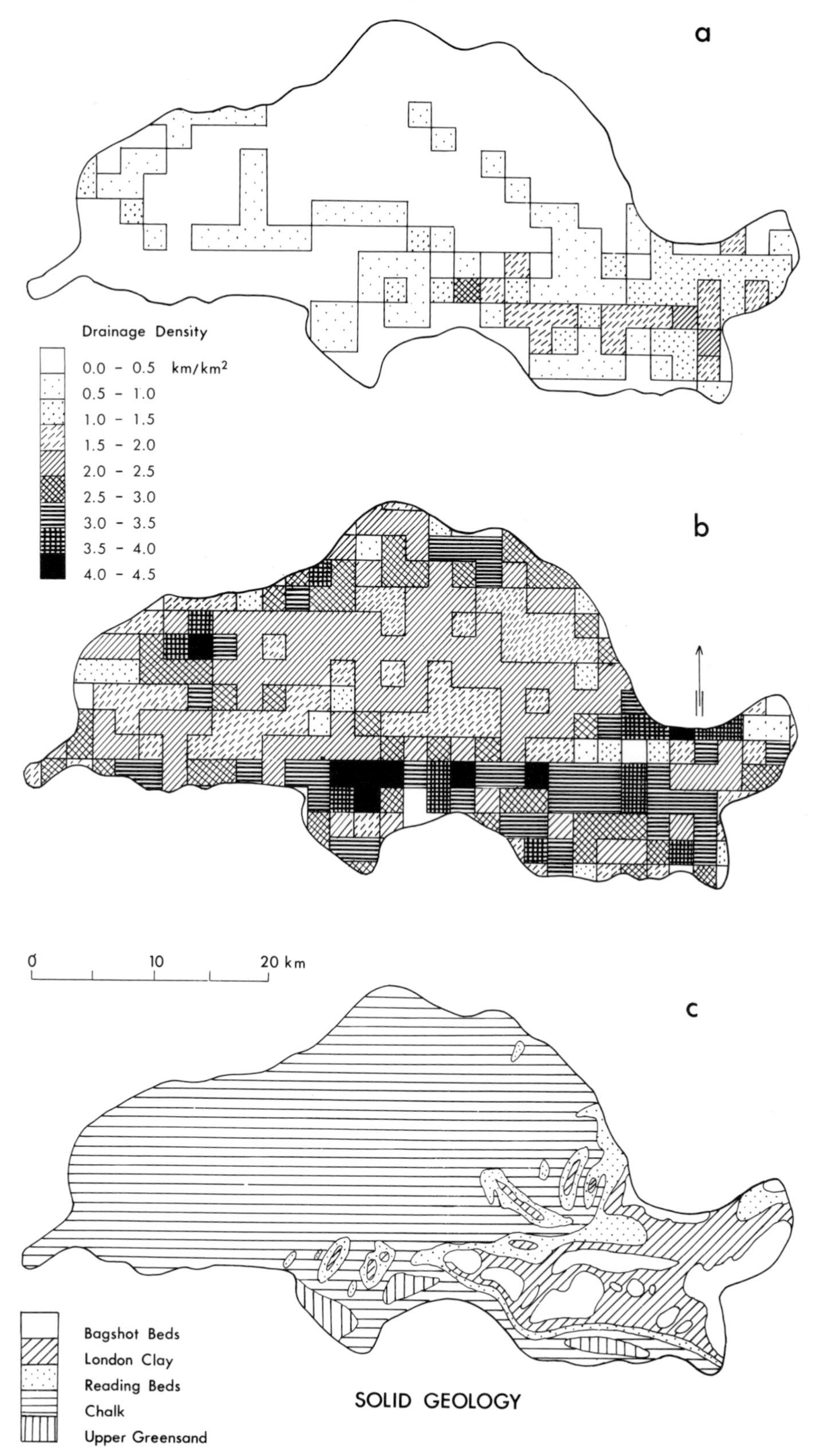


*Figure 11.* The Kennet catchment downstream to Theale: (a) channel densities (b) valley axis densities (c) solid geology.

are distinct indications that the dry valleys of the Kennet catchment area were part of an active channel network at the time of the deposition of the Beenham Grange Terrace on account of their frequent gradation to this level rather than to the present-day river. This is consistent with Gregory (1971) who suggested that many of the dry valleys of south-west England are inherited features of a former hydrological regime.

Further indication of an inherited network in the Kennet Valley can be seen by comparing the spatial distribution of valley axis densities with lithology. Higher valley axis densities as compared to channel densities throughout the catchment indicate the widespread occurrence of dry valleys on all the lithologies, including the London Clay. Valley axis densities are also spatially more uniform than channel densities. Although the valley axis densities tend to be slightly higher on the south side of the catchment area, probably as a result of the more steeply dipping limb of the syncline which bounds the catchment area divide along the south, the influence of lithology is much more subdued, and often apparently non-existent. Such evidence clearly gives weight to the suggestion that many of the dry valleys developed in a period when the effects of lithology were overridden, such as the incidence of permafrost which inhibited infiltration and produced one relatively uniform and widespread substratum.

## Palaeodischarge

Discharge and drainage density are dynamically related at any point in time and the two vary with each other in a definable way.

Gregory and Walling (1968) asserted that discharge is proportional to the square of drainage density. Thus, for the Kennet catchment downstream to Theale, the mean valley axis density of 2·34 km km$^{-2}$ as compared with the mean 'blue-line' density of 0·45 km km$^{-2}$ would imply a discharge increase of approximately 27 times if all the dry valleys were active. However, these discharge relationships need to be considered in absolute terms in connection with present-day hydrological records since the two types of density need not necessarily refer to flow conditions of the same recurrence. It is therefore necessary to employ an empirical relationship between discharge of a specified recurrence and drainage density. Ideally, such a relationship ought to include a precipitation input term, such as the formula proposed by Rodda (1969). However, in the present instance to avoid having to speculate on an unknown quantity, the more direct relationship of Carlston (1963) based on fifteen basins in eastern USA is considered preferable. This assumes, of course, that it is valid for application to an inferred late Quaternary permafrost environment. This relationship in metric form is:

$$q = 0{\cdot}0368\, D^2$$

where $q$ is the discharge of the mean annual flood per km$^2$ in m$^3$ s$^{-1}$, and $D$ is drainage density in km km$^{-2}$. In this relationship, Carlston assumed discharge to be proportional to the square of drainage density. A re-examination of his raw data by least-squares regression analysis, however, shows that the relationship more precisely should be:

$$q = 0{\cdot}063\, D^{1{\cdot}59},$$
$$\text{where } r = 0{\cdot}903,\ \text{SE} = 0{\cdot}1094\ \text{m}^3\text{s}^{-1} \text{ in } \log_{10} \text{ units.}$$

The application of this revised relationship to the drainage system of the Kennet upstream of Theale is summarized in Table III. A channel density of 0·45 km km$^{-2}$ would thus indicate a mean value of 18·4 m$^3$ s$^{-1}$ for the mean annual flood. Gauging records at Theale give a value of 40 m$^3$ s$^{-1}$. Even the upper limit of the 95 per cent probability range is appreciably less than the gauging station value. This discrepancy most probably reflects the fact that the 'blue-line' network on the topographic maps is related to drainage densities associated with much lower discharges than the mean annual flood.

TABLE III

*Palaeodischarge calculations based on drainage density downstream to Theale*

| Drainage area (km$^2$) | Channel density (km km$^{-2}$) | Calculated present-day $Q_{2\cdot33}$ (m$^3$s$^{-1}$) | | Actual $Q_{2\cdot33}$ (Theale gauging records 1961-73) (m$^3$s$^{-1}$) | Valley axis density (km km$^{-2}$) | Calculated palaeo-discharge $Q_{2\cdot33}$ (m$^3$s$^{-1}$) | |
|---|---|---|---|---|---|---|---|
| | | Mean | 95% probability range (± 2 SE) | | | Mean | 95% probability range (± 2 SE) |
| 1037·7 | 0·45 | 18 | 11<br>30 | 40 | 2·34 | 253 | 153<br>418 |

The highest discharge of the River Kennet recorded at Theale in the 13 year period, 1961-73, is 70·8 m$^3$ s$^{-1}$ at 03.00 h on 11 June, 1971. Although the recurrence interval of this discharge level cannot be established beyond the period of existing records, it is important to note discharge estimates based on the valley axis densities are considerably greater. It is not unreasonable to assume, therefore, that the dry valleys were part of an active channel network during the deposition of the Beenham Grange Terrace. Accordingly, there is a 95 per cent probability that the discharge of the mean annual flood at this time was between 153 m$^3$s$^{-1}$ and 418 m$^3$s$^{-1}$ as compared to 40 m$^3$s$^{-1}$ for the modern Kennet, i.e. between 4 and 10 times greater than at present. Furthermore, the upper value of this range is considerably less than Dury's (1965) minimum discharge ratio. He estimated that since Devensian times, there has been at least a 20 times decrease in bankfull discharge for rivers in lowland Britain.

## Conclusions

On the basis of sedimentological features indicative of a braided stream environment, it is concluded that, within a 95 per cent probability, the bankfull discharge of the River Kennet was at least two times greater than the present-day Kennet when the Thatcham and Beenham Grange Terraces were deposited.

More specific reconstruction is possible for the Beenham Grange Terrace because of preserved palaeochannels and the functional relationship of many extensive dry valley networks. Reconstruction of palaeochannel dynamics showed that bankfull discharge was insufficient to create bedload motion, an essential feature of braided streams to maintain their form. Bed shear stresses approximately double those for bankfull stage appeared to be necessary. However, assuming that all the dry valleys formed part of an active channel network at the Beenham Grange stage, there is a 95 per cent probability that the discharge of the mean annual flood was between 4 and 10 times greater than the modern Kennet. This range is compatible with an estimate of minimum bankfull discharge based on the incidence of braiding, as adjusted for an increase for the creation of bedload motion in the palaeochannels.

Dury (1965) estimated that there has been at least a twenty times decrease in bankfull discharge for rivers in lowland Britain since Devensian times. The main point arising from the present study is that discharge range of the roughly comparable mean annual flood, even within a 95 per cent probability, differs markedly from Dury's conclusions. The upper limit of this range falls considerably short of Dury's minimum value. The results of the present study obtained by a completely independent methodology, however, largely support Tinkler's (1971) revised interpretation of underfit streams which is corrected for differences in flow recurrence.

## Acknowledgements

This research project was undertaken during the tenure of a University of Reading Postgraduate Studentship. Grateful thanks for their guidance and helpful criticism are due to Dr P. Worsley, Professor J. R. L. Allen, Mr I. M. Fenwick and Dr J. B. Thornes, and to the Kennet Valley Research Committee for providing financial support to undertake aerial survey.

## References

Cant, D. J. and Walker, R. G. (1976). Development of a braided-fluvial facies model for the Devonian Battery Point Sandstone, Quebec, *Can. J. Earth Sci.* **13**, 102-119.

Carlston, C. W. (1963). Drainage density and streamflow, *U.S. Geol. Surv. Prof. Paper* **422-C**, 8 pp.

Chartres, C. J. (1975). "Soil development on the terraces of the River Kennet", unpublished Ph.D. thesis, University of Reading.

Cheetham, G. H. (1975). "Late Quaternary palaeohydrology, with reference to the Kennet Valley", unpublished Ph.D. thesis, University of Reading.

Cheetham, G. H. (1976). Palaeohydrological investigations of river terrace gravels, *in*, Davidson, D. A. and Shackley, M. (eds), "Geo-archaeology: Earth Science and the Past" pp. 335-344. Duckworths, London.

Cheetham, G. H. (1980). Flow competence in relation to stream channel form and braiding, *Geol. Soc. Amer., Bull.* (in press).

Dury, G. H. (1964). Principles of underfit streams, *U.S. Geol. Surv., Prof. Paper* **452-A**, 67 pp.

Dury, G. H. (1965). Theoretical implications of underfit streams, *U.S. Geol. Surv., Prof. Paper* **452-C**, 43 pp.

Gardiner, V. (1974). Drainage basin morphometry, *Br. Geomorphol. Res. Group, Tech. Bull.* **14**, 48 pp.

Gregory, K. J. (1971). Drainage density changes in south-west England, *in*, Gregory, K. J. and Ravenhill, W. L. D. (eds), "Exeter Essays in Geography", pp. 33-53. University of Exeter Press.

Gregory, K. J. and Walling, D. E. (1968). The variation of drainage density within a catchment, *Int. Ass. Sci. Hydrol., Bull.* **13**, 61-68.

Kirkby, M. J. (1972). Alluvial and non-alluvial meanders, *Area* **4**, 284-288.

Leopold, L. B. and Wolman, M. G. (1957). River channel patterns—braided meandering and straight, *U.S. Geol. Surv., Prof. Paper* **282-B**, 39-85.

McDonald, B. C. and Banerjee, I. (1971). Sediments and bedforms on a braided outwash plain, *Can. J. Earth Sci.* **8**, 1282-1301.

Meyer-Peter, E. and Müller, R. (1948). Formulas for bed-load transport. *Int. Ass. Hydraul. Str. Res.*, 2nd meeting, Stockholm, pp. 39-64.

Neill, C. R. (1968). A re-examination of the beginning of movement for coarse granular bed materials, *Hydraulic Research Station, Wallingford, Int. Rep.* **68**, 37 pp.

Osborne White, H. J. (1907). The Geology of the country around Hungerford and Newbury, *Mem. Geol. Surv., Engl. and Wales.* H.M.S.O., London.

Rodda, J. C. (1969). The significance of characteristics of basin rainfall and morphometry in a study of floods in the United Kingdom. UNESCO Symposium on Floods and their compilation, *Int. Ass. Sci. Hydrol., UNESCO-WMO* **2**, 834-845.

Smith, N. D. (1974). Sedimentology and bar formation in the Upper Kicking Horse River, a braided outwash stream, *J. Geol.* **82**, 205-223.

Thiel, G. H. (1932). Giant current ripples in coarse fluvial gravel, *J. Geol.* **40**, 452-458.

Thomas, M. F. (1957). "River terraces and drainage patterns in the Reading area", unpublished M.A. thesis, University of Reading.

Thomas, M. F. (1961). River terraces and drainage development in the Reading area, *Geol. Ass. (London), Proc.* **72**, 415-436.

Tinkler, K. J. (1971). Active valley meanders in south central Texas and their wider implications, *Geol. Soc. Amer., Bull.* **81**, 1783-1799.

Wooldridge, S. W. and Linton, D. L. (1939). Structure, surface and drainage in south-east England, *Inst. Br. Geogr.* **10**, 1-176.

Wymer, J. (1968). "Lower Palaeolithic Archaeology in Britain". Baker, London.

NINE

# The weathering and erosion of chalk under periglacial conditions

R. B. G. WILLIAMS

*The Geography Laboratory, University of Sussex*

## Introduction

The chalk landscapes of southern and eastern England bear the clear imprint of periglacial processes. Almost everywhere one goes, unmistakable evidence can be found of former frost climates. For example, many frost disturbances are preserved in the South Downs, together with deposits due to mass wasting under cold conditions. Involutions are particularly abundant, and they can be seen, for example, at many places along the top of the chalk cliffs between Brighton and Eastbourne. The total length of the cliffs is about 22 km if cliffs where Tertiary deposits overlie the Chalk are excluded. Involutions are present for some 9 km, or 40 per cent of that length. Solifluction and sheetwash deposits, also, are well exposed where the cliffs cut through the floors of dry valleys.

Periglacial features are even more common in an area such as the Breckland (south-west Norfolk and north-west Suffolk). Chalkland polygons and stripes are very widely preserved; some individual spreads extend for 5 km or more in one direction. The patterns occupy an estimated 90 per cent of the area of certain 10 km grid squares. Shallow depressions resulting from the melting out of bodies of ground-ice are also abundant, particularly towards the Fenland edge. In some 10 km grid squares they occupy over 50 per cent of the land surface.

The abundance of periglacial features makes it tempting to infer that the chalk landscape was mainly fashioned under periglacial conditions, but this would be to ignore the fact that many periglacial features are little more than surface ornamentation, and are not indicative of any profound degree of landform modification. The awkward truth is that it is very difficult to judge the extent to which chalk landforms are the result of periglacial processes.

## Amounts of frost weathering

No clear picture has yet emerged concerning the weathering of chalk in periglacial times. Frost weathering of the chalk may have been rapid, but the field evidence is rather unsatisfactory. Beneath the surface soil the chalk is often severely shattered, consisting of angular fragments about 0·01 to 0·1 m long, generally set in a matrix of fines, but also existing as loose pieces with air spaces in between. The size of the fragments increases with depth, and there is a gradual transition to unshattered bedrock. Where shallow deposits overlie the shattered chalk, frost heaving has often produced involutions or chalkland polygons and stripes, but where there are no such deposits the chalk, though shattered, is often not particularly disturbed. The fragments usually fit together like a three-dimensional jig-saw puzzle.

The shattered zone is usually attributed to frost weathering, but definite proof is wanting. Where the chalk has a relatively high insoluble residue (greater than two per cent) there is a possibility that the shattering is due simply to repeated wetting and drying, for experiments have shown that such chalk often breaks quite readily after only a few cycles of wetting and drying. But where the chalk is relatively pure (with an insoluble residue of less than two per cent) this cause can seemingly be ruled out, because chalk of this kind appears to be immune from damage (Williams, unpublished). Even if it be assumed that frost weathering was mainly responsible for the shattered zone, this still leaves the question of when it developed. Frost penetrates up to one metre below the surface in very hard winters at present, and no doubt helps to break up the bedrock. However, since the shattering often extends to depths of more than a metre, and is often associated with periglacial disturbances such as involutions, one can safely conclude that it mostly developed under periglacial conditions and is a relict feature.

However carefully one examines the frost-shattered zone one cannot deduce how many years it took to form. The evidence is simply not there. One must resort to laboratory experiments in order to derive some idea of the rate of weathering of chalk under periglacial conditions. It is important in such experiments to try to duplicate the conditions that existed in periglacial times as closely as possible.

Tricart (1956) experimented on some samples of chalk from two sites in northern France (departments of Pas de Calais and Marne). He found that the samples disintegrated rapidly after only a few cycles of freezing and thawing with −30 °C as the minimum temperature. In addition, investigators at the CNRS laboratory at Caen have carried out a long series of experiments on Upper Chalk from two sites in the department of Seine Maritime (Coutard *et al.*, 1970; Lautridou, 1970; Aubry and Lautridou, 1974). They too found the chalk to be very susceptible to frost.

No one has published any results of experiments on British chalk. A series of experiments was carried out to investigate the weathering of British chalk using (1) samples from sites in Cambridgeshire where the chalk forms only very subdued relief, and (2) samples from sites in Sussex where the chalk forms more impressive relief and is conceivably more resistant to weathering. Large blocks of apparently unweathered (or only slightly weathered) chalk were obtained from quarries and other sites. Cubes measuring 7·5 cm (3 in.) on each edge were cut from the centres of the blocks and were then oven-dried for 3 days at 60 °C. After cooling, the cubes were weighed, and were left to soak in tap water for a week before being weighed once again to determine the amount of water absorption. The temperature of the laboratory in which this took place was a constant + 15 °C.

Freezing was carried out using a specially modified commercial deep-freeze. The cubes were placed in the freezing compartment in separate plastic containers which were 11 cm in height and 10 cm × 10 cm in cross-section. The containers were open at the top, but were surrounded on the sides and base by an insulating jacket of polystyrene. They were kept filled with enough water to cover the tops of the cubes. The purpose of the water was to prevent the cubes drying out during freezing. Previous investigators have found that rock specimens become unsaturated after a series of freeze-thaw cycles, whilst the walls of the freezing compartment become covered with condensation.

The water around the cubes could not crush them on freezing because the containers had very flexible walls which took up the expansion. In addition, the ice had an easy escape upwards because the containers were open at the top.

The polystyrene was placed round the containers to try to ensure that the specimens froze from the top downwards (as would happen under natural conditions) rather than inwards from all faces simultaneously. It was only partly successful in this respect.

In one set of experiments, the temperature inside the freezing compartment was pre-set at − 30 °C (see cycle A in Fig. 1). The containers with their cubes were put straight into the freezing compartment, and there was no attempt to lower the temperature of the cubes gradually. The cubes briefly warmed the compartment before the refrigeration equipment was able to restore the temperature to − 30 °C. The cubes started to freeze within one hour of being placed in the freezing compartment. After the cubes had been in the freezing compartment for some 57 h, the refrigeration equipment was switched off, and the cubes were allowed to start thawing. The freezing compartment was warmed using an electric light bulb until an air temperature of + 15 °C was attained. The cubes and their containers were returned to the deep-freeze at the end of five days to begin a new cycle of freezing and thawing. After six cycles the frost-shattered debris was wet-sieved and the various size fractions weighed.

Table I shows the amount of disintegration of a set of 15 cubes from five sites in Cambridgeshire. Many of the cubes were badly damaged by frost despite being subjected to only six cycles. The immediate question is whether the ultra-low temperatures employed in the experiments were commensurate with those experienced in periglacial Britain. Precise quantitative information is difficult to obtain, but there seem to have been times when average temperatures in the coldest month (January or February) were below − 20 °C and in the warmest month (July) were at or just below 10 °C (Coope, 1975, 1977; Williams, 1975). It is quite possible in

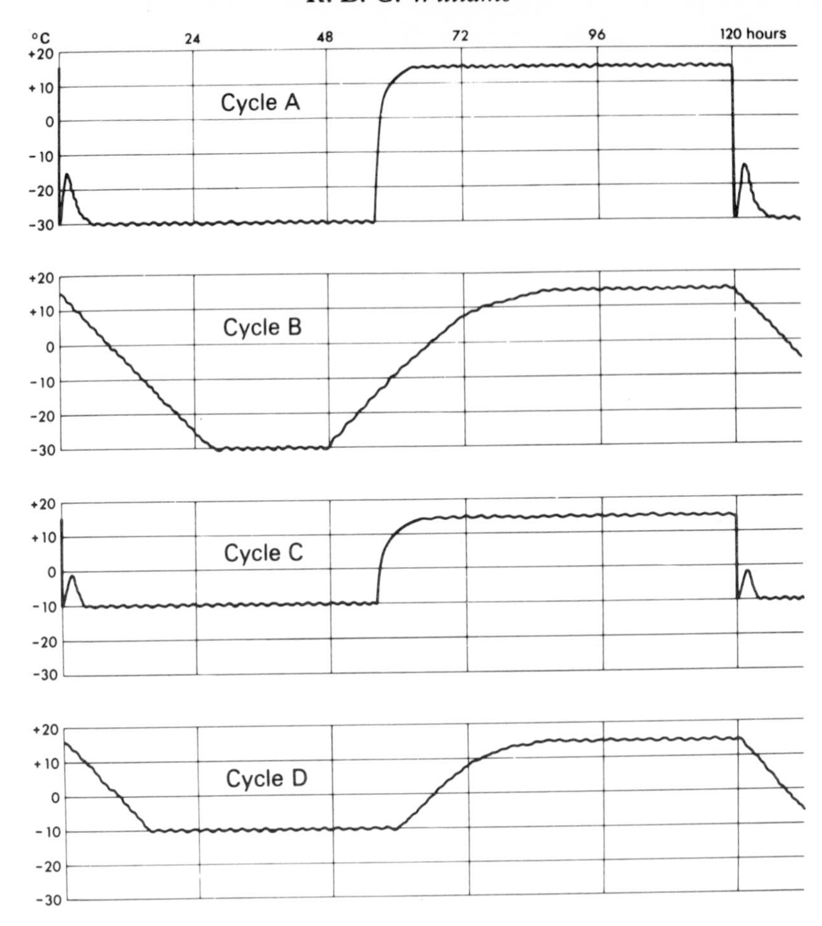


*Figure 1.* Curves showing the various cycles used to test chalk samples for frost-susceptibility.

fact that temperatures rose above + 15 °C on hot summer days and dropped to below − 30 °C on the coldest days in winter. The temperature limits employed in the experiments, would appear therefore to have been fairly realistic. However, the rate of freezing was almost certainly too rapid. Each temperature cycle telescoped into five days changes which presumably took a year under periglacial conditions.

Some research workers have claimed that fast freezing leads to accelerated disintegration of rocks. Thomas (1938; p. 93), for example, was convinced that slow freezing is accompanied "by only small extensional strains. Rapid rates of freezing, on the other hand, are usually accompanied by large extensional strains, which are liable to cause much more damage". It appears, however, that this view is incorrect. Frozen food firms have found that fast freezing causes less damage to plant and animal tissues than slow freezing because only small crystals of ice can form. Slow freezing allows the growth of large ice crystals which destroy the tissues. It seems likely that slow freezing also promotes the destruction of chalk,

TABLE I

*Disintegration produced by six "A" cycles*

| Location | Fossil zone | Specimen number | Dry weight (g) | Weight when saturated (g) | Density (dry) | Water absorption as % of dry weight | Size of debris after frost weathering: Percentage of total wet weight in different size classes | | | | | | Size of debris after frost weathering: Wet weight of largest fragment as % of total |
|---|---|---|---|---|---|---|---|---|---|---|---|---|---|
| | | | | | | | >20·0 (mm) | 20·0-9·5 | 9·5-3·4 | 3·4-1·7 | 1·7-0·08 | 0·08-0·00 | |
| Burwell Town Quarry, Burwell, Cambs. TL591663 | *Holaster subglobosus* (Totternhoe Stone) | 1 | 744·0 | 863·8 | 1·83 | 16·1 | 45·0 | 15·1 | 18·2 | 8·4 | 8·2 | 5·2 | 10·5 |
| | | 2 | 824·5 | 958·1 | 1·88 | 16·2 | 53·6 | 18·1 | 16·1 | 4·4 | 4·8 | 3·1 | 20·0 |
| | | 3 | 826·5 | 979·4 | 1·80 | 18·5 | 39·0 | 13·0 | 29·1 | 6·1 | 7·8 | 5·0 | 5·3 |
| Quarry on Gog Magog Hills, Cambs. TL 484539 | *Holaster subglobosus* (above Totternhoe Stone) | 1 | 886·2 | 1069·6 | 1·61 | 20·7 | 53·4 | 11·2 | 15·1 | 6·2 | 10·1 | 3·9 | 52·8 |
| | | 2 | 770·5 | 953·1 | 1·66 | 23·7 | 12·7 | 10·1 | 40·3 | 15·7 | 15·4 | 5·9 | 1·9 |
| | | 3 | 758·0 | 921·7 | 1·72 | 21·6 | 35·3 | 25·6 | 37·7 | 6·1 | 10·2 | 3·9 | 18·9 |
| Quarry on Cherry Hinton Hill, Cambs. TL485556 | *Inoceramus labiatus* (Melbourn Rock) | 1 | 673·6 | 844·6 | 1·56 | 25·4 | 16·6 | 17·6 | 23·6 | 12·9 | 18·1 | 11·2 | 3·8 |
| | | 2 | 642·5 | 821·9 | 1·44 | 27·9 | 19·0 | 10·2 | 28·3 | 11·8 | 19·0 | 11·8 | 5·0 |
| | | 3 | 743·8 | 896·2 | 1·68 | 20·5 | 86·3 | 3·5 | 3·9 | 1·8 | 2·8 | 1·8 | 73·6 |
| Quarry near West Wratting Farm, Cambs. TL565545 | *Terebratulina lata* | 1 | 735·6 | 918·0 | 1·48 | 24·8 | 88·2 | 1·5 | 5·0 | 1·2 | 2·8 | 1·3 | 82·2 |
| | | 2 | 686·9 | 873·1 | 1·61 | 27·1 | 91·0 | 1·1 | 2·8 | 1·4 | 2·5 | 1·2 | 87·5 |
| | | 3 | 749·2 | 923·8 | 1·65 | 23·3 | 98·3 | 0·2 | 0·2 | 0·2 | 0·8 | 0·4 | 98·3 |
| Quarry near Underwood Hall, Cambs. TL613572 | *Holaster planus* | 1 | 680·5 | 865·6 | 1·47 | 27·2 | 39·0 | 2·8 | 9·5 | 12·7 | 27·2 | 8·6 | 24·1 |
| | | 2 | 700·9 | 880·4 | 1·52 | 25·6 | 49·3 | 2·8 | 6·8 | 8·6 | 24·8 | 7·9 | 27·9 |
| | | 3 | 719·9 | 897·6 | 1·58 | 24·7 | 77·9 | 4·6 | 5·0 | 1·9 | 5·3 | 1·7 | 70·5 |

but this has yet to be proved conclusively. A second set of experiments was carried out in which the air temperature inside the freezing compartment was lowered from +15 °C to −30 °C over a period of 28 h (see cycle B in Fig. 1). After a period of 20 h at −30 °C, the temperature was slowly raised to +15 °C and then held constant. The entire cycle took 5 days to complete, the same time as the cycle previously described which involved a rapid fall in temperature. The same freezing and thawing periods were selected (57 h and 63 h respectively). The first four days of the cycle were an imitation of the four-day 'Siberian' cycle used by Tricart (1956), Wiman (1963) and Potts (1970). An extra day was allowed at the end, however, to ensure that the cubes thawed completely.

After six cycles of freezing and thawing the debris was wet-sieved and weighed (Table III). The amount of disintegration was greater than with the fast freezing cycle, but not so great as to rule out the possibility of sampling error. Analysis of variance (discussed in greater detail later) indicates that there is a chance of more than 0·05 that the difference in the amount of disintegration is an accident of sampling. The experiments are therefore somewhat inconclusive, although they do tend to support the findings of frozen food firms that slow freezing causes more damage than fast freezing.

It is not suggested that the rate of freezing in cycle B was slow enough to reproduce periglacial conditions accurately, but at least it was slower than in the previous set of experiments. Quite possibly the rate of weathering was even greater in periglacial times than Table II suggests.

It is likely that temperatures in periglacial Britain fluctuated around the freezing point in early and late summer. In order to discover the effect of comparatively minor fluctuations in temperature, freezing experiments were carried out using five-day cycles with a minimum temperature of only −10 °C. In one set of experiments the cubes were put straight into the freezing compartment which was pre-set at −10 °C (cycle C in Fig. 1). In a second set of experiments the temperature was lowered gradually (cycle D in Fig. 1).

Tables III and IV record the amounts of disintegration obtained with different cubes. Table V presents a summary of the four sets of experiments. Cycles A and B (the −30 °C cycles) produced 4·9 and 3·5 times more fines on average than cycles C and D (the −10 °C cycles). Each cycle involving slow freezing (cycle B and cycle D) produced more fines than its fast freezing counterpart (cycle A or cycle C). Cycle B produced 1·2 times more fines on average than cycle A and cycle D produced 1·7 times more fines on average than cycle C.

Analysis of variance of the data suggests that minimum temperature is an important factor determining the percentage of fines (the variance ratio $F$ = 47·7 with 1 and 49 df; $P < 0{\cdot}0005$). Rate of freezing is not clearly identifiable as a factor ($F$ = 2·0 with 1 and 49 df; $P > 0{\cdot}05$). Since the experience of the frozen food trade suggests that rate of freezing probably is a factor, it would seem that more cubes ought to be tested.

Geographical location is apparently not as important a factor as minimum temperature ($F$ = 6·5 with 4 and 49 df; $P < 0{\cdot}0005$). Interestingly, there appears to be

TABLE II

*Disintegration produced by six "B" cycles*

| Location | Fossil zone | Specimen number | Dry weight (g) | Weight when saturated (g) | Density (dry) | Water absorption as % of dry weight | Size of debris after frost weathering: Percentage of total wet weight in different size classes | | | | | | Wet weight of largest fragment as % of total |
|---|---|---|---|---|---|---|---|---|---|---|---|---|---|
| | | | | | | | >20·0 (mm) | 20·0-9·5 | 9·5-3·4 | 3·4-1·7 | 1·7-0·08 | 0·08-0·00 | |
| Burwell Town Quarry, Burwell, Cambs. TL591663 | *Holaster subglobosus* (Totternhoe Stone) | 1 | 798·8 | 916·6 | 1·74 | 14·7 | 7·8 | 14·2 | 36·3 | 15·5 | 16·9 | 9·3 | 1·5 |
| | | 2 | 827·4 | 961·2 | 1·70 | 16·2 | 48·6 | 16·9 | 20·0 | 5·1 | 6·1 | 3·3 | 5·7 |
| | | 3 | 822·1 | 958·8 | 1·73 | 16·6 | 62 | 9·2 | 16·7 | 4·5 | 5·0 | 2·7 | 10·8 |
| Quarry on Gog Magog Hills, Cambs. TL484539 | *Holaster subglobosus* (above Totternhoe Stone) | 1 | 699·0 | 892·1 | 1·64 | 26·6 | 0 | 5·0 | 64·0 | 15·0 | 9·9 | 6·1 | 0·4 |
| | | 2 | 672·5 | 852·9 | 1·59 | 26·8 | 0 | 11·4 | 57·2 | 11·2 | 12·6 | 7·7 | 0·5 |
| | | 3 | 767·2 | 950·0 | 1·63 | 23·8 | 20·8 | 14·1 | 32·3 | 12·4 | 12·0 | 7·4 | 13·0 |
| Quarry on Cherry Hinton Hill, Cambs. TL485556 | *Inoceramus labiatus* (Melbourn Rock) | 1 | 718·9 | 867·4 | 1·60 | 20·6 | 74·6 | 5·1 | 7·6 | 3·6 | 5·3 | 3·9 | 69·7 |
| | | 2 | 686·2 | 854·2 | 1·57 | 24·5 | 12·5 | 9·1 | 23·0 | 14·4 | 23·5 | 17·4 | 2·0 |
| | | 3 | 673·6 | 830·5 | 1·51 | 23·3 | 23·8 | 13·8 | 28·4 | 8·6 | 14·6 | 10·8 | 8·5 |
| Quarry near West Wratting Farm, Cambs. TL565545 | *Terebratulina lata* | 1 | 707·6 | 882·4 | 1·53 | 24·7 | 79·3 | 3·7 | 8·2 | 2·6 | 4.3 | 2·0 | 50·4 |
| | | 2 | 701·0 | 865·4 | 1·60 | 23·5 | 94·8 | 0·5 | 1·7 | 0·9 | 1·5 | 0·7 | 92·3 |
| | | 3 | 686·9 | 859·2 | 1·56 | 25·1 | 94·3 | 0·4 | 1·8 | 0·8 | 2·0 | 0·9 | 62·5 |
| Quarry near Underwood Hall, Cambs. TL61 3572 | *Holaster planus* | 1 | 600·4 | 781·9 | 1·44 | 30·2 | 26·9 | 5·2 | 8·1 | 12·9 | 26·1 | 20·9 | 10·4 |
| | | 2 | 696·8 | 850·9 | 1·60 | 22·1 | 68·6 | 3·9 | 5·9 | 3·9 | 9·8 | 7·9 | 32·7 |
| | | 3 | 667·8 | 837·6 | 1·50 | 25·4 | 38·8 | 9·2 | 14·9 | 8·8 | 15·7 | 12·6 | 8·5 |

## TABLE III

*Disintegration produced by six "C" cycles*

| Location | Fossil zone | Specimen number | Dry weight (g) | Weight when saturated (g) | Density (dry) | Water absorption as % of dry weight | Size of debris after frost weathering | | | | | | |
|---|---|---|---|---|---|---|---|---|---|---|---|---|---|
| | | | | | | | Percentage of total wet weight in different size classes | | | | | | Wet weight of largest fragment as % of total |
| | | | | | | | >20·0 (mm) | 20·0-9·5 | 9·5-3·4 | 3·4-1·7 | 1·7-0·08 | 0·08-0·00 | |
| Burwell Town Quarry, Burwell, Cambs. TL591663 | *Holaster subglobosus* (Totternhoe Stone) | 1 | 808·1 | 1061·9 | 1·76 | 16·0 | 50·5 | 28·2 | 14·3 | 2·2 | 3·3 | 1·4 | 14·1 |
| | | 2 | 761·4 | 892·4 | 1·82 | 17·2 | 66·3 | 19·5 | 8·6 | 1·7 | 2·3 | 1·5 | 10·6 |
| | | 3 | 798·1 | 932·2 | 1·80 | 16·8 | 91·3 | 4·0 | 2·8 | 0·7 | 0·9 | 0·4 | 84·7 |
| Quarry on Gog Magog Hills, Cambs. TL484539 | *Holaster subglobosus* (above Totternhoe Stone) | 1 | 725·9 | 890·0 | 1·65 | 22·6 | 75·8 | 12·8 | 6·8 | 1·1 | 2·8 | 0·7 | 34·2 |
| | | 2 | 717·8 | 885·0 | 1·65 | 23·3 | 54·6 | 13·3 | 23·1 | 2·6 | 4·8 | 1·6 | 30·1 |
| | | 3 | 767·2 | 945·2 | 1·63 | 23·2 | 59·2 | 19·4 | 16·9 | 0·7 | 2·6 | 0·9 | 39·7 |
| Quarry on Cherry Hinton Hill, Cambs. TL485556 | *Inoceramus labiatus* (Melbourn Rock) | 1 | 740·0 | 890·2 | 1·61 | 20·3 | 99·5 | 0 | 0 | 0 | 0·2 | 0·3 | 99·5 |
| | | 2 | 673·2 | 826·0 | 1·60 | 22·7 | 99·2 | 0 | 0 | 0 | 0·3 | 0·5 | 99·2 |
| | | 3 | 704·0 | 860·3 | 1·61 | 22·2 | 98·7 | 0·2 | 0·2 | 0 | 0·3 | 0·6 | 97·4 |
| Quarry near West Wratting Farm, Cambs. TL565545 | *Terebratulina lata* | 1 | 679·9 | 858·0 | 1·57 | 26·2 | 96·5 | 0·6 | 0·7 | 0·2 | 0·8 | 1·2 | 96·3 |
| | | 2 | 675·5 | 868·7 | 1·57 | 28·6 | 88·8 | 3·0 | 3·3 | 0·8 | 1·5 | 2·5 | 42·2 |
| | | 3 | 701·1 | 888·3 | 1·57 | 26·7 | 97·6 | 1·1 | 0·1 | 0·4 | 0·3 | 0·4 | 95·6 |
| Quarry near Underwood Hall, Cambs. TL613572 | *Holaster planus* | 1 | 762·4 | 930·9 | 1·69 | 22·1 | 74·0 | 10·3 | 8·3 | 1·6 | 2·9 | 3·0 | 67·2 |
| | | 2 | 715·1 | 883·9 | 1·55 | 23·6 | 92·9 | 1·8 | 2·8 | 0·6 | 0·9 | 0·9 | 52·5 |
| | | 3 | 657·5 | 836·3 | 1·43 | 27·2 | 50·2 | 18·4 | 15·8 | 2·8 | 6·2 | 6·5 | 13·8 |

TABLE IV

*Disintegration produced by six "D" cycles*

| Location | Fossil zone | Specimen number | Dry weight (g) | Weight when saturated (g) | Density (dry) | Water absorption as % of dry weight | Size of debris after frost weathering: Percentage of total wet weight in different size classes >20·0 (mm) | 20·0-9·5 | 9·5-3·4 | 3·4-1·7 | 1·7-0·08 | 0·08-0·00 | Wet weight of largest fragment as % of total |
|---|---|---|---|---|---|---|---|---|---|---|---|---|---|
| Burwell Town Quarry, Burwell, Cambs. TL591663 | *Holaster subglobosus* (Totternhoe Stone) | 1 | 866·8 | 1010·7 | 1·80 | 16·6 | 77·5 | 10·7 | 7·6 | 1·4 | 2·0 | 0·7 | 46·3 |
| | | 2 | 721·5 | 836·9 | 1·75 | 16·0 | 36·0 | 26·3 | 23·3 | 5·7 | 6·4 | 2·3 | 11·1 |
| | | 3 | 783·2 | 904·6 | 1·77 | 15·5 | 70·6 | 13·7 | 8·3 | 2·9 | 3·3 | 1·2 | 31·4 |
| Quarry on Gog Magog Hills, Cambs. TL484539 | *Holaster subglobosus* (above Totternhoe Stone) | 1 | 743·2 | 850·2 | 1·76 | 14·4 | 23·3 | 24·9 | 34·3 | 6·9 | 8·6 | 2·0 | 13·2 |
| | | 2 | 681·4 | 925·3 | 1·57 | 35·8 | 79·2 | 8·7 | 7·5 | 1·7 | 2·3 | 0·5 | 53·7 |
| | | 3 | 679·4 | 853·3 | 1·60 | 25·6 | 7·5 | 26·7 | 43·5 | 10·0 | 9·9 | 2·3 | 1·3 |
| Quarry on Cherry Hinton Hill, Cambs. TL485556 | *Inoceramus labiatus* (Melbourn Rock) | 1 | 741·6 | 874·3 | 1·71 | 17·9 | 89·9 | 4·7 | 3·3 | 0·7 | 0·7 | 0·7 | 52·2 |
| | | 2 | 670·3 | 807·0 | 1·65 | 20·4 | 98·7 | 0·5 | 0·4 | 0·1 | 0·1 | 0·1 | 96·7 |
| | | 3 | 715·5 | 874·3 | 1·59 | 22·2 | 97·0 | 0·6 | 1·3 | 0·2 | 0·4 | 0·4 | 70·6 |
| Quarry near West Wratting Farm, Cambs. TL565545 | *Terebratulina lata* | 1 | 686·6 | 883·7 | 1·55 | 28·7 | 87·6 | 4·0 | 5·4 | 1·0 | 1·5 | 0·6 | 35·8 |
| | | 2 | 724·6 | 908·6 | 1·59 | 25·4 | 98·8 | 0·4 | 0·4 | 0 | 0·3 | 0·1 | 98·8 |
| | | 3 | 686·8 | 868·1 | 1·52 | 26·4 | 97·0 | 1·4 | 0·9 | 0 | 0·4 | 0·2 | 96·5 |
| Quarry near Underwood Hall, Cambs. TL613572 | *Holaster planus* | 1 | 634·5 | 828·0 | 1·41 | 30·5 | 24·5 | 21·8 | 23·6 | 6·5 | 15·3 | 8·2 | 3·9 |
| | | 2 | 725·9 | 903·7 | 1·55 | 24·5 | 80·4 | 10·1 | 5·0 | 1·0 | 2·3 | 1·2 | 42·8 |
| | | 3 | 805·2 | 949·3 | 1·70 | 17·9 | 95·2 | 1·9 | 1·8 | 0·3 | 0·5 | 0·3 | 93·7 |

an interaction between minimum temperature and location ( $F = 35$ with 4 and 49 df; $P < 0{\cdot}025$). The samples of Melbourn Rock, for instance, proved very resistant to temperatures of − 10 °C, but were relatively easily disintegrated by temperatures of − 30 °C. It would seem that chalks can be ranked in order of resistance to frost weathering only if a particular minimum temperature be specified. The rank order apparently varies with temperature.

Analysis of variance assumes random sampling, but actually the blocks of chalk used for making the cubes were not selected at random. Only blocks that were a

TABLE V

*Percentage of total wet weight of cubes that turned into fines (particles of less than 3·4 mm diameter) after six cycles of freezing and thawing*

| Average of three cubes of: | | Type of cycle | | | |
|---|---|---|---|---|---|
| | | A | B | C | D |
| Totternhoe Stone | Burwell Town Quarry | 17·7 | 22·8 | 4·8 | 8·6 |
| *subglobosus* Chalk | Quarry on Gog Magog Hills | 25·8 | 31·4 | 5·9 | 14·7 |
| Melbourn Rock | Quarry on Cherry Hinton Hill | 30·4 | 34·0 | 0·7 | 1·2 |
| *Terebratulina* Chalk | Quarry near West Wratting Farm | 3·9 | 5·2 | 2·7 | 1·4 |
| *planus* Chalk | Quarry near Underwood Hall | 32·9 | 39·5 | 8·5 | 11·9 |
| Totals | | 110·7 | 132·9 | 22·6 | 37·8 |

convenient size and readily accessible were selected, and they had to be free from fractures, flints, or fossils. Random sampling would doubtless have increased the residual or error variance, but nevertheless it is thought that the results of the analysis of variance are broadly correct.

Some subsidiary experiments were conducted in which the debris was sieved at the end of each cycle and then replaced in the containers. These experiments showed that the amount of disintegration accelerates with each cycle. The percentage of fines tends to increase rapidly, while the percentage of large fragments correspondingly declines. Although some cubes resist frost weathering for a number of cycles and then suddenly weaken, most start to disintegrate with the first cycle.

It seems safe to conclude that chalk is a highly frost-susceptible material when saturated. In periglacial times disintegration must have been rapid if the surface layers of the chalk were saturated. Even in a single winter measurable amounts of weathering would have occurred. Over a period of a few hundred years the entire active layer would have been pulverized. Even on gentle slopes the weathered chalk would have been easily removed by sheetwash and solifluction.

One possible limitation of the experiments needs to be discussed. All the chalk cubes were saturated with water before freezing and kept in containers filled with water. Had the cubes been completely dry they would almost certainly have been unaffected by the changing temperatures. Preliminary experiments were conducted with cubes of dry chalk, and after six cycles no damage was visible. Dry chalk may eventually break as a result of repeated thermal contraction and expansion, but no experiments have been long enough to show this.

Probably only chalk that is saturated, or nearly saturated, can be damaged by frost, but this has yet to be proved. If the water in the pores is limited, and much of the pore space is filled with air, the water will compress the air when it freezes and not damage the rock. Experiments with rocks other than chalk have shown that there is a critical moisture content for a given type of rock which must be exceeded if significant damage is to occur. Some rocks break if the pore spaces are only 70 per cent filled with water, but most rocks break only if the pores are at least 80 per cent filled with water (Hirschwald, 1908, 1912). However, experiments have not yet been performed to determine the minimum moisture content that will cause chalk to break.

In periglacial times the surface layers of the chalk would undoubtedly have been saturated, or nearly saturated, at the foot of escarpments and on the floors of the deeper valleys. However, it is less easy to be certain about the chalk of upland areas and the heads of the valleys. One might imagine that the chalk was relatively dry near the surface and weathering proceeded relatively slowly. This might help to explain why ancient erosion surfaces and deposits are often preserved (although it is still surprising that they have managed to survive continuous solutional lowering of the ground surface). If the chalk was dry this would also explain why it forms high ground in many areas even though it is one of the most frost-susceptible of rocks when saturated.

But ought one to assume the chalk of upland areas was fairly dry? The answer would appear to be: no. The fractures and joints of the chalk are often dry at the present time, but this is hardly likely to have been the case under periglacial conditions. Rain and meltwater from snowbanks, percolating downwards into the chalk, would have frozen at depth preventing further infiltration. The surface layers of the chalk would have become saturated, at least during spring thaws. There is some field evidence to suport these ideas. Watt *et al.* (1966) note that involutions in chalk sometimes exhibit a peculiar layering which appears to be the result of the growth of semi-horizontal layers or lenses of ice in periglacial conditions. Such layers or lenses develop when saturated ground freezes and water is drawn up from depth towards the freezing front. The excess water crystallizes as layers or lenses of ice, and when the ground finally thaws it becomes temporarily supersaturated, and turns into a slurry. The laminations observed by Watt *et al.*, are a common feature of chalk involutions, even in upland areas. Apparently there was plenty of water around at the time the involutions were forming.

Tables I to IV suggest that chalk varies greatly in frost resistance from site to site. Each site in the tables belongs to a different stratigraphic or fossil zone, and it seems likely that the zones differ in frost resistance but more data are needed to prove this. The blocks of chalk used in the experiments were not necessarily representative of the selected sites, and the sites themselves were not necessarily representative of the fossil zones. One would need to investigate several sites within each fossil zone to prove that the zones differ in frost resistance. It is very difficult to believe, however, that the zones are equally resistant since the porosity of the chalk and its strength of bonding varies a good deal from zone to zone and these factors must partly determine its frost resistance. Some of the differences between the sites given in the tables can be safely assumed to be due to differences between the fossil zones, but nevertheless considerable caution is needed in interpreting the tables.

Some additional experiments that have not yet been completed show that chalks from different sites within the same fossil zone often vary considerably in their susceptibility to frost. This is not surprising, since the chalk of each fossil zone is far from homogeneous. As Hancock (1975) has pointed out, hard-grounds representing breaks in deposition recur at frequent intervals separated by beds of much softer rock. It would be necessary to test a large number of specimens of chalk in order to establish the average frost-susceptibility of a particular zone accurately.

Many relief features in chalk areas have been traditionally explained in lithological terms. The resistance of the chalk to differential weathering processes (including frost disintegration) is alleged to vary from one fossil zone to another. Yet, as Sparks (1971) has observed, the arguments tend to be circular. The chalk of certain fossil zones forms relatively high ground and the explanation is advanced that the chalk possesses greater resistance to weathering processes. Very often, however, the only evidence that the chalk is relatively resistant is the fact that it forms high ground. Experimental evidence is needed to confirm that the chalk really is resistant and to identify the precise process of weathering.

The most striking relief feature attributed to differential weathering and erosion is the secondary escarpment of the South Downs and Hampshire Basin. The main escarpment is capped by chalk belonging to the *Micraster* fossil zones, and this same chalk underlies the dip-slope to the south. The dip-slope merges into a discontinuous strike-vale underlain by chalk of the *Marsupites* and *Offaster* fossil zones. Southwards the ground rises again to form the secondary escarpment, which is not as high as the main escarpment, but is nevertheless a conspicuous relief feature. The lower part of the secondary escarpment is developed in *Offaster* chalk, but the upper part is capped by *Gonioteuthis (Actinocamax)* chalk.

The *Micraster* and *Gonioteuthis* chalks would appear to be much more resistant to weathering than the *Marsupites* and *Offaster* chalk, to judge from the differences in relief. Yet when one examines the chalks in the field they appear to be very similar, and it is difficult to account for the supposed differences in resistance. Sparks (1949) carried out some analyses which suggested that the *Marsupites* and *Offaster* chalks contain rather more insoluble matter than the *Micraster* and *Gonioteuthis* chalks (3 to 4 per cent as opposed to 1 to 2 per cent). Published descriptions and field observations convinced him that the *Marsupites* and *Offaster* chalks are more frequently interrupted by marl seams. He thought it likely that the *Marsupites* and *Offaster* chalks are less permeable than the chalks above and below and have been subject to more run-off and more erosion. He was puzzled, however, by the fact that the lithological differences are very slight compared with the considerable differences in relief.

Small and Fisher (1970) carried out additional analyses of the insoluble residues of the different chalks and came to the conclusion that the chalks are even more alike than Sparks supposed. They were inclined to attribute the greater resistance to weathering of the *Micraster* and *Gonioteuthis* chalks to the presence of greater numbers of nodular flints. The evidence they collected suggests that the *Marsupites* chalk, and the chalk of the lower part of the *Offaster* zone contain fewer flints. There is also a possibility (although the evidence is not strong) that the *Gonioteuthis* chalk may be better jointed and thus more permeable than the chalks immediately beneath.

Sparks (1971; p. 358) does not believe that flints contribute appreciably to the resistance of the chalk, but he is willing to accept the possibility that the *Gonioteuthis* chalk is better jointed than the *Marsupites* and *Offaster* chalks. Hodgson *et al.* (1974) consider that none of the existing explanations is satisfactory. They point out that "the summits of the secondary escarpment are invariably capped by Clay-with-Flints", and suggest that the Clay-with-Flints has protected the *Gonioteuthis* chalk from weathering and erosion.

One possibility that has not been given serious consideration is that the chalks of the main and secondary escarpments differ appreciably in frost resistance. Table VI records the results of some experiments on chalks from four sites in West Sussex, where the secondary escarpment is particularly well developed. Cubes of *Marsupites* and *Offaster* chalk tended to be severely damaged after only six cycles of freezing and thawing (cycle B). The result was that on average some 21·9 per cent of the original wet weight of the cubes came to consist of particles less than 3·4 mm in diameter. Cubes of *Micraster* and *Gonioteuthis* chalk, on the other hand, suffered comparatively little damage. Particles less than 3·4 mm in diameter accounted on average for only 2·6 per cent of the original wet weight of the cubes.

These results suggest that differences in frost resistance may have been a major factor in the development of the secondary escarpment. At first glance the differences are much more striking than the slight differences in the number of marl seams, flint layers, etc. noted by Sparks and Small and Fisher, which seem wholly inadequate to explain the size of the secondary escarpment. Too much reliance ought not to be placed on the frost weathering experiments, however, since only four sites have been investigated, and they may be very unrepresentative. Chalk from many more sites will need to be investigated to determine whether the fossil zones really differ in frost resistance. The results shown in Table VI could be an accident of sampling. It is also important to bear in mind that the experiments were concerned only with minimum temperatures of −30 °C, and the relative resistance of the fossil zones may be different at −10 °C or −20 °C.

Some additional experiments have already been carried out, but the results are difficult to interpret. Chalks were tested from a variety of sites at the eastern end of the South Downs, the minimum temperature again being −30 °C. The *Marsupites* and *Offaster* chalks proved much more resistant than the *Gonioteuthis* chalk, in complete reverse of the situation shown in Table VI. However, the *Marsupites* and *Offaster* chalks were collected at sites where there is no secondary escarpment, so

## TABLE VI

*Disintegration produced by six "B" cycles*

| Location | Fossil zone | Specimen number | Dry weight (g) | Weight when saturated (g) | Density (dry) | Water absorption as % of dry weight | Size of debris after frost weathering: Percentage of total wet weight in different size classes | | | | | | Wet weight of largest fragment as % of total |
|---|---|---|---|---|---|---|---|---|---|---|---|---|---|
| | | | | | | | >20·0 (mm) | 20·0-9·5 | 9·5-3·4 | 3·4-1·7 | 1·7-0·08 | 0·08-0·00 | |
| River cliff, | *Micraster* | 1 | 749·5 | 914·0 | 1·60 | 21·9 | 96·5 | 0·3 | 0·7 | 0·8 | 1·4 | 0·4 | 87·3 |
| South Stoke, | *coranguinum* | 2 | 783·5 | 950·2 | 1·60 | 21·3 | 99·2 | 0·6 | 0·1 | 0·1 | 0 | 0 | 97·1 |
| near Arundel, | | 3 | 782·6 | 948·7 | 1·60 | 21·3 | 99·7 | 0·3 | 0 | 0 | 0 | 0 | 99·7 |
| Sussex, | | 4 | 757·9 | 929·7 | 1·52 | 23·0 | 80·5 | 9·6 | 6·3 | 2·0 | 1·3 | 0·3 | 69·9 |
| TQ 020100 | | 5 | 780·4 | 950·3 | 1·60 | 21·9 | 96·8 | 1·2 | 1·1 | 0·5 | 0·3 | 0·1 | 96·8 |
| River cliff, | *Marsupites* | 1 | 670·7 | 786·2 | - | 17·2 | 44·3 | 13·8 | 23·9 | 8·5 | 6·2 | 3·3 | 14·3 |
| Burpham, | *testudinarius* | 2 | 721·8 | 874·1 | - | 21·1 | 39·8 | 13·2 | 24·1 | 10·0 | 8·4 | 4·4 | 26·5 |
| near Arundel, | | 3 | 720·4 | 868·8 | - | 20·6 | 40·5 | 6·4 | 27·1 | 11·3 | 9·6 | 5·1 | 35·8 |
| Sussex, | | 4 | 627·7 | 758·3 | - | 20·8 | 39·2 | 16·9 | 25·0 | 8·9 | 6·6 | 3·5 | 22·2 |
| TQ 038086 | | 5 | 670·4 | 821·9 | - | 22·6 | 16·7 | 8·7 | 33·8 | 18·4 | 14·7 | 7·8 | 8·1 |
| Quarry at | *Offaster* | 1 | 790·5 | 953·3 | 1·64 | 20·7 | 46·2 | 19·5 | 18·3 | 4·9 | 3·9 | 7·2 | 6·6 |
| Offham, near | *pilula* | 2 | 775·4 | 938·4 | 1·68 | 20·8 | 88·7 | 5·1 | 3·4 | 0·8 | 0·7 | 1·3 | 87·5 |
| Arundel, | | 3 | 757·8 | 922·1 | 1·65 | 21·8 | 7·1 | 13·1 | 44·9 | 10·6 | 8·5 | 15·9 | 1·5 |
| Sussex, | | 4 | 752·5 | 913·4 | 1·62 | 21·0 | 75·2 | 8·0 | 8·5 | 8·4 | 2·1 | 3·9 | 73·1 |
| TQ 024085 | | 5 | 838·7 | 1021·4 | 1·64 | 21·9 | 17·6 | 29·2 | 28·9 | 7·1 | 6·0 | 11·2 | 2·0 |
| Quarry, | *Gonioteuthis* | 1 | 675·6 | 832·9 | - | 23·3 | 90·2 | 2·1 | 4·8 | 1·4 | 1·2 | 0·3 | 67·4 |
| summit of | *quadrata* | 2 | 784·6 | 964·2 | - | 22·9 | 85·8 | 2·9 | 5·2 | 2·6 | 2·5 | 0·7 | 45·3 |
| Warning- | | 3 | 798·7 | 976·0 | - | 22·2 | 87·7 | 1·5 | 8·2 | 1·3 | 1·1 | 0·3 | 87·7 |
| Camp Hill, | | 4 | 755·0 | 914·2 | - | 21·1 | 92·2 | 2·2 | 3·7 | 0·7 | 0·8 | 0·2 | 90·6 |
| near Burpham, | | 5 | 796·3 | 976·7 | - | 22·7 | 83·8 | 5·5 | 5·9 | 2·3 | 2·0 | 0·5 | 75·2 |
| Sussex, | | | | | | | | | | | | | |
| TQ 047084 | | | | | | | | | | | | | |

The *Marsupites* and *Gonioteuthis* samples were sieved after each cycle, but the sieving is not thought to have contributed to the disintegration.

perhaps one would expect them to be resistant. The *Gonioteuthis* chalk was collected from two quarries well south of the crest of the secondary escarpment where it reaches its easternmost limit. There was thus no *a priori* reason to suppose that the chalk would prove to be particularly resistant. It is clearly important that in future, specimens of chalk be collected only from areas where the secondary escarpment is well developed, and only from crest-top sites and sites at the base of the escarpment. Unfortunately, as Small and Fisher observe, suitable exposures are comparatively rare, and it may be doubted whether enough exposures are available for a proper statistical analysis.

The origin of the escarpment will not be finally settled, even if further experiments confirm that the *Micraster* and *Gonioteuthis* chalks are more frost resistant than the *Marsupites* and *Offaster* chalks in areas where the secondary escarpment occurs. There is an obvious danger of circular argument. It will not be enough to show that the beds at the base of the escarpment are appreciably weaker than those at the top. Strictly speaking, it will also be necessary to prove that the weakness is original and owes nothing to the existence of the escarpment. This will not be easy. The chalk at the base of the escarpment may be much damper, for instance, than the chalk at the top of the escarpment merely because of the form of the relief. For all that is known, dampness over long periods may seriously weaken a rock like chalk, causing it to become more susceptible to frost.

## Amounts of periglacial erosion

The freeze-thaw experiments suggest that the surface layers of the chalk disintegrated rapidly with the onset of periglacial conditions. Apparently great quantities of shattered debris were produced annually. The surface of the ground must in places have been quickly reduced to a finely-divided sludge. Given its size composition one can readily imagine that the weathered material was stripped from slopes by solifluction and sheetwash almost as quickly as it formed, exposing the bedrock to almost continuous weathering. Only on very gentle interfluves is the weathered material likely to have remained *in situ*, thus protecting the bedrock beneath from further weathering. If these ideas are correct, the chalk landscape must have undergone fairly considerable modification in periglacial times.

All this is hypothetical, of course, and it is necessary to consider whether any supporting evidence shows extensive periglacial erosion. It must be said at the outset that there is a serious lack of agreement as to the effect of periglacial processes on chalk landscapes. Some investigators have suggested that there was considerable erosion, and that chalk landscapes are almost entirely periglacial in origin. Others have been more cautious and have postulated only very limited erosion. A crucial question for many investigators is the origin of the dry valleys. If the valleys are periglacial in origin, one can only conclude that a very considerable amount of erosion took place in periglacial times. Indeed, it would seem safe to regard most of the chalk landscape as having been fashioned under periglacial conditions. Not everyone is agreed, however, that the dry valleys are entirely periglacial, and some would say that they are largely inter-glacial or even post-glacial in origin.

Reid (1887) was apparently the first writer to champion a periglacial origin. He suggested that the chalk was once deeply frozen and thus impervious to water. Rain falling in summer was unable to sink into the ground and collected as surface streams which carried away the debris loosened by the frost and so carved out the valleys. With the passing of periglacial conditions the valleys became dry and all excavation ceased.

The streams that excavated the valleys were likened by Reid to mountain torrents possessed of "enormous scouring and transporting power", but nowadays it is thought that the streams were not particularly violent, and that excavation of the valleys proceeded comparatively slowly. As a matter of fact, there is no evidence that permanent stream courses have ever existed in the upper sections of the valleys. Periglacial deposits towards the heads of the valleys are poorly sorted and appear to be the result of sheetwash and solifluction, rather than stream action. The fossil Mollusca that are found in the deposits are exclusively land species (see, for example, Kerney, 1963, and Williams, 1971). Freshwater species are found only in deposits in the lower sections of the valleys and at the base of the chalk escarpments where there were springs. If streams ever flowed in the upper sections of the valleys they could only have been very shallow and transitory.

Reid did not attempt to explain why chalk dry valleys vary greatly in form, though this in fact is a major problem. Later writers have tried to distinguish different types of dry valleys and have pointed out that they may not all have had the same origin. Steep-sided escarpment valleys with narrow, flat floors and abrupt ends have received especial attention. They are often rectilinear in plan with curious right-angled bends. While some of the valleys may have been excavated by rain-fed streams running over frozen ground or meltwater issuing from snow-banks, others (see, for example, Kerney *et al.* 1964) may have been excavated during the Post-glacial Period by springs sapping headwards along major joint lines in the chalk (see, for example, Sparks and Lewis, 1957).

Remarkable armchair-like hollows occur in the face of the escarpment of the South Downs near Eastbourne. Bull (1940) thought the hollows were occupied by bodies of snow or ice when the climate was periglacial in character. Debris produced by frost shattering was sludged out of the hollows and accumulated in front of the escarpment. Kerney (1963) has investigated the deposits in one hollow in some detail and suggests that solifluction and meltwater were active during the Late-glacial and earlier on in the Devensian.

Shallow, paddle-shaped or bowl-like dry valleys occur on the dip-slopes of the chalk escarpments, often in great numbers, and these too may be periglacial in origin. They are not properly integrated with the main network of dip-slope valleys. French (1976) considers that they probably developed through meltwater erosion beneath snowdrifts when the chalk was either permanently or seasonally frozen. He refers to them as "dells".

The theory that some or all dry valleys have a periglacial origin meets strong competition. It is an undoubted fact that the clay vales surrounding the chalk uplands have undergone considerable erosion in Quaternary times. The excavation of the clay vales has certainly lowered the water table in the chalk and could explain why the valleys are now dry. It is likely also that the excavation of the dry valleys has itself contributed to the lowering of the water-table and the desiccation of the valleys. Pinchemel (1954) has aptly termed this process 'auto-assèchement'.

It seems likely that only a minority of chalk dry valleys originated under periglacial conditions, and that Reid overstated his case. The majority of valleys would appear to have been initiated before the onset of periglacial conditions, perhaps on a cover of Tertiary deposits. During the early Quaternary, the water-table in the chalk would have been much higher than at present and the valleys not as deeply excavated. Interglacial erosion may in these early stages have been fully as important as periglacial erosion. During the late stages of the Quaternary, however, the valleys may have evolved mainly through periglacial erosion. Interglacial and Post-glacial erosion is likely to have been negligible, owing to the low level of the water-table.

It is rarely possible to determine how much periglacial erosion has occurred from the volume of soliflUctional and sheetwash deposits that have been left behind. Often there is no indication of the precise area from which the deposits have been derived. A further difficulty is that the deposits are often thin or even totally absent. It would seem that in periglacial times great quantities of weathered material were washed away in streams and finally deposited somewhere out to sea. The volume of deposits in the immediate vicinity of chalk outcrops tends to be seriously misleading and only where the valleys descending the chalk uplands reached an extensive plain was the weathered and eroded material deposited in anything approaching full measure. Exceptional amounts of material were deposited, on the relatively flat and undissected coastal plain of Sussex and Hampshire and this allows of estimates of the amount of erosion in the chalk uplands bordering the plain. For instance, the volume of deposits resting on the plain in the Portsmouth area shows that at least 10·5 m (35 ft) was stripped from the southern side of Portsdown in periglacial times (Edmunds, 1929).

The Portsmouth deposits have not been dated precisely. They may have accumulated during the whole of the Devensian cold period or only during a sub-stage. The truth seems to be that very few periglacial deposits can be dated accurately. A remarkable exception is provided by the fans of debris at the foot of the chalk escarpment near Brook in Kent (Kerney *et al.*, 1964). The fans overlie radio-carbon dated marsh deposits belonging to Zone II of the Late Glacial Period. Evidently the fans formed during Zone III (9350 to 8500 B.C.). The debris forming the fans came from a series of steep-sided dry valleys which are cut deep into the face of the escarpment.

To judge from the volume of debris in the fans about 33 per cent of the erosion of the valleys was accomplished during Zone III. As the authors point out, even this astounding figure may be a serious underestimate because a considerable amount of debris may have been swept away by streams and not deposited on the fans. Before Zone III the valleys are likely to have been very much smaller and shallower than they are now. Indeed there can be no certainty that they even existed.

What importance ought one to attach to the Brook site? It is clear that considerable modification of chalk landscapes must have occurred during Zone III if erosion was everywhere as rapid as at Brook. But there may have been peculiar local conditions that led to greatly accelerated erosion. Certainly, debris fans as extensive as those at Brook are not usually found at the foot of chalk escarpments. Thus, although large fans occur below the Chilterns escarpment (near Wallingford, for example) there are no large fans below the South Downs escarpment, or below the North Downs escarpment except in the vicinity of Brook. At the moment it seems wisest to treat the Brook site as a special case and not to base generalizations upon it.

One must also bear in mind that special climatic conditions are likely to have existed during Zone III. As Kerney *et al.* remark, the temperature may have oscillated continually around the freezing point, thus promoting exceptionally rapid frost weathering. There was probably much more water around than during the full Devensian, and hence erosion is likely to have been unusually rapid. The annual rate of erosion during the full Devensian was probably only a fraction of the rate during Zone III.

Although most estimates of periglacial erosion have been based on computations of the volume of deposits left by the erosion, there is an alternative method that can be sometimes employed with success. It involves reconstructing the changes that took place in the form of the landsurface during the period of periglacial erosion. The strengths and weaknesses of the method can be illustrated with reference to valleys that have been rendered asymmetrical as a result of periglacial erosion.

Dry valleys cut in the chalk are often strikingly asymmetrical in cross-section. The asymmetry in many places has a simple structural cause: one side of the valley is cut against the dip and has a steeper slope than the other side which is cut with the dip. Sometimes the asymmetry has a lithological cause: one side of the valley is cut into harder chalk than the other and tends therefore to be steeper. Asymmetry also occurs at valley bends, the slope on the outside of the bend being characteristically much steeper than the slope on the inside owing to undercutting by former streams. In addition, there is a systematic tendency for slopes facing west and south to be steeper than slopes facing north and east. This 'regional asymmetry' is especially noticeable in areas where the valleys are aligned in a north-west-south-east direction. Where the valleys have a different alignment, or a complex dendritic pattern, the asymmetry tends to be less obvious. Differences in structure and lithology also serve to obscure or even eliminate regional asymmetry.

Areas where regional asymmetry is well developed include the Chiltern Hills (Ollier and Thomasson, 1957; French, 1972), the North Downs between Canterbury and Folkestone (Smart *et al.*, 1966), the North Wiltshire Uplands and the North Dorset Downs (French, 1973, 1976). As French points out, the asymmetry is often confined to the middle and lower sections of the valleys; the shallow heads of the valleys are generally symmetrical. In the asymmetric sections of the valleys the gentler slopes facing north and east are the longer in length and often dissected by small, shallow symmetrical valleys (dells) which are tributary to the main valleys. The steeper and shorter slopes facing south and west tend to be relatively undissected.

The asymmetry of slopes is usually accompanied by an asymmetry of periglacial deposits. In the Chilterns, for example, the upper parts of the steep south- and west-facing slopes are usually bare of deposits, whereas the lower slopes and the valley floors are covered with chalky, clayey, solifluction deposits that are full of flints. The gentle north- and east-facing slopes tend to be covered for their entire length in flinty clay that is largely non-calcareous and derived by solifluction from Clay-with-Flints on the hill crests.

In the area around Canterbury and Folkestone asymmetry of deposits is particularly marked. Slopes facing north and east are covered with 'head-brickearth' composed of fine sand and silt with only minor amounts of chalk. The valley floors are underlain by chalky solifluction debris, as in the Chilterns. Deposits are generally lacking on the steeper slopes facing south and west. The asymmetry of the slope deposits is very apparent on the geological maps of the area (Geological Survey sheets 289 and 290).

The occurrence of periglacial deposits in asymmetrical valleys does not prove that the asymmetry is periglacial in origin for the volume of deposits is not great compared with the size of the valleys. Nevertheless, the asymmetry is usually thought to be periglacial, if only because there is no adequate alternative explanation. During the warm periods of the Quaternary the forest vegetation is likely to have been similar on opposite sides of the valleys. It is difficult to believe that there were major differences in the rate of soil creep or solutional weathering. Slope asymmetry does not seem to be actively developing in humid temperate regions at the present time, but there is ample evidence that it is developing in arctic regions (see, for example, Kennedy and Melton, 1972). One can readily imagine that in periglacial Britain slopes facing in different directions were subject to very different rates of weathering and erosion. Slopes facing south and west were sunnier than slopes facing north and east, and probably thawed to greater depths. They would have had less snow cover if the prevailing wind was south-westerly. Slopes facing north and east would have been on the lee side of the hills and would possibly have had a thicker and longer-lasting snow cover. Solifluction on these slopes may have been very much greater than on the slopes facing south and west.

French (1972) has collected evidence showing that the asymmetrical valleys were once symmetrical and less deeply incised. Downcutting by streams during periglacial episodes was accompanied by a lateral movement of the stream channel in a north-easterly direction, thus promoting an increase in the steepness of the south- and west-facing slopes. The cause of this lateral movement is unclear. French suggests that the south- and west-facing slopes were more easily eroded by the stream because they thawed more deeply. An alternative explanation is that greater solifluction on the north- and east-facing slopes pushed the streams across the valley floors.

Minimum figures for the amount of periglacial erosion can be obtained by measuring cross-sections of the asymmetrical valleys. Figure 2 is a cross-section of the valley at Little Hampden in the Chilterns, the most closely studied of all the asymmetrical valleys in the English chalk (see Ollier and Thomasson, 1957, and French, 1973 for previous discussion). Assuming the valley was symmetrical, the maximum depth it could have attained before periglacial erosion began is 45 m (150 ft). Periglacial erosion has removed a triangular 'wedge' of chalk from the eastern side of the valley at least as large as that indicated on the figure. The cross-sectional area of the figured wedge is approximately 23 per cent of the area of cross-section of the present valley. Hence at least a fifth of the excavation of the valley would seem to be periglacial. The figure of 23 per cent does not allow for any erosion of the western side of the valley, and the presence of shallow solifluctional deposits on this side of the valley indicates that there was some erosion, though not necessarily very much.

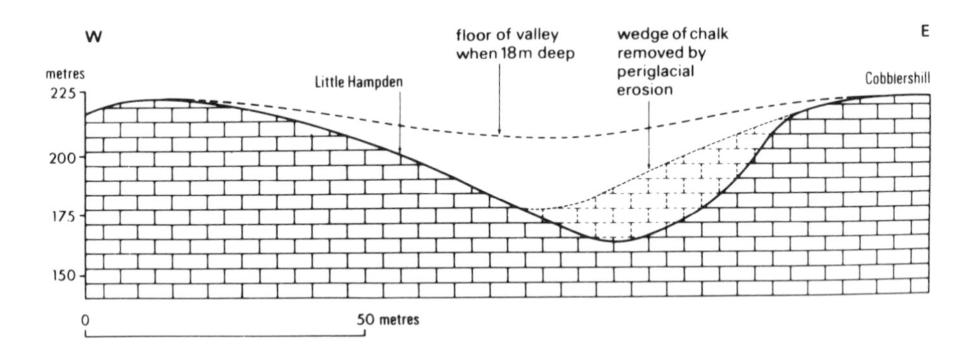


*Figure 2.* A cross-section of the valley at Little Hampden in the Chilterns.

If the erosion of the valley during its symmetrical phase was also partly or wholly periglacial, then the figure of 23 per cent may need to be greatly increased. French believes that asymmetry began to develop when the Chiltern valleys were only 14 to 18 m (45 to 60 ft) deep. If periglacial erosion began when the Little Hampden valley was only 18 m deep, some 63 per cent of the excavation of the valley at the line of cross-section is periglacial.

Examination of a number of other asymmetrical valleys suggests that the Little Hampden figures are fairly typical (Table VII). Periglacial downcutting is responsible for at least 15 per cent, sometimes 30 per cent or more, of the depth of the valleys in their middle and lower sections. The amount of chalk removed from the valleys is generally much greater than can be inferred from the shallow and often patchy deposits that have been left behind.

There is no evidence that symmetrical valleys were subject to any less erosion than the valleys that display regional asymmetry. As already mentioned, it appears likely that the processes responsible for the asymmetry operated throughout the chalk outcrop even though the asymmetry often did not develop (or is difficult to discern) because of lithological variations, changes in valley direction, etc. Accord-

TABLE VII

*Minimum amounts of erosion required to produce valley asymmetry (as % of cross-sectional area of each valley)*

| | |
|---|---|
| *Chiltern Hills* | |
| Valley at Little Hampden, 867037 | 23% |
| Bryants Bottom, 856994 | 18% |
| Chesham Vale, 964040 | 28% |
| Pednor Bottom, 928033 | 15% |
| Valley south of Bottom Farm, 813967 | 16% |
| Valley NW of Chawley Manor Farm, 805964 | 17% |
| *Marlborough Downs* | |
| Wick Bottom, 135739 | 36% |
| *North Dorset Downs* | |
| Devil's Brook near Dewlish, 773975 | 38% |
| Devil's Brook near Water Barn, 775953 | 25% |
| Valley north of Druce Farm, 960742 | 23% |

ing to this view, all dry valleys would be asymmetric were it not for special local effects. Estimates of periglacial erosion based on asymmetrical valleys, as far as one can tell, are thus valid estimates for dry valleys in general.

Estimates of periglacial erosion given in an earlier paper (Williams, 1968) are roughly comparable with those given here. It is hard to escape the conclusion that amounts of erosion were quite large in many places and the chalk landscape underwent major modification. The rapidity of weathering suggested by the freeze-thaw experiments thus receives some support from the calculations of the erosion. The amounts of erosion would undoubtedly have been much less had the chalk been more resistant to frost weathering.

Slopes were probably subject to differing amounts of periglacial erosion according to their steepness. Where slopes were steep considerable solifluction and sheetwash may have taken place. On gentle slopes the downslope component of gravity was less and amounts of erosion may therefore have been small. This may be the reason why ancient erosion surfaces and deposits like the Lenham Beds are preserved on broad, flat interfluves. Alternatively, erosion may have been slow because the chalk was relatively dry and weathered only slowly.

The distribution of patterned ground provides some indication of rates of erosion. It is noticeable for instance, that chalkland stripes are restricted to slopes of under four or five degrees. The underground structure of the stripes (Williams, 1968, 1973) is never seriously deformed in a downslope direction and often is not deformed at all. One can only conclude that the slopes on which the stripes formed were more or less stable. Moreover, it would seem that on slopes greater than about five degrees solifluction was so rapid it prevented the formation of stripes, or eliminated them as quickly as they formed. A similar argument can be advanced for involutions. On slopes of four degrees or less involutions show little sign of downslope movement, but on steeper slopes they tend to be severely distorted. The steeper the slope the less frequent they become. No involutions are found on slopes steeper than nine degrees, presumably because the slopes were subject to such rapid solifluction that involutions could not form or were destroyed.

The available evidence, then, would seem to indicate that periglacial erosion was highly selective. On broad, flat interfluves erosion was often slight, yet considerable erosion took place in the valleys. The sides of many valleys remained steep despite massive solifluction because the valleys were subject to active downcutting by streams.

One difficulty remains and this is the steepness of the chalk escarpments. The escarpments have evidently been subject to periglacial erosion since solifluctional and other deposits are found at their base. What is surprising is that the erosion did not succeed in reducing the escarpments to a low angle. There are many places on the escarpments where the slope is 25-35° and the surface soil is visibly unstable. Under periglacial conditions the chalk ought to have been rapidly carried away and the slope lowered.

The steepness of the escarpments does not seem to be a recent phenomenon. The slopes cannot have changed much in inclination since Late-glacial times because deposits of this age are found mantling parts of the scarps that are identical in angle to parts now bare. Nor is there much evidence of retreat since the Late-glacial.

The escarpments remained steep despite periglacial erosion. One possible explanation is that the chalk at the base of the escarpments was damper and underwent more rapid frost weathering than the chalk at the top of the escarpments. Surface water, derived perhaps from the melting of snow drifts, may have transported the weathered debris away, thus helping to maintain the steep face of the escarpments. There are important objections to this explanation, however. Permafrost at depth ought to have ensured that the surface layers of the chalk were saturated not just at the base of the escarpments but also at the top. And weathered debris moving down the escarpments ought to have protected the bedrock at the base from over-rapid weathering.

Another possibility is that the chalk capping the escarpments is particularly frost resistant. But it is difficult to see this as the sole explanation for the steepness of the escarpments since the chalk that forms the capping belongs to a whole series of different fossil zones. The fossil zone forming the capping in one place is found low down the escarpment in another.

One further possibility is that the processes of erosion were different from those on gentle slopes. Because the escarpments were high and steep they would have tended to trap snow in large amounts. Snow avalanching and nivation may have helped to maintain the escarpments at a steep angle despite rapid solifluction.

The idea of former snow action on the face of the escarpment is a very old one. It has been proposed from time to time in connection with various types of dry valleys that cut into the face of the escarpments. But the features most suggestive of snow action appear never to have been commented on. These are giant 'flutes' which run down the face of the escarpments particularly in western chalk areas, giving the escarpments a resemblance from a distance to a steeply-pitching roof of corrugated iron. Each 'flute' is shallow (no more than 3 or 4 m deep) and smoothly concave in horizontal cross-section, being separated by convex ridges from its neighbours. The 'flutes' are rhythmically spaced about 50 to 75 m apart. The floors of the flutes are roughly parallel with the face of the escarpment, and at the base of each flute there is usually a subdued fan, presumably made of debris rather than bedrock.

Because of the shallowness of the 'flutes' they do not appear on contour maps and they are hard to see even in the field, except when the sun is casting long shadows. They are found not only on the escarpments but also on the sides of some of the deeper dry valleys. Bratton in Wiltshire is one of the best places to see them.

The flutes are apparently former avalanche chutes similar to those described by Rapp (1959) from the mountains of northern Scandinavia. Their form is strikingly similar, and they are often occupied by snowdrifts in present severe winters. The presence of the flutes appears to indicate that snow avalanching occurred on the face of the escarpments in periglacial times, and this may have helped maintain a fairly steep angle of slope. The flutes have yet to be studied in detail, however, and any ideas about them must necessarily be tentative. The problem of the steepness of the chalk escarpments cannot be regarded as solved at present, and much detailed research is clearly needed.

## References

Aubry, M. P. and Lautridou, J. P. (1974). Relations entre proprietes physiques, gēlivitē et caracteres microstructuraux dans divers types de roches: craies, calcaires crayeux, calcaire sublithographique et silex, *Bull. Centre de Gēomorphologie C.N.R.S. Caen* **19**, 7-16.

Bull, A. J. (1940). Cold conditions and land forms in the South Downs, *Proc. Geol. Ass.* **51**, 63-71.

Coope, G. R. (1975). Climatic fluctuations in northwest Europe since the Last Interglacial indicated by fossil assemblages of Coleoptera, *in* Wright, A. E. and Moseley, F. (eds), "Ice Ages: ancient and modern", pp. 153-168. *Geol. J. Sp. Issue* **6**.

Coope, G. R. (1977). Fossil coleopteran assemblages as sensitive indicators of climatic changes during the Devensian (Last) cold stage, *Phil. Trans. Roy. Soc. B* **280**, 313-340.

Coutard, J.-P., Helluin, M., Lautridou, J.-P. and Pellerin, J. (1970). Gēlifraction expēri-mentale des calcaires de la Campagne de Caen; comparison avec quelques dēpôts pēri-glaciaires de cette rēgion, *Bull. Centre de Geomorphologie C.N.R.S. Caen* **6**, 7-44.

Edmunds, F. H. (1929). The Coombe Rock of the Hampshire and Sussex Coast, *Summ. Prog. Geol. Surv.*, pp. 63-68.

French, H. M. (1972). Asymmetrical slope development in the Chiltern Hills, *Biul. Peryglacjalny* **21**, 51-73.

French, H. M. (1973). Cryopediments on the Chalk of Southern England, *Biul. Peryglacjalny* **22**, 149-156.

French, H. M. (1976). "The periglacial environment". Longman, London.

Hancock, J. M. (1975). The petrology of the Chalk, *Proc. Geol. Ass.* **86**, 499-536.

Hirschwald, J. (1908). "Die Prüfung der Näturlichen Bausteine auf ihre Wetterbeständigkeit". Borntraeger, Berlin.

Hirschwald, J. (1912). "Handbuch der Bautechnischen Gesteinsprüfung". Borntraeger, Berlin.

Hodgson, J. M., Rayner, J. H. and Catt, J. A. (1974). The geomorphological significance of Clay-with-Flints on the South Downs, *Trans. Inst. Br. Geogr.* **61,** 119-129.

Kennedy, B. A. and Melton, M. A. (1972). Valley asymmetry and slope forms of a permafrost area in the Northwest Territories, Canada; *in* Price, R. J. and Sugden, D. E. (eds), "Polar Geomorphology", pp. 107-22. *Inst. Br. Geogr. Sp. Pub.* **4**.

Kerney, M. P. (1963). Late-glacial deposits on the Chalk of South-East England, *Phil. Trans. Roy. Soc. B* **246**, 203-254.

Kerney, M. P., Brown, E. G. and Chandler, T. J. (1964). The Late-glacial and Post-glacial history of the Chalk escarpment near Brook, Kent, *Phil. Trans. Roy. Soc. B* **248,** 135-204.

Lautridou, J.-P. (1970). Gélivité de la Craie de Tancarville; le head de l'estuaire de la Seine, *Bull. Centre de Géomorphologie C.N.R.S. Caen* **6,** 47-61.

Ollier, C. D. and Thomasson, A. J. (1957). Asymmetrical valleys of the Chiltern Hills, *Geog. J.* **123,** 71-80.

Pinchemel, P. (1954). "Les plaines de craie du nord-ouest du bassin parisien et du sud-est du bassin de Londres et leur bordures: *étude de géomorphologie*". Colin, Paris.

Potts, A. S. (1970). Frost action in rocks: some experimental data, *Trans. Inst. Br. Geogr.* **49**, 109-124.

Rapp, A. (1959). Avalanche boulder tongues in Lappland, *Geogr. Annaler* **41,** 34-48.

Reid, C. (1887). On the origin of dry valleys and of coombe rock, *Q. J. Geol. Soc.* **43,** 364-73.

Small, R. J. and Fisher, G. C. (1970). The origin of the secondary escarpment of the South Down. *Trans. Inst. Br. Geogr.* **49,** 97-107.

Smart, J. G. O., Bisson, G. and Worssam, B. C. (1966). "Geology of the Country around Canterbury and Folkestone". HMSO, London.

Sparks, B. W. (1949). The denundation chronology of the dip-slope of the South Downs, *Proc. Geol. Ass.* **60,** 165-215.

Sparks, B. W. (1971). "Rocks and relief". Longman, London.

Sparks, B. W. and Lewis, W. V. (1957). Escarpment dry valleys near Pegsdon, Hertfordshire, *Proc. Geol. Ass.* **68,** 26-38.

Thomas, W. N. (1938). Experiments on the freezing of certain building materials, *Building Res. Station Tech. Paper* 17.

Tricart, J. (1956). Etudes expérimentales due problème de la gélivation, *Biul. Peryglacjalny* **4,** 282-318.

Watt, A. S., Perrin, R. M. S. and West, R. G. (1966). Patterned ground in Breckland: structure and composition; *J. Ecol.* **54,** 239-258.

Williams, R. B. G. (1968). Some estimates of periglacial erosion in Southern and Eastern England, *Biul. Peryglacjalny* **17,** 311-335.

Williams, R. B. G. (1971). Aspects of the geomorphology of the South Downs, *in* R. B. G. Williams (ed.), "Guide to Sussex Excursions". Inst. Brit. Geogr. Conference.

Williams, R. B. G. (1973). Frost and the works of man, *Antiquity* **47,** 19-31.

Williams, R. B. G. (1975). The British climate during the Last Glaciation: an interpretation based on periglacial phenomena *in* Wright, A. E. and Moseley, F. (eds), "Ice Ages: ancient and modern", pp. 95-120. *Geol. J. Sp. Issue* **6**.

Wiman, S. (1963). A preliminary study of experimental frost weathering, *Geogr. Annaler* **45,** 113-120.

TEN

# The shape of the future

C. P. GREEN
*Department of Geography, Bedford College, London.*

## Introduction

What shall we discover in the future about the shaping of southern Britain? Foreseeing the future is difficult, but it relies on skills similar to those required for reconstructing the past. To succeed in either case, we must recognize the factors that are likely to be operative at a time other than the present. Such factors can be recognized only from a careful study of the components and mechanisms of the system in which we are interested. The first part of this chapter therefore examines the study of landform development in southern Britain to identify from what disciplines students of landform development come, where studies of landform development fit into the broader geomorphological domain, and how such studies have evolved within British geomorphology as a whole in the past. In the second part of the chapter, an attempt is made to evaluate the possible impact of recent developments on the course of future progress.

To understand the study of landform development, it is necessary first to recognize that the boundaries of geomorphology are not closely related to the boundaries of other traditional disciplines within the natural sciences. Direct contributions to the study of landform development have come both from Geography and Geology, and indirect contributions from a wide range of earth and life sciences. Recent discussion of the future of British geomorphology (Stoddart, 1975; Dury, 1978) has stressed the importance of this diversity as a potential stimulus in the future growth of the subject.

Direct contributions to the study of landform development can be separated into three fairly distinct classes. Firstly, study of the spatial organization of landforms is, by defintion, the concern of geomorphologists and seems logically to belong in the broader field of Geography (Sugden, 1979). Secondly, the relative lack of interest in the Quaternary among British geologists has encouraged geographical geomorphologists to extend their work into this field. Such geomorphologists have often, because of their geographical backgrounds, sought interpretations of Quaternary evidence based primarily on geographical distributions rather than on the principles of stratigraphy. Thirdly, the contribution of geologists has usually been through the stratigraphical interpretation of the depositional evidence.

In the study of landform development, progress since the late nineteenth century has been made, not by the discovery of fundamental geomorphological principles or laws, but by the discovery of geomorphologically relevant evidence and by its accommodation in a general environmental or palaeoenvironmental framework. This framework can be seen to have evolved gradually as the various disciplines that contribute to its formulation have themselves evolved. In the presentation of studies of landform development, the formulation of the environmental framework has usually been implicit rather than explicit, but has primarily reflected the contemporary state of understanding in three main fields: the geomorphological system itself; the fluctuations of the broader environmental system; and the dating of the past.

## The geomorphological domain

Geomorphology in Britain, and elsewhere, has developed along three major axes which can be described as functional, spatial and temporal. The functional axis defines the nature of the geomorphological system, and can be viewed in terms of three elements: (1) Geomorphological processes—both the instantaneous processes of erosion, deposition and chemical reaction, and the evolutionary processes of morphological change. (2) The independent environmental variables that govern and regulate process—specifically, the behaviour of climate; geological conditions, both structural and lithological; and the behaviour of base-level. (3) The landforms and materials that result from the operation of geomorphological processes, but which may themselves exert, through feedback relationships, a considerable influence upon the activity of those processes. These elements are susceptible to precise definition and their measurement and characterization have been an important branch of geomorphology during the last twenty years. The relationships between these main elements of the geomorphological system are the other essential component of that system, defining the ways in which the effects of changes in any part of the system are transmitted along the functional axis.

The spatial axis defines the geographical distribution of environmental conditions and, therefore, of geomorphological systems and of relief. The temporal axis defines the evolutionary nature of geomorphological systems, and along the temporal axis it is possible to trace the pattern of change produced by those processes that are not instantaneous in their effect. It is also possible to trace the occurrence and significance of relic forms as components in the spatial and functional axes. These are forms that are the product of past environmental conditions, but which persist, and influence the nature of the contemporary geomorphological system. And finally it is possible, on the temporal axis, to reconstruct past geomorphological systems and to evaluate the functional relationships associated with environmental change in the past, which may have been of a kind or an intensity that no longer occurs.

## The development of British geomorphology

A brief examination of the history of geomorphology in Britain, in terms of this triaxial model, is useful as it facilitates the identification of several well-marked developmental stages, and reveals how the course of development has been influenced by the aims, traditions and limitations of the main exponents.

In the nineteenth century, landforms were recognized by geologists as an important class of evidence in the study and explanation of Tertiary and Quaternary events (Foster and Topley, 1865). This interest among geologists in landform has of course continued down to the present day and represents an important contribution to studies in the temporal dimension of geomorphology. In the nineteenth century, however, geographical studies of relief were mainly descriptive. It was not until the first half of the present century, that the study of landforms as a branch of Physical Geography was transformed, under the pervasive influence of W. M. Davis, from a descriptive procedure into the forerunner of the modern geomorphological sub-discipline. In this transformation, S. W. Wooldridge played a leading and influential role. At the same time and subsequently, the geological and geographical aspects of geomorphology developed side by side, and the relationship between the two aspects, often one of hardly disguised tension, has been an important influence in the development of the subject.

### The supremacy of denudation chronology

Prior to 1960, despite the formalization of geomorphology within Geography, and perhaps because of the subject's early geological affinities, its development occurred, with rare exceptions, through the investigation of its temporal axis. This focus is clearly illustrated by the frequent use during this period of the term 'denudation chronology' (Wooldridge and Kirkaldy, 1936; Sparks, 1949). The essential aim of such chronological studies was to recognize in the landscape, elements or facets representing evidence of successive stages of development. The key facets were base-level surfaces, notionally associated with stable environmental conditions, and separated from one another by valley-side slopes, indicative of phases of dissection and notionally associated with environmental instability. This largely geographical work was linked to, and proceeded alongside, studies of the depositional evidence, which tended to rely on stratigraphical principles and to build on biological and archaeological schemes of relative dating (King and Oakley, 1936; Arkell, 1943).

The interpretation of landform development at this time tended to assume a satisfactory understanding of the functional axis of geomorphology, both in terms of the geomorphological processes and of the environmental factors regulating them. This assumption appears now to have been unwarranted, although the influential ideas were not so much wrong, as partial in their scope. The geomorphological processes themselves were, for all practical purposes, *never* the subject of discussion, except in terms of a few relatively uncomplicated 'rules'. The diversity of the environmental variables and the complexity of the relationships among them had not been fully recognized, nor the complexity of the relationships between geomorphological process and relief. Consequently in the reconstruction of past geomorphological systems, relatively few alternative palaeoenvironmental states were visualized and evidence could therefore be accommodated in an uncomplicated sequence of clearly defined episodes. Thus the influential consideration was not the processes, which were taken for granted, but the environmental 'rules', of which three have been particularly important. Firstly, understanding of the climatic fluctuations of the Quaternary was completely dominated for many years in Britain by the four cycle Alpine model of Penck and Bruckner (1909). Secondly, understanding of eustatic effects relied heavily on the Mediterranean evidence in which five major high sea-level episodes appeared to be associated with interglacial conditions (Depéret, 1918); a scheme that could itself be fairly easily related to the Alpine model of climatic fluctuation. Thirdly, a fundamental assumption about the context of landform development in Britain was that major tectonic effects, those associated with the Alpine orogenic episode, could be neatly contained in a mid-Tertiary episode, before and after which, landform development proceeded without tiresome tectonic complications.

## The measurement revolution in geomorphology

As it became increasingly evident after 1960 that no satisfactory understanding of geomorphological processes existed, the focus of study in Britain changed fairly rapidly to the functional axis of geomorphology, concentrating there on studies connected with the frequency and magnitude of erosion. In its initial form this approach was a simple measurement of erosion (Young, 1960). In its more elaborate forms it attempted to relate measured temporal variations in erosional intensity to measured variations of the independent environmental variables (Bridges and Harding, 1971). In such studies the spatial dimension of the subject was often not of paramount interest, and the influence of erosion in terms of morphological change was presented, if at all, as a generalized measurement of regional denudation (Young, 1969).

The temporal dimension of the subject was also somewhat neglected at this time. It was almost inevitably neglected in short-term studies of process that examined relationships monitored during periods of, at most, a few years. The significance of the temporal axis was much more seriously challenged by students of the functional axis who argued that the geomorphological system was not in any real sense evolutionary (Hack, 1960), that present-day processes could account entirely for present-day landforms, and that, in effect, the temporal dimension of the subject did not exist at all.

Despite such obvious limitations in the treatment of the geomorphological domain, some geographers began, by the early 1970s, to suggest that an effective understanding of the functional axis of geomorphology had been largely achieved and that predictive models of the geomorphological system could, therefore, be constructed and put to use (Chorley and Kennedy, 1971). At the same time it was argued that the temporal dimension of geomorphology might be split off from the functional along a boundary between Geography and the Earth Sciences. Although these views might be appropriate in the limited temporal context of certain socially relevant problems, they are wholly unsatisfactory for geomorphology as a scientific discipline, and particularly for problems involving the longer spans of time within which significant environmental change has occurred in the past and may occur again in the future.

Fortunately the idea that geomorphologists might, with advantage, dismantle their subject has not had a wide intellectual appeal, as events of the last ten years show. Nevertheless, the survival of a coherent geomorphological sub-discipline and the proper disciplinary context for geomorphology are matters which continue to exercise the minds of geomorphologists (Brown, 1975; Dury, 1978; Worsley, 1979).

## Towards a comprehensive study of geomorphology

While the focus of investigation was on the measurement of environmental variables and of geomorphological processes, relatively little attention was paid to the exact nature of the relationships by which the effects of change are transmitted between the elements of the system. Uncomplicated relationships were generally assumed or implicit. The most recent developments in the study of geomorphological process have been concerned less with the characterization of the geomorphologically significant environmental variables, geomorphological processes, and landforms, and more with the way in which the effects of variability are transmitted along the functional axis. Hence the recent interest in the identification of thresholds (Schumm, 1979), and the emerging appeal of catastrophe theory (Brunsden and Thornes, 1979). The development of this interest in the nature of change within the geomorphological system inevitably requires a renewal of interest in the temporal axis of the subject, for it is along that axis that the effects of change must be traced (Thornes and Brunsden, 1977).

Alongside this reappraisal of change within the geomorphological system, studies of landform development in Britain, which seemed to have reached a very low ebb between 1965 and 1970, have developed a new impetus. This impetus is based on the application to geomorphological and related environmental evidence of a wide variety of techniques of measurement and classification which provide a more precise basis for correlation and differentiation. As is suggested by the content of this volume, these techniques have been applied mainly to sediments and soils. Quantitative appraisal of the morphological evidence has also been attempted (Thornes and Jones, 1969) but has been pursued rather less enthusiastically. This increased emphasis on the provision of a scientific basis for the interpretation of landform development reinforces the convergence of interest between the temporal and functional axes of the subject.

The other main influences upon the study of landform development in the last ten years have been:

(1) An increasing awareness of the complexity of the environmental system of

which the geomorphological system forms a part. This development is evident in the study of Pleistocene biology, where attention has tended to be transferred from consideration of the chronological significance of the evidence (West, 1963) to consideration of its palaeoenvironmental significance (Coope, 1977).

(2) A recognition of the need for radical changes in the palaeoenvironmental framework (Bowen, 1979) due to the emergence of new global models of climatic, eustatic and tectonic change, based on the oceanic record of Cenozoic stratigraphy and on the theory of plate tectonics.

(3) The development of absolute dating techniques to replace or supplement the various schemes of relative dating based on biological evidence.

It is difficult across the relatively short perspective of the last few years to generalize the effects of these influences on studies of landform development in southern Britain. The position seems, however, to be one in which geomorphologists have recognized the far-reaching implications of the new concepts and have broken away from the constraints of earlier palaeoenvironmental frameworks (Jones, 1974), while rarely extending interpretative discussion to include explicit reference to the new concepts. It will perhaps be easier, therefore, to examine these concepts more fully in terms of the foreseeable progress of the subject.

## Future progress

### Sources of new evidence

The discovery of previously unrecognized landforms or materials in southern Britain is now an infrequent occurrence. Elsewhere in this volume Green and McGregor note that almost all the key Quaternary localities in the Thames Basin were known before 1914. Much new evidence comes, therefore, from the application of new analytical techniques to landforms, sediments, and soils which have been considered before, sometimes on numerous occasions. Nevertheless, certain circumstances do increase the chances of completely new raw material coming to light. The continuing extension and improvement of the motorway and trunk road networks has served geomorphologists well (e.g. Briggs *et al.*, 1975) and will undoubtedly offer further opportunities for the examination of the depositional evidence. However, an even richer source of new information lies off the coasts of southern Britain in the sediments, structures and submarine relief of the English Channel and the North Sea. Evidence from these sources coming mainly from geological investigations is already being deployed in geomorphological studies (Wood, 1976) but the relationship of this evidence to the terrestrial record has yet to be worked out. As well as providing new information about the off-shore area itself, the submarine evidence may also provide a more direct link between Britain and the mainland of Europe, making possible the more effective integration into the British model of landform development, of evidence from such sources as the Tertiary sediments of the Paris Basin, and the Quaternary terrace sequences of the rivers Rhine, Seine and Somme.

## The shaping of ideas

Because completely new discoveries are comparatively rare, the re-interpretation of evidence must in future be an important source of progress in the study of landform development. The process of re-interpretation depends on the evolution of the palaeoenvironmental framework, which is here considered under the three headings identified at the beginning of this account.

### *The geomorphological system*

A further convergence of studies of the temporal and functional axes of geomorphology seems to offer considerable potential for progress. Both R. B. G. Williams and G. H. Cheetham (this volume, pp. 225 and 203 respectively) show that the study of contemporary process continues to improve our insight into the nature of past geomorphological episodes. In this general context two areas of investigation are of particular significance—the evolution of contemporary floodplains (Lewin and Manton, 1975) and the processes of till emplacement (Boulton, 1975). These two areas promise to contribute in the short term more fundamentally than any others to an understanding of the geomorphological implications of Quaternary sediments. Another critical area of study, but one in which there seems less likelihood of development in the short term, is in the understanding of coastal processes. Particularly in need of elucidation are the processes of erosion that seem to be required for the formation of the extensive wave-cut platforms that characterize some Quaternary shorelines (Keen, 1978), and may, or may not, be an important element of Tertiary relief (George, 1974).

Improved understanding of the temporal axis of geomorphology has important implications for the development of studies of the functional axis itself. The most far-reaching implication arises from the recognition that almost all landforms are relics and have not been shaped only, or even largely, by present-day processes. In other words, a powerful variable in the present-day geomorphological system is the relief inherited from the past and often shaped in environmental conditions very different from those of the present. Given a more integrated appreciation of the temporal, functional and spatial dimensions of the geomorphological domain, models, such as that of Boulton *et al.* (1977), representing past geomorphological systems should become increasingly realistic.

The juxtaposition of long- and short-term studies of landform development highlights one problem to which geomorphologists will have to address themselves in the future. Measured erosion rates (Young, 1974; McArthur, 1977) suggest that present-day processes, if sustained in the past, could have created the present relief of Britain in, at most, a few million years, and over the full span of the Cenozoic could have reduced the whole landmass of Britain much closer to base-level than it is. Nevertheless landforms apparently of early Tertiary age are present even in the landscape of lowland Britain at levels far above present base-level. The solution to this enigma may lie in a better understanding of the structural history of Britain, or of the long term continuity of geomorphological processes. The importance of the latter point is stressed by R. B. G. Williams in this volume (p. 225) where he shows that the significance of periglacial erosion of the Chalk depends on the duration of the erosional episodes, and that this is almost always unknown and usually very difficult to find out.

*The palaeoenvironmental framework*

Reappraisal of the interpretative framework of long-term landform development is more explicit now than at most times in the past. The importance attached to the interpretative framework is clearly seen in the present volume in which several attempts are made to model past geomorphological events in terms of specific environmental constraints.

*Climatic effects.* The record of Quaternary climatic fluctuations derived from deep ocean sediments suggests a sequence of between 15 and 25 climatic cycles of glacial-interglacial intensity. The terrestrial evidence in southern Britain can be extended, at best, to 10 or 12 cycles, with unequivocal indications of only two glacial episodes, Devensian and Anglian, and equivocal indications of perhaps two or three more. This discrepancy obviously calls for early explanation, either in terms of later erosional events obliterating the evidence of earlier events, or, as seems more likely, in terms of global climatic fluctuations of differing intensity and geographical incidence, some of which did not lead to identifiable transformations of the geomorphological system in southern Britain. The outstanding problem in either case is the dating of the identifiable episodes, which is considered below.
latter point is stressed by R. B. G. Williams in this volume (p. 225) where he shows
*Eustatic effects.* Possible limitations of the classical Mediterranean scheme of sea-level behaviour are now recognized (Hey, 1978), and at the same time the evidence of the oceanic record suggests the need for a complete re-examination of our views on Quaternary sea-levels. The oceanic record indicates (Shackleton and Opdyke, 1973) that the amount of water in the oceans has fluctuated within the same range throughout the Quaternary. The maximum volume in the latter part of the Quaternary can be identified during the last interglacial in isotope stage 5e. The terrestrial record in southern Britain suggests that at that time sea-level did not rise above 8 m which would therefore be the maximum sea-level for the Middle and Upper Pleistocene, with fluctuations occurring between +8 m O.D. and well below O.D. Such a model of sea-level behaviour leaves the geomorphologist with two outstanding problems: (a) the problem of tracing the effects of numerous changes of sea-level within the range below +8 m O.D.; (b) the problem of explaining the evidence of sea-levels above +8 m O.D. Failing a satisfactory structural explanation of high Quaternary sea-levels, a global cause, superimposed on the glacio-eustatic fluctuations of the oceanic record must be considered, possibly as a manifestation of global tectonics during the Quaternary (Flemming and Roberts, 1973).

*Tectonic effects.* Many aspects of geological explanation have been radically changed by the emergence, during the last twenty years, of plate tectonic theory. The influence of the theory on explanations of landform development in southern Britain is likely to be felt in two ways. Firstly, the structural context of landform development can now be understood in terms of two patterns of stress. Hitherto effects of Alpine orogenesis have been of primary concern to geomorphologists. Now the consequences of Britain's position on the eastern margin of the North Atlantic Rift add a new dimension to structural explanations, and may be particularly helpful in resolving problems of regional uplift and subsidence, and the complementary problems of long-term rates of denudation. Secondly, the plate tectonic model is consistent with a less episodic history of structural events, in which the former emphasis on a 'mid-Tertiary Alpine storm' has no place. Consequently hypotheses based on long-term structural stability require re-examination.

*Chronology and stratigraphy*

If, as seems probable, a global stratigraphy for the Quaternary can be based on the oceanic evidence, a major problem for the geomorphologist is to relate the less complete terrestrial geomorphological record to this stratigraphy. Only by way of an absolute chronology can a completely satisfactory link be achieved. For the Cenozoic in southern Britain, suitable dating techniques have been lacking in the past, except for the very short period covered by the radio-carbon assay of biological material.

In the last ten years, however, several new methods of dating have been applied for the first time to British material and the initial results give real indications of important developments in the near future. For the Tertiary, the radio-isotopic assay of glauconite in marine sediments (Odin, Curry and Hunziker, 1978), and of a small number of newly discovered volcanic ash bands (Knox and Ellison, 1979) has begun to provide an absolute chronological framework for the sedimentary sequences in the London and Hampshire Basins. For the Quaternary, uranium series (Szabo and Collins, 1975) and amino-acid (Miller, Hollin and Andrews, 1979) methods aplied to biological material, and uranium/thorium assay of calcite speleothems (Atkinson *et al.*, 1978) are beginning to produce significant results. But perhaps the greatest potential lies with the application of thermoluminescence dating to mineral sediments (Wintle and Huntley, 1979).

Although important progress is likely using absolute dating techniques, there is also scope for closer comparison between stratigraphic successions, both within Britain, and between Britain and neighbouring areas of Europe. Such comparisons, while not resolving problems of absolute age, ought to provide insights into the relative continuity or otherwise of the various successions.

## Conclusion

In the study of landform development in southern Britain, the scope for rapid advancement of understanding is evident. Firstly, an interdisciplinary approach to the reconstruction of past environments can make possible a better understanding of the stages of landform development, while at the same time the geomorphological evidence can contribute significantly to the environmental reconstruction itself. Secondly, the development of dating techniques can place the recognizable stages of landform development within a global stratigraphic context and can provide an accurate basis for the elucidation of long-term rates of geomorphological change. Thus a more scientific appraisal of long-term landform development can contribute to a comprehensive understanding of the geomorphological system in all its dimensions.

Whether the potential is realized or not, depends not so much on the institutional organization of geomorphology, as some exponents of the subject have suggested (Worsley, 1979), but on the ability of individuals to recognize the full extent and unifying principles of the geomorphological domain, to communicate effectively within that domain, and to encourage the flow of ideas into geomorphology from related disciplines.

## References

Arkell, W. J. (1943). The Pleistocene rocks at Trebetherick Point, north Cornwall; their interpretation and correlation, *Proc. Geol. Ass.* **54**, 141-170.

Atkinson, T. C., Harmon, R. S., Smart, P. L. and Waltham, A. C. (1978). Palaeoclimatic and geomorphic implications of $^{230}$Th/$^{234}$U dates on speleothems from Britain, *Nature, Lond.* **272**, 24-28.

Boulton, G. S. (1975). Processes and patterns of sub-glacial sedimentation: a theoretical approach, *in* Wright, A. E. and Moseley, F. (eds), "Ice ages: ancient and modern" pp. 7-42. Seel House *Geol. J.*, Special issue No. 6.

Boulton, G. S., Jones, A. S., Clayton, K. M. and Kenning, M. J. (1977). A British ice-sheet model and patterns of glacial erosion and deposition in Britain, *in* Shotton, F. W. (ed.) "British Quaternary studies", pp. 231-46. Oxford University Press.

Bowen, D. Q. (1979). Geographical perspective on the Quaternary, *Progress in Physical Geography* **3**, 167-86.

Bridges, E. M. and Harding, D. M. (1971). Micro-erosion processes and factors affecting slope development in the Lower Swansea Valley, *in* Brunsden, D. (ed.), "Slopes form and process", pp. 65-80. *Inst. Br. Geogr. Sp. Pub.* **3**.

Briggs, D. J., Gilbertson, D. D., Goudie, A. S., Osborne, P. J., Osmaston, H. A., Pettitt, M. E., Shotton, F. W. and Stuart, A. J. (1975). New interglacial site at Sugworth, *Nature, Lond.* **257**, 477-79.

Brown, E. H. (1975). The content and relationships of physical geography, *Geog. J.* **141**, 35-40.

Brunsden, D. and Thornes, J. B. (1979). Landscape sensitivity and change, *Trans. Inst. Br. Geogr.* (New Series) **4**, 463-84.

Chorley, R. J. and Kennedy, B. (1971). "Physical geography, a systems approach". Prentice-Hall, Englewood Cliffs.

Coope, G. R. (1977). Fossil coleopteran assemblages as sensitive indicators of climatic changes during the Devensian (Last) cold stage, *Phil. Trans. Roy. Soc. Lond.* **B280**, 313-40.

Depēret, C. (1918). Essai de coordination chronologique des temps quaternaires, *Comptes rendus hebdomadaires de l'Acadēmie des Sciences Paris* **166**, 480-86.

Dury, G. H. (1978). The future of geomorphology, *in* Embleton, C., Brunsden, D. and Jones, D. K. C. (eds), "Geomorphology". Oxford University Press.

Flemming, N. C. and Roberts, D. G. (1973). Tectono-eustatic changes in sea level and sea-floor spreading, *Nature, Lond.* **243**, 19-22.

Foster, C. Le N. and Topley, W. (1865). On the superficial deposits of the valley of the Medway, with remarks on the denudation of the Weald, *Q. J. Geol. Soc. Lond.* **21**, 443-74.

George, T. N. (1974). Prologue to a geomorphology of Britain, *in* Brown, E. H. and Waters, R. S. (eds), "Progress in geomorphology", pp. 113-126. *Inst. Br. Geogr. Sp. Pub.* **7**.

Hack, J. T. (1960). Interpretation of erosional topography in humid temperate regions, *Am. J. Sci.* **258-A**, 80-97.

Hey, R. W. (1978). Horizontal Quaternary shorelines of the Mediterranean, *Quat. Res.* **10**, 197-203.

Jones, D. K. C. (1974). The influence of the Calabrian transgression on the drainage evolution of south-east England, *in* Brown, E. H. and Waters, R. S. (eds), "Progress in geomorphology", pp. 139-61. *Inst. Br. Geogr. Sp. Pub.* **7**.

Keen, D. H. (1978). The Pleistocene deposits of the Channel Islands, *Rep. Inst. Geol. Sci.* No. 78/26.

King, W. B. R. and Oakley, K. P. (1936). The Pleistocene succession in the lower part of the Thames valley, *Proc. Prehist. Soc.* **2**, 52-76.

Knox, R. W. O'B. and Ellison, R. A. (1979). A Lower Eocene ash sequence in SE England, *J. Geol. Soc. Lond.* **136**, 251-3.

Lewin, J. and Manton, M. (1975). Welsh floodplain studies: the nature of floodplain geometry, *J. Hydrol.* **25**, 37-50.

McArthur, J. L. (1977). Quaternary erosion in the upper Derwent basin and its bearing on the age of the surface features in the southern Pennines, *Trans. Inst. Br. Geogr.* (New Series) **2**, 490-97.

Miller, G. H., Hollin, J. T. and Andrews, J. T. (1979). Animostratigraphy of U.K. Pleistocene deposits, *Nature, Lond.* **281**, 539-43.

Odin, G. S., Curry, D. and Hunziker, J. C. (1978). Radiometric dates from NW European glauconites and the Palaeogene time-scale, *J. Geol. Soc. Lond.* **135**, 481-497.

Penck, A. and Bruckner, E. (1909). "Die Alpen im Eiszeitalter". Tauchnitz, Leipzig.

Schumm, S. A. (1979). Geomorphic thresholds: the concept and its applications, *Trans. Inst. Br. Geogr.* (New Series) **4**, 485-515.

Shackleton, N. J. and Opdyke, N. D. (1975). Oxygen isotope and palaeomagnetic stratigraphy of Equatorial Pacific core V28-238: oxygen isotope temperatures and ice volumes on a $10^5$ year and $10^6$ year scale, *Quat. Res.* **3**, 39-55.

Sparks, B. W. (1949). The denudation chronology of the dipslope of the South Downs, *Proc. Geol. Ass.* **60**, 165-215.

Stoddart, D. R. (1975). In discussion of Brown, E. H. (1975). The content and relationships of physical geography, *Geogr. J.* **141**, 35-40.

Sugden, D. E. (1979). Whither geomorphology?, *Area* **11**, 309-12.

Szabo, B. J. and Collins, D. (1975). Age of fossil bones from British interglacial sites, *Nature, Lond.* **254**, 680-81.

Thornes, J. B. and Brunsden, D. (1977). "Geomorphology and time". Methuen, London.

Thornes, J. B. and Jones, D. K. C. (1969). Regional and local components in the physiography of the Sussex Weald, *Area* **2**, 13-21.

West, R. G. (1963). Problems of the British Quaternary, *Proc. Geol. Ass.* **74**, 147-86.

Wintle, A. G. and Huntley, D. J. (1979). Thermoluminescence dating of a deep-sea sediment core, *Nature, Lond.* **279**, 710-12.

Wood, A. (1976). Successive regressions and transgressions in the Neogene, *Marine Geology* **22**, M23-30.

Wooldridge, S. W. and Kirkaldy, J. F. (1936). River profiles and denudation chronology in southern England, *Geol. Mag.* **73**, 1-16.

Worsley, P. (1979). Whither geomorphology?, *Area* **11**, 97-101.

Young, A. (1960). Soil movement by denudational processes on slopes, *Nature, Lond.* **188**, 120-22.

Young, A. (1969). Present rate of land erosion, *Nature, Lond.* **224**, 851-52.

Young, A. (1974). The rate of slope retreat, *in* Brown, E. H. and Waters, R. S. (eds), "Progress in geomorphology", pp. 65-78. *Inst. Br. Geogr. Sp. Pub.* **7**.

# SUBJECT INDEX

D

J



K



L



M

Q



R

## T

U



V



W